MAGIC, REASON AND EXPERIENCE

MAGIC
REASON
AND
EXPERIENCE

Studies in the Origins and Development of Greek Science

G.E.R. Lloyd

Published in the U.K. by
Gerald Duckworth & Co. Ltd.
London

Published in North America by
Hackett Publishing Company, Inc.
Indianapolis/Cambridge

First published in 1979 by Cambridge University Press

This edition published in 1999 by
Bristol Classical Press
an imprint of
Gerald Duckworth & Co. Ltd
61 Frith Street
London W1V 5TA
and by
Hackett Publishing Company, Inc.
P.O. Box 44937
Indianapolis, Indiana 46244-0937

© 1979 by G.E.R. Lloyd

U.K. edition:
ISBN 1-85399-602-5
A CIP catalogue record for this book
is available from the British Library

U. S. edition
Paper ISBN 0-87220-528-2
Cloth ISBN 0-87220-529-0
**Library of Congress Catalog
card number: 99-75596**

The paper used in this publication meets the minimum
requirements of American National Standard for Information
Sciences – Permanence of Paper for Printed Library
Materials, ANSI Z39.48-1984

Printed in the United States of America

For Ji

CONTENTS

ACKNOWLEDGEMENTS

Many friends and colleagues have been kind enough to read and comment, from different points of view, on parts or the whole of this book in draft: Myles Burnyeat, Moses Finley, Edward Hussey, Gilbert Lewis, Jeremy Mynott, Martha Nussbaum, G. E. L. Owen, Robert Parker, Malcolm Schofield. What I have learnt from them both on the strategy of my arguments and on points of detail far outreaches the warmest thanks I can here express. But it is a pleasure to record, however inadequately, my debts to their stimulating criticisms and advice. They need, of course, no absolution for the shortcomings that remain.

I wish also to thank Michael Loewe and Denis Twitchett for help on Chinese questions, Peter Khoroche and Piers Vitebsky on points of comparative religion, and Jack Goody, Alan Macfarlane and S.J. Tambiah on anthropological matters. Sections of this work were delivered as seminars at the École des Hautes Etudes in the spring of 1978 and I wish to thank Mary Dallos for the French version of my text. I am most grateful for the comments and criticisms made on those occasions, especially by Jacques Brunschwig, Marcel Detienne, Nicole Loraux and Jean-Pierre Vernant. I wish, too, to record how much I have benefited from many instructive conversations with Pierre Vidal-Naquet. In addition to the detailed observations I have received from Jeremy Mynott, I owe much to the Officers of the Press, and especially to Pauline Hire, for their patient and constructive help. Finally I thank my family for their continuing tolerance, and my son Adam in particular for his perceptive criticisms of my ideas. The book is dedicated, with love and thanks, to my wife.

Cambridge 1978   G. E. R. L.

TEXTS AND ABBREVIATIONS

Except where otherwise stated, the fragments of the Presocratic philosophers are quoted according to the edition of Diels, revised by Kranz, *Die Fragmente der Vorsokratiker* (6th ed., 1951–2) (referred to as DK), the works of Plato according to Burnet's Oxford text, the treatises of Aristotle according to Bekker's Berlin edition and the fragments of Aristotle according to the numbering in V. Rose's edition (1886). Greek medical texts are cited, for preference, according to the *Corpus Medicorum Graecorum (CMG)* editions. For those Hippocratic treatises not edited in *CMG*, I use E. Littré, *Oeuvres complètes d'Hippocrate*, 10 vols., Paris, 1839–61 (L), with two exceptions. For *On the Sacred Disease* I use H. Grensemann's Ars Medica edition (Berlin, 1968) (G), cited by chapter and paragraph, and for *On Sevens* W. H. Roscher, Paderborn, 1913. Galen is cited according to *CMG* and Teubner editions (where these exist), but the reference is also given to the edition of C. G. Kühn (Leipzig, 1821–33) (K), which is also used for works not in *CMG* or Teubner. Euclid's *Elements* are cited according to the edition of Heiberg, revised by Stamatis, 4 vols., Leipzig, 1969–73 (HS), and the works of Archimedes according to Heiberg, revised by Stamatis, 2 vols., Leipzig, 1972 (HS). Ptolemy's *Syntaxis* is cited according to the two volume edition of Heiberg (1898, 1903) (cited as ii and iii to distinguish them from the other volumes of the Teubner edition of Ptolemy: thus vol. ii contains *Opera Astronomica Minora*, edited by Heiberg, 1907). Ptolemy's *Harmonics* are cited according to I. Düring, Göteborg, 1930, and his *Optics* according to A. Lejeune, Louvain, 1956.

Otherwise Greek authors are cited according to the editions named in the *Greek–English Lexicon* of H. G. Liddell and R. Scott, revised by H. S. Jones, with Supplement (1968) (LSJ), though, where relevant, references are also provided to more recent editions, and Latin authors are cited according to the editions named in the new *Oxford Latin Dictionary* (*OLD*), supplemented, where necessary, from Lewis and Short. Abbreviations are those in LSJ and *OLD*, again supplemented, where necessary, from Lewis and Short.

Full details of modern works referred to will be found in the bibliography on pp. 268ff. They are cited in my text by author's name and publication date or dates. A double date is used to distinguish, where this has seemed relevant, the original publication from the revised or reprinted version used. Such works are listed in the bibliography by the first date, but cited according to the second. Thus Owen (1960) 1975 refers to the 1975 reprint (with additions) of an article originally published in 1960.

INTRODUCTION

This study, like my *Polarity and Analogy*, is a contribution to what might be thought to be, by now, a very hoary problem, namely the relationship between what may be called 'traditional' and 'scientific' patterns of thought. It is not hard to suggest reasons for the dominant role of this issue in early anthropological writings, or for the way in which it was often represented as a matter of a polar contrast between 'primitive' and 'civilised' societies or between two distinct mentalities, the one 'pre-logical' or 'pre-scientific' and the other 'logical' or 'scientific'.[1] But the manifest unacceptability of the terms in which some aspects of the problem were debated in the nineteenth and early twentieth centuries does not mean either that there was no problem or that its resolution has now been agreed. Indeed after a period of comparative neglect, the issues have recently been revived by both anthropologists and philosophers.

Four main lively areas of debate have generated work that has modified the way in which certain fundamental questions now present themselves. First there is the major philosophical issue of understanding alien societies,[2] that is of the commensurability or incommensurability of the modes of thought, beliefs and values of different societies. To translate the concepts of any given society into those of any other is to interpret them, and – so it has been argued – in so doing inevitably to distort them, in particular by prejudging certain key issues relating both to the nature of truth and to that of rationality. There are, then, some have said,[3] no culture-independent

---

[1] Lloyd 1966, pp. 3ff, outlines Lévy-Bruhl's hypothesis and mentions some of the classical scholars who were directly or indirectly influenced by his ideas.

[2] Among important recent studies of the sociological aspects of this problem are Winch 1958, Horton 1967, the papers of Gellner (1962), Winch (1964), MacIntyre (1967) and Lukes (1957) in B. R. Wilson 1970, S. B. Barnes 1972, the contributions of Gellner, Lukes and S. B. Barnes in Horton and Finnegan 1973, S. B. Barnes 1974 and Skorupski 1976. Of the more purely philosophical discussions, Quine's work (1953 and 1960, especially ch. 2) is fundamental.

[3] See, for example, Mannheim 1936, pp. 239ff, 262ff. The idea that language not merely influences but determines thought has been much debated in connection with the ideas of Sapir and Whorf (e.g. Feuer 1953, Hoijer 1954, Black 1959, Penn 1972).

criteria that can be used as the basis of 'objective' judgements concerning other societies, and a society can only be understood from 'within', that is by the actors themselves, not by outside observers. However it was appreciated early in this debate that many of the points that had been expressed as difficulties concerning the understanding of one society by another apply also to the mutual understanding of different groups within the same society even when those groups use the same natural language. While one of the salutary effects of this controversy has been to emphasise both the crucial importance of the distinction between 'actor' and 'observer' categories and the ever-present danger of distortion in the application of our own concepts to another culture, I shall not here attempt to justify what after all must always be assumed in any discussion of the ancient world, namely that some progress towards understanding is possible, even if at a quite modest level and subject to the reservations implied by the problems of interpretation I have mentioned. Meanwhile one of the aims of my inquiry is to analyse within Greek thought both the conditions under which confrontations between contrasting belief-systems were possible and the nature and limits of such confrontations as occurred.

Secondly, there have been important anthropological studies devoted to the interpretation of the complex of phenomena loosely categorised as 'magic' and directed, in particular, to a critique of the old idea which saw magic as, broadly speaking, failed applied science.[4] Evans-Pritchard's classic monograph on Zande religion employed what now seem simple-minded distinctions between 'mystical', 'common-sense' and 'scientific' notions, defining the first as 'patterns of thought that attribute to phenomena supra-sensible qualities which, or part of which, are not derived from observation or cannot be logically inferred from it, and which they do not possess', and taking science and logic to be the 'sole arbiters of what are mystical, common-sense, and scientific notions'.[5] Any attempt to contrast magic as a whole directly with science is now seen to be liable to distort the nature and aims of the former. Magic, so it has been forcefully and in part, at least, surely rightly argued, should be seen less as attempting to be efficacious, than as affective, expressive or symbolic. The criteria that are relevant to judging magical behaviour are not whether it achieves practical results but whether it has been carried out appropriately or not. Thus Tambiah has

---

[4] See especially Frazer 1911–15, I pp. 220ff and cf. Jarvie and Agassi (1967) 1970.
[5] Evans-Pritchard 1937, p. 12. I made what I now see as an uncritical use of Evans-Pritchard's categories in my 1966, pp. 177–9.

focused on the performative or illocutionary nature of magical actions and insisted that they should be assessed from the point of view of their felicity or infelicity, not from that of their practical effectiveness.[6] The distance that many anthropologists have moved from Evans-Pritchard (let alone from earlier writers) can be illustrated by a passage in Mary Douglas:[7] 'Once when a band of !Kung Bushmen had performed their rain rituals, a small cloud appeared on the horizon, grew and darkened. Then rain fell. But the anthropologists who asked if the Bushmen reckoned the rite had produced the rain, were laughed out of court.'

The third main area of discussion where recent contributions have far-reaching implications for the understanding of the early development of science is the philosophy and sociology of science itself, where the work of Popper, Kuhn, Feyerabend and Lakatos on the demarcation between scientific and other forms of knowledge and on the growth of science has been especially influential.[8] In this debate Kuhn has argued for a fundamental distinction between puzzle-solving 'normal' science and periods of crisis when the shared assumptions, or paradigms, of a scientific community are at issue and when a 'gestalt switch' of paradigms may occur. Many aspects of this thesis remain controversial or obscure or both, and the key notion of the 'paradigm' itself – used both globally of the 'constellation of group commitments' and much more specifically of certain 'shared examples' – has been acknowledged by Kuhn himself to have been unclear.[9] Yet the effect of his work has certainly been to draw

---

[6] Tambiah 1973, pp. 220ff (where he refers explicitly to Austin 1962a). At pp. 227–9, however, he expresses reservations concerning the applicability of this point to the whole of magic. Cf. also Tambiah 1968.

[7] Douglas 1966, p. 73, with a reference to Marshall 1957. What Marshall wrote was: 'One night when the Rain dance was being beautifully danced with a fine precision and vigour of clapping, singing, and stamping, which to us suggested fervour, we were watching it so intently that we had not noticed the sky. The first storm of the season had crept up behind us and suddenly burst over us like a bomb. We asked Gao if he believed the dance had brought the rain. He said that the rain was due to come. The dance had not brought it' (Marshall 1957, p. 238). Marshall cited this as evidence that the purpose of the Rain dance is not to control the weather: it is, rather, one of a number of 'medicine' dances 'danced during the medicine men's ceremony to cure the sick and protect the people and drive away any of the...spirits of the dead who might be lurking to bring some evil upon the people' (p. 238). However the same article of Marshall's contains other suggestions about how the Bushmen do believe they can control the weather, for example by cutting the throats of particular animals to bring on or to stop rain (p. 239).

[8] Apart from Popper 1959, 1963, and 1972, Kuhn (1962) 1970 and 1970b, Feyerabend 1962 and 1975, Lakatos 1978a and 1978b, see also the papers in Lakatos and Musgrave 1970, and Hesse 1974. Several of these writings are also concerned with, and refer directly to, the debates on the problems of commensurability and translation mentioned above, pp. 1f.

[9] See Kuhn 1974 (written before, but published after, 1970a and 1970b) as well as 1970a,

attention to the role of the consensus of the scientific group and to that of their shared implicit or explicit assumptions, and this in turn has made it easier to see the important similarities between the scientific, and other, communities, and between science itself and other belief-systems. Thus while Evans-Pritchard's category of the 'mystical' has been subject to drastic revision from the side of the anthropologists, the same is equally true of his notion of the 'scientific' from the side of the sociologists and philosophers of science.

Fourthly, there have been fundamental studies of the development of literacy carried out by Goody and others, dealing, especially, with the effects of changes in the technology of communication – including the availability of written records of various types – on the nature and complexity of what is communicated. Most recently, in his significantly entitled *The Domestication of the Savage Mind*, Goody has argued forcefully for a recasting of the problem of the 'Grand Dichotomy' between 'primitive' and 'advanced', 'traditional' and 'modern', 'cold' and 'hot', societies, and emphasised the importance of complex and multiple changes in the means of communication, focusing attention particularly on those that took place in the ancient Near East. '"Traditional" societies', he maintained, 'are marked not so much by the absence of reflective thinking as by the absence of the proper tools for constructive rumination',[10] and he put it that 'the problem can be partly resolved, the Grand Dichotomy refined, by examining the suggested differences in cultural style or achievement as the possible outcome of changes in the means of communication, an outcome that will always depend for its realisation on a set of socio-cultural factors.'[11]

So far as the interpretation of early Greek thought is concerned, the debates of the anthropologists and philosophers have been, at most, intermittently influential. In the discussion of many aspects of the relationship between 'magic' and 'science' in the ancient world Dodds' classic *The Greeks and the Irrational* is still the starting-point.[12] Thanks very largely to this study, certain points may now be taken as generally agreed. Thus it is abundantly clear that the 'irrational' in one or other of its complex and diffuse forms is to be found at every period of Greek thought for which there is any evidence. Magical beliefs and practices of a wide variety of kinds can be documented

---

pp. 174ff, and 1970*b*, pp. 231ff: and cf. Shapere 1964, Scheffler 1967, ch. 4 and Masterman 1970.

[10] Goody 1977, p. 44.

[11] Goody 1977, p. 147. I return to consider the bearing of these theses for the understanding of early Greek thought below, ch. 4, pp. 239f.

[12] Dodds 1951.

from Homer to the end of antiquity and on into the middle ages. This is not to deny that we can talk about certain fluctuations in the extent of particular beliefs, or even in 'magic' as a whole, at particular periods. Dodds himself argued,[13] with some plausibility, that after the fifth-century B.C. 'enlightenment' there was something of a reversion or reaction in the fourth century. Again the evidence both from literary sources and from papyri[14] suggests an increase in magical beliefs and practices – notably in the form known as theurgy – from about the second century A.D., though this is, to be sure, impossible to quantify. But the principal conclusion, not only from Dodds but from many other studies, is clear, namely that magic and the irrational can be documented from the very earliest to the very latest times.

This by itself gives the lie to any classical scholar who might still be tempted to suppose that in ancient Greece 'science' supplanted 'magic', or 'reason' 'myth'. Moreover, as we shall be illustrating later,[15] several of those who were prominent in the development of Greek cosmology and science combined an interest and belief in magic with their other work in the 'inquiry into nature'. To mention just the most obvious single example here, it is well known that most ancient, like most medieval and Renaissance, astronomers were also practising astrologers.[16]

But if there can be no doubt about the continuous importance of myth and magic throughout antiquity, it is also agreed on all sides, at the broadest and most general level, that inquiries that are recognisable as science and philosophy were developed in the ancient world. However much scholars differ in their detailed interpretations, they acknowledge that certain significant changes or developments occurred during the period from the sixth to the fourth centuries B.C. But just how those changes are to be described – let alone explained – is problematic. Those who have proposed general accounts of the growth of early Greek speculative thought have advanced various theses on such questions as what the Greeks owed to their ancient Near Eastern neighbours, on the debt of science and philosophy to religion, and on the influence of social, political, economic and technological factors.[17] Yet although

---

[13] Dodds, 1951, ch. 6 ('Rationalism and reaction in the classical age').

[14] See especially Preisendanz 1973–4.

[15] See, for example, below, pp. 33ff, on Empedocles.

[16] This is true of both Hipparchus and Ptolemy, for instance: see further below, p. 180 n. 292.

[17] Among the most important contributions to these debates have been those of Burnet (1892) 1948, Cornford 1912 and 1952, Farrington (1944–9) 1961, Vlastos (1947) 1970, pp. 56ff, (1952) 1970, pp. 92ff, (1955) 1970, pp. 42ff and 1975, G. Thomson 1954 and

references to a 'revolution in thought' are common in works on the Presocratic period, the prior question of what exactly that revolution consisted in has often been dealt with somewhat schematically. In particular the force of the obvious point that Greek science and philosophy developed in the continuing presence of traditional patterns of thought has tended to be underestimated, as also sometimes have the complexities, heterogeneities and limitations of what we know as Greek science and philosophy. One of the chief difficulties that this inquiry faces is to do justice to the differences *between* and indeed *within* the various distinguishable, if overlapping, strands of early Greek speculative thought – cosmology, 'natural science', medicine and mathematics – a problem that is exacerbated by the disparities in the evidence that is available to us for each of them.[18] Moreover, so far from speculative thought being totally hostile and injurious to all aspects of the irrational, there are certain ways in which the development of magic itself may be said to have depended on, and followed the model of, philosophy and science. The development of 'temple medicine' may well owe a good deal to – certainly it often imitates – rationalistic medicine.[19] The systematisation of astrology does not happen before the third century B.C.,[20] nor that of 'alchemy'[21] before the first century B.C., and in both cases it is possible to argue that the very success of natural science and philosophy contributed to the systematisation of these other 'sciences'.[22]

The primary task is to delineate as carefully as possible the nature of the developments that took place in early Greek thought during the crucial period when science and philosophy were emerging as recognisable inquiries. We must juxtapose Greek and non-Greek material – whether from the ancient Near East or from modern non-literate societies – in order to isolate such distinguishing features as the former exhibits. We have, too, to ask how far the new inquiries involved or consisted in new aims and methods, and at what point

---

1955, Vernant (1957) 1965, pp. 285ff, 1962 and (1963) 1965, pp. 145ff, Popper (1958–9) 1963, pp. 136ff, Vidal-Naquet 1967 and Hussey 1972.

[18] See below, pp. 8f.

[19] See further below, pp. 40f.

[20] Although Herodotus (II 82), for instance, attributes to the Egyptians the belief that a man's future can be foretold according to the day of his birth and says that Greek poets used these ideas, astrological beliefs do not become prominent in Greece until the fourth century B.C., and it is not until later still that astrology was turned into a universal system: see Neugebauer 1957, pp. 170ff, and 1975, II pp. 613f, and cf. further below, p. 180 and n. 292.

[21] On the limitations of 'alchemy' in the Greco-Roman world, see especially J. Needham 1954–, V, 2.

[22] As Préaux 1973, for example, has suggested in connection with astrology.

those aims and methods themselves became the object of self-conscious reflection. Only then can we pose the question, finally, of how such developments as we can identify occurred, and analyse the social, political and other conditions that may be thought to have stimulated or permitted them.

The problems as outlined relate to early Greek thought: yet they clearly have general implications. In a contribution to the recent anthropological debate on magic, Jarvie and Agassi concluded: 'The problem is not, then, "how on earth can they [primitive peoples] believe in magic?"; it is rather "can people with inefficient magical beliefs come to be critical of them, under what conditions and to what extent?" To us this seems the really urgent sociological problem posed by magic.'[23] But that very way of stating the problem indicates its diachronic character. What is required is an examination of the beliefs and practices of any given society over a considerable period of time. Yet it is just evidence of that kind – on changes in beliefs and practices over an extended period – that the anthropologist studying a non-literate society often finds it hard or impossible to obtain. Whilst the data for early Greek thought suffer from their own severe limitations,[24] our sources do provide some information on the vital issue of changes in the beliefs and attitudes of at least certain groups within Greek society over a period when science and philosophy were first developing.

We must, to be sure, bear in mind Tambiah's caveats concerning the relevance of the European experience to the study of magic as a whole. Having remarked that 'there is no denying that in Europe there is some kind of developmental sequence by which out of more "primitive" notions and "magical" practices more "scientific" notions and experimentation were born',[25] Tambiah went on: 'it may very well be that the Western experience is a *privileged* case of transition from "magic" to "science".'[26] When he thus insists on the possibly exceptional nature of aspects of the Greek situation, Tambiah's points are well taken.[27] But if caution is certainly in order in any attempt to apply conclusions based on the study of the Greeks to other societies, we may still recognise that the evidence for early

[23] Jarvie and Agassi (1967) 1970, p. 193. Cf. Gellner 1973, pp. 162ff, who deliberately 'thinks away' the 'middle ground' in posing the problem of what he calls the 'Great Divide' between 'the Savage and the Modern mind', and contrast Goody 1977, p. 147, quoted above, p. 4 n. 11.
[24] See below, pp. 8f.
[25] Tambiah 1973, p. 227.
[26] Tambiah 1973, p. 228, his italics.
[27] Indeed from a different point of view, classical scholars, for their part, have often in the past been too ready to see the Greek experience as exceptional.

Greek thought offers a quite special opportunity[28] to examine the background circumstances, and the precise form and limitations, of the emergence of science.

THE EVIDENCE

The chief evidence for this study is, quite simply, the sum total of extant Greek literature from the period that principally concerns us – roughly from Homer to the end of the fourth century B.C. – together with such later texts as bear on our problems. To this we can add the inscriptional and papyrological data relevant especially to Greek magic, medicine and religion. This evidence suffers from two major shortcomings. First we are dealing, in the main, with literary texts. The effect of this limitation is clear: in discussing the ancient world we must forego any extravagant ambition to reconstruct the beliefs, attitudes and practices of ordinary Greeks *as such,* and confine ourselves largely to analysing those beliefs *as they are represented by particular individuals or groups.* The question of how typical or deviant particular views and reactions are will, of course, occupy us in considering the social background and conditions of those developments, but here the richness and variety of our sources can, to some extent, compensate for their mainly literary character. Outside the early philosophers and scientists themselves there is important direct and indirect testimony relevant to our concerns, in, for example, the dramatists, the orators and the historians.

Secondly, the evidence is fragmentary and uneven, particular individuals and groups being much better represented in our extant sources than others. The lack of reliable information concerning the Milesians is a notable handicap, and so too is that for the early Pythagoreans and for the atomists. While we can reconstruct certain aspects of early Greek mathematics from Euclid and later commentators, the almost complete loss of original Greek mathematical texts from the fifth and early fourth centuries B.C. severely restricts our discussion of the development of techniques of argument and of the notion of proof in that field. On the other hand we are fortunate enough to have, for example, more than fifty complete medical treatises from the period in which we are chiefly interested. The limitations, and the possible distorting effects, of our evidence must be acknowledged and borne in mind: yet we have material enough

[28] Perhaps the only comparable opportunity is that presented by the rich documentation for the developments that took place in Chinese thought. Conscious that these would require a full length study in themselves – and that by a specialist – I have paid only very limited attention to them in the course of this inquiry.

to undertake an exploratory, if at points necessarily speculative, investigation of the problems we have posed.

The natural starting-point of our inquiry is the texts in which what are represented as magical beliefs and practices are explicitly criticised and rejected. These will enable us to analyse in some detail the precise character of the criticisms advanced, and the nature, and grounds, of the ideas that were put in place of the beliefs that were rejected. Our study of this evidence will, in turn, suggest further more general topics for investigation relating to the growth – and limitations – of science and philosophy in the sixth to the fourth centuries B.C. Two related, even at points overlapping, aspects of the methodology of early Greek speculative thought appear to be of fundamental importance, namely (1) the development of techniques of argumentation, in refutation, persuasion and demonstration, and (2) that of observation and research. In chapter 2 we shall outline the development, and assess the significance, of the theory and the practice of argument in early Greek science, including for example the influence of rhetoric on the growth of natural science, and the development of the notions of an axiomatic system and of an exact science. Similarly chapter 3 will study the development of empirical research, where the question of the relationship between theory and practice, and the reasons for the variable performance of the Greeks in different domains of science, will occupy us, and where we shall extend the discussion beyond the fourth century B.C. in order to put the achievements of earlier and later periods into perspective. No claim is or can be made that the two topics thus chosen for the two principal sections of the inquiry exhaust the problem: nor is it possible to discuss more than a small proportion of the extensive material available to us under each head. Similar reservations apply with even greater force to our final study which follows up aspects of the picture of early Greek science that emerges from our investigations and confronts the issue of the social background to the developments described. While many of the questions surrounding the problem of why the Greeks produced the natural science they did are, admittedly, not ones to which definitive answers can be given, an attempt will be made to assess various hypotheses that may be advanced on the possible relations between Greek science and Greek society.

1

THE CRITICISM OF MAGIC AND THE INQUIRY CONCERNING NATURE

THE PLURALISM OF GREEK RELIGIOUS BELIEFS

The first time in extant Greek – indeed in extant Western – literature that an attempt was made explicitly to refute a set of what the writer himself called magical[1] beliefs and practices is in a work that dates from the later part of the fifth or the early fourth century B.C. But the attack on magic – including, especially, any claim to be able forcibly to manipulate the divine or the supernatural – must be understood against the background of the pluralism of Greek religious beliefs; so we must first consider briefly the development of critical attitudes towards certain aspects of Greek traditional notions concerning the gods. This begins already with Hesiod, if not with the Homeric poems themselves.[2] Although the extent of the originality of Hesiod's *Theogony* is hard to estimate, it represents at the very least a systematisation of a group of stories about the origins of the gods. Although he invokes the Muses at the start of the poem, it is the 'fine song' that they taught *him* (and he identifies himself by name)[3] that he recounts. Hesiod stands at the head of a line of writers of theological cosmogonies – the group Aristotle refers to as the θεολόγοι.[4] These include such men as Pherecydes and Epimenides – and we can now add Alcman on the evidence of the recently discovered fragment of his theogony – as well as a number of other more shadowy figures. Our sources for their ideas are often, to be sure, late and unreliable, but Alcman in the seventh,[5] and Pherecydes in the sixth, century, at least, evidently introduced a number of new theogonical myths, based partly on earlier Greek and partly, it may be, on non-Greek material.[6] Again another recent discovery, the so-called Derveni

---

[1] On the origin and application of the terms μάγοι and μαγεία, see below, p. 13 n. 20.

[2] The extent to which the Homeric poems introduced new religious conceptions has been much debated. See, for example, Guthrie 1950, pp. 117ff, and Finley (1954) 1977, pp. 135ff (and the works listed in his bibliographical essay, pp. 183f), and Burkert 1977, pp. 191ff.

[3] *Th.* 1ff, 22ff.   [4] E.g. *Metaph.* 1000a9f, 1071b26f, 1075b26 and 1091a33ff.

[5] The interpretation of Alcman's theogony (Fr. 5) is notoriously controversial. See, for example, Page 1959, pp. 20f, Fränkel (1962) 1975, pp. 164 and 253f, Burkert 1963a, West 1963, 1967 and 1971, pp. 206ff, Vernant 1970 and Penwill 1974.

[6] For an account of Pherecydes' myths, see, for example, Kirk and Raven 1957, pp. 48–

papyrus, provides valuable evidence that is independent of Plato for Orphic theogonical speculation.[7]

Moreover the first natural philosophers, the Milesians, may also be thought of as innovators in this area in two respects. First they attempted naturalistic explanations of phenomena such as earthquakes, lightning and thunder, which had often been ascribed to the gods.[8] Secondly, there is evidence that they considered their principles – that is, what the world comes from – to be divine,[9] and in that, admittedly very limited, sense they may be seen as putting forward a new or 'reformed' theology.[10] Again although the precise nature of Pythagoras' religious teaching is disputed,[11] we have good early evidence that he held that the soul is immortal and transmigrates from one species of living being to another.[12]

Whilst a number of seventh-, sixth- and early fifth-century writers may be represented as religious innovators, the two outstanding early explicit critics of certain traditional Greek religious notions are Xenophanes (*c.* 570–470 B.C.) and Heraclitus (active at the turn of the sixth and fifth centuries). Xenophanes inveighed against the conception of the gods in Homer and Hesiod first on moral grounds. 'Homer and Hesiod have ascribed to the gods everything that is shameful and a reproach among men, thieving, adultery and deceiving each other' (Fr. 11, cf. Fr. 12). But he also satirised anthropomorphism more generally. 'But men', he says in Fr. 14, 'think that gods are born and that they have clothes and voices and shapes like their own.' In the first extant text to bring to bear knowledge of what other societies believed about the gods he says: 'the Ethiopians say their own gods are snub-nosed and black, the

---

72. The degree of dependence on non-Greek ideas has recently been rather exaggerated in West's discussion (West 1971, chh. 1 and 2).

7 See especially Burkert 1968 and 1970. The papyrus itself dates from the second half of the fourth century, but the commentary on Orphic ideas it contains is thought to be a product of 400 B.C. or shortly afterwards.

8 See further below, p. 32.

9 Thus Aristotle suggests that Anaximander described the Boundless as immortal and imperishable (*Ph.* 203b13ff, DK 12A15). Our late sources report that Anaximenes held his principle, air, to be divine (e.g. Aetius I 7.13, Cicero, *N.D.* I 10.26, both in DK13A10, and cf. Aet. I 3.4, DK 13B2 and Hippolytus, *Haer.* I 7.1, DK 13A7). Even Thales, too, may have considered his principle, water, to be divine, though the precise application of the dictum that 'all things are full of gods', ascribed to him by Aristotle (*de An.* 411a8, DK 11A22; cf. Plato, *Lg.* 899b, where, however, there is no mention of the author of the saying), is controversial (see Lloyd 1966, pp. 233ff).

10 Different versions of this line of interpretation can be found in, for example, Jaeger 1947 and Hussey 1972.

11 See especially Burkert 1972a, ch. 2.

12 Xenophanes Fr. 7 is quoted by Diogenes Laertius, VIII 36, as referring to Pythagoras. Even if that were incorrect, the fragment is good early evidence of the belief in transmigration.

Thracians say theirs are blue-eyed and red-haired' (Fr. 16). Another fragment (15) attempts to reduce anthropomorphism to absurdity by drawing an analogy with animals: 'If oxen and horses and lions had hands and could draw with their hands and produce works of art like men, horses would draw the forms of the gods like horses, and oxen like oxen, and they would make their bodies such as each of them had themselves.' In place of the crude anthropomorphism he rejects, he puts forward an idea of god as the divine Mind (Frr. 24–6), a notion that is, to be sure, still influenced by a human model, even if his god is said to be 'not like mortals either in shape [form] or in thought' (Fr. 23).

With Heraclitus, the range of religious notions and practices under attack is extended.[13] Thus in one fragment (5) he condemns ritual purifications after murder and praying to statues: 'They purify themselves polluting themselves further[14] with blood, as if a man who had stepped into mud were to wash it off in mud: he would be thought mad[15] if anyone remarked him doing this. And they pray to these statues, as if someone were to converse with houses, not knowing at all who the gods and heroes are.' In another passage (Fr. 15) he refers to the Dionysiac religion in particular: 'For if it were not for Dionysus that they were holding processions and singing the hymn to the phalli,[16] it would be a most shameless act: but Hades and Dionysus, in whose honour they go mad and perform bacchic rites, are the same.' Here and elsewhere it may be that it is not the acts themselves that he objects to, so much as performing them in ignorance of their true significance, that is of the true nature of the god or gods to whom they are addressed.[17] In a third fragment (14), the extent and authenticity of which are unfortunately in doubt,[18] he is again reported as criticising the mystery religions ('what are

---

[13] Heraclitus often expresses his contempt both for the ordinary mass of mankind (e.g. Frr. 1, 2, 17, 29, 34, 104) and for most of the rest (including Xenophanes himself) who passed as 'wise men' (e.g. Frr. 35, 40, 42, 56, 57, 106).

[14] Reading ἄλλῳ with DK (and Marcovich). Alternatively, reading ἄλλως, 'they purify themselves in vain, polluting themselves with blood'.

[15] Here, as so often elsewhere in the fragments, there is a calculated play on words – μιαινόμενοι (translated 'polluting') and μαίνεσθαι (translated 'mad') – which cannot be captured in English.

[16] Again there is a play on words. The term translated 'phalli' is αἰδοίοισιν (lit. 'shameful parts'), which is immediately followed by ἀναιδέστατα ('most shamelessly').

[17] Cf. Guthrie 1962, pp. 475f, who also refers to Fr. 69.

[18] Our source is Clement of Alexandria (who is also responsible for Fr. 15). Clement is not a very reliable witness at the best of times, since his own chief purpose, in the *Protrepticus*, is to expose all heathen religions (and especially the Greek mysteries). But there is an additional reason to be cautious about the first part of what appears in DK as fragment 14: it does not form a grammatical sentence, but consists simply of a list of the types of people whom Clement represents Heraclitus as 'prophesying against'. The dangers of such a list being subject to interpolation and corruption are obvious.

deemed to be mysteries among men are unholy mysteries') and as 'prophesying against'[19] 'night-roamers, "mages" (μάγοι), bacchants, maenads, initiates'. If μάγοι here is part of the original quotation and not – as is quite possible – an addition by our source, this is the first reference in extant Greek literature to these men: our earliest extensive authority, Herodotus, represents them as a Median tribe who – or members of which – acted as priests and the interpreters of signs and dreams.[20] Like Xenophanes, Heraclitus' remarks about the gods are not merely destructive and critical, for he has his own quite different, if in parts obscure, conception of the divine to propose, one that is linked with his central philosophical doctrine of the unity of opposites. Thus we are told in Fr. 67 that 'god is day night, winter summer, war peace, satiety hunger', while another fragment (102) says that 'to god all things are beautiful and good and just, but men have thought that some things are unjust, others just'.[21]

These texts show that in the sixth and early fifth centuries it was, within broad limits, perfectly possible both to criticise existing

---

19 This is Clement's term (μαντεύεται) and Clement held (incorrectly, as is now generally thought) that Heraclitus, like the Stoics much later, believed that the world is periodically destroyed by fire (the doctrine of ἐκπύρωσις). Yet Clement's misinterpretation of Heraclitus on that point does not, by itself, undermine the value of this testimony as a whole, since it is still possible that it reflects some statement of Heraclitus criticising some at least of the types of person that Clement mentions.

20 See Herodotus I 101, 107, 120, 128, 132, 140, VII 19, 37, 43. It is clear that for Herodotus the μάγοι were a distinct tribe (the doubtful accuracy of his reports does not affect their value as evidence of what was believed about the μάγοι in Greece). But already in the fifth century μάγος and its derivatives came to be used pejoratively – often in association with such other words for vagabonds, tricksters and charlatans as ἀγύρτης, γόης and ἀλαζών – for deception, imposture and fraudulent claims for special knowledge. This is so not only in *Morb. Sacr.* (on which see below), but also in Gorgias' *Helen* (para 10, cf. para. 14), Sophocles, *OT* 387ff and Euripides, *Or.* 1496ff (cf. Aristotle Fr. 36, which however exonerates the μάγοι themselves of the practice of γοητική μαγεία). Thus these texts already exhibit what was to remain a prominent feature of words from the μαγ- root (and of their Latin equivalents, *magus, magicus* etc.). They were never clearly defined in terms of particular beliefs or practices, but were commonly used of such activities or claims to special knowledge as any particular author or speaker suspected of trickery or fraudulence. Pliny, for instance, attacks the 'magical art' at length in *Nat.* xxx especially (as often elsewhere, e.g. xxiv 1.4f, xxvi 9.18ff, xxviii 23.85f). But that does not prevent him from including in his work a mass of homeopathic and sympathetic remedies, amulets and the like, which he is half inclined himself to believe to be efficacious: he often mentions, for example, the special, ritualistic procedures to be used in their collection and preparation (see e.g. xxiii 71.137ff, xxvi 62.95ff, xxvii 43.66, xxviii 23.77ff, xxix 32.98ff). See further Hubert 1904, Thorndike 1923–58, Hopfner 1928, Bidez and Cumont 1938 and Nock 1972, I pp. 308ff, especially.

21 Cf. such other, often cryptic, fragments on god and the divine as Frr. 24, 32, 53, 62, 86, 114 and notably those that emphasise the contrast between divine knowledge and human ignorance, e.g. Frr. 78, 79, 83. Even though his statements on soul and on immortality are exceptionally obscure, it is fairly clear that he believed in some form of after-life, see, e.g., Frr. 63, 77, 88 and cf. Frr. 27, 36, 45, 98, 115.

religious ideas and practices and to introduce new ones.[22] To put it negatively, there was no dogmatic or systematic religious orthodoxy.[23] Although there were certain widespread and deeply held beliefs, there was no common sacred book,[24] no one true religion, represented by universally recognised spokesmen – priests or prophets – and backed by an organised religious authority such as a church. The expression of new and quite individualistic views on god and the divine was, as our examples show, not only possible but quite common, and by the end of the fifth century we have evidence[25] of a series of rationalistic accounts of the origin of religion.[26] First Democritus explained belief in the gods as in part a mistaken inference from terrifying natural phenomena,[27] although he evidently did not dismiss notions of the gods entirely, for he is also reported to have related certain such ideas to 'images' that appear to men.[28] Secondly, Prodicus is said to have accounted for beliefs in the gods in terms of man's gratitude for the benefits he derives from such things as bread, water, wine and fire. Thus Sextus reports that

Prodicus of Ceos says: 'The ancients considered as gods the sun and moon and rivers and springs and in general everything that aids our life because of the benefit from them, just as the Egyptians consider the Nile a god.' He adds that for this

---

[22] Modifications to religious practices and the introduction of new ones appear to continue throughout the sixth and fifth centuries – especially, though not exclusively, in connection with the growth of the mystery religions.

[23] Thus Herodotus, II 3, puts it that all men have equal knowledge – or ignorance – of the gods. We shall be discussing later the significance of trials for impiety, see below, p. 255 and n. 129, p. 257 and n. 138.

[24] Such 'sacred stories', ἱεροὶ λόγοι, as the Greek possessed were associated with particular exclusive cults, such as the mysteries: see, for example, Burkert 1972a, pp. 178ff, 219ff, 1977, pp. 414f.

[25] Admittedly much of our most striking evidence derives from a single source, Sextus Empiricus, who sets out in *M.* IX 13ff to show the doubtfulness of the inquiry concerning gods. But it is clear that by the end of the fifth century rationalising speculations about the gods were common in two contexts in particular, etymologising on the gods' names and allegorical interpretations of incidents in Homer. The Derveni papyrus reflects the former interest: for the latter, see Richardson 1975, pp. 66f, 70ff.

[26] Conjectures concerning the possible origins of particular religious beliefs and customs begin already in Herodotus. Thus at II 43ff (especially 50) he speculates on the Egyptian origin of the Greek names of the gods. See also II 81 on the prohibition concerning the use of wool (on the problems posed by the alternative readings in this text, see, for example, Burkert 1972a, pp. 127f), II 104 on circumcision and II 123 on the Egyptian origin of the belief in immortality. Cf. also, for example, Euripides, *Hec.* 799ff, where the gods themselves are said to be subject to νόμος, custom or convention.

[27] Sextus, *M.* IX 24, DK 68 A 75, mentioning thunder, lightning, the conjunctions of stars and eclipses of the sun and moon among the 'happenings in the upper regions' for which men in the past thought the gods responsible.

[28] Democritus Fr. 166 (some of these images are beneficent, others harmful: he is reported to have wished for the former kind himself), cf. also e.g. Plutarch, *Quaest. Conv.* VIII 10.2, 734Ff (DK 68 A 77). In Fr. 30 Democritus was, presumably, being ironical in referring to those who, gesturing towards the air, spoke of Zeus as 'king of all'.

reason bread was worshipped as Demeter, wine as Dionysus, water as Poseidon, fire as Hephaestus, and so on with each of the things that are good for use.[29]

Thirdly, and far more radically, a text from Critias' *Sisyphus* represents the gods as a human invention for the purposes of moral control:

> Then when the laws prevented them [men] from committing open deeds of violence, but they continued to do them in secret, it seems to me that a man of clever and cunning wit first invented for men fear of the gods, so that there might be something to frighten the wicked, even if they do or say or think something in secret. Hence he introduced the divine, saying that there is a deity [daimon] who enjoys immortal life, hearing and seeing with his mind, thinking of everything and caring about these things, and possessing a divine nature, who will hear everything said among mortals and be able to see everything that is done...The place he said the gods lived in was one by the mention of which he could most frighten men – from which he knew came fears for mortals and rewards for their miserable life – the upper circuit, where he remarked lightnings and fearful claps of thunder, and the starry frame of heaven, the beautiful workmanship of the cunning craftsman Time...With such fears he surrounded men...and quenched lawlessness by his ordinances...So I think did someone first persuade men there is a race of deities.[30]

THE CRITICISM OF MAGIC

In addition to this evidence for the rational criticism of religious beliefs and customs in the philosophers and sophists, we have first-hand information relating to the rejection and refutation of certain magical notions. For this we have to turn to the medical writers. Our chief text is the treatise *On the Sacred Disease*,[31] the date of which cannot be fixed at all precisely but which is generally thought to belong to the end of the fifth or the beginning of the fourth century B.C.[32] The principal aims of this work are (1) to establish that the 'sacred disease' – that is, epilepsy[33] – is, as the author puts it, 'no

---

[29] Sextus, *M.* ix 18 (cf. 52). On the important evidence in Philodemus, *Piet.* (*PHerc.* 1428) see most recently Henrichs 1975, pp. 107ff. Cf. also Cicero, *N.D.* i 42.118, Themistius, *Or.* xxx 349ab (Hardouin), ii 183.1ff (Schenkl, Downey, Norman) (DK 84 B 5).

[30] Critias, Fr. 25.9ff: see, for example, Burkert 1977, p. 465 and cf. ch. 7, pp. 452ff, in general on the topic of philosophical criticism of religious beliefs.

[31] I follow Grensemann's edition (1968, cited by chapter and paragraph) except where otherwise indicated. My translations are adapted from those in Chadwick and Mann (1978) who follow the chapter divisions in W. H. S. Jones, 1923–31, ii (J) rather than those in Littré (L) and Grensemann (G). Some aspects of this material are discussed in Lloyd 1975c.

[32] See, for example, W. H. S. Jones 1923–31, ii p. 134, Pohlenz 1938, p. 35, Heinimann 1945, pp. 170ff, especially 206–9, Bourgey 1953, pp. 75f, Grensemann 1968, pp. 7–31. The philosopher Diogenes of Apollonia, whose floruit is usually assigned to about 430 B.C., provides a probable *terminus post quem*, but we have no reliable means of determining how long after Diogenes the treatise was written.

[33] On the identification of this disease, see especially Temkin 1933a, and b, and 1971, pp. 15ff, for example 19: 'various diseases were called "sacred disease" in Antiquity', but 'in the great majority of cases "the sacred disease" meant epilepsy for physicians as well as laymen'.

more sacred than any other disease' and that it has a natural cause like all other diseases, and (2) to expose as frauds those who claimed to be able to cure the disease by purifications, incantations and other ritual means. The work begins:

> I do not believe that the sacred disease is any more divine or sacred than any other disease but, on the contrary, just as other diseases have a nature from which they arise, so this one has a nature (φύσις) and a definite cause (πρόφασις). Nevertheless, because it is completely different from other diseases, it has been regarded as a divine visitation by those who, being only human, view it with ignorance and astonishment.[34]

Shortly afterwards the writer makes a suggestion about why the disease came to be considered 'sacred':

> It is my opinion that those who first called this disease 'sacred' were the sort of people we now call mages (μάγοι), purifiers (καθαρταί), vagabonds (ἀγύρται) and charlatans (ἀλαζόνες). These are exactly the people who pretend to be very pious and to be particularly wise. By invoking a divine element they were able to screen their own failures to give suitable treatment and so called this a 'sacred' malady to conceal their ignorance of its nature. By picking their phrases carefully, prescribing purifications and incantations along with abstinence from baths and from many foods unsuitable for the sick, they ensured that their therapeutic measures were safe for themselves.[35]

The writer's criticisms of his opponents[36] take various forms. He accuses them not only of ignorance, but also of deceit and fraudulence, of inconsistency and indeed of impiety.[37] In opposition to the views he attacks he puts forward his own naturalistic doctrines about diseases in general and about the sacred disease in particular, during the course of which he produces some fairly detailed anatomical and physiological theories. Several of the criticisms he advances can be paralleled either from anthropologists' reports concerning attitudes towards witchdoctors and magic in non-literate societies, or from the accounts of historians of witchcraft, such as Keith Thomas' celebrated study of sixteenth- and seventeenth-century England, *Religion*

---

[34] Ch. 1 paras. 2f (G). Cf. the rather different texts of Littré, VI 352.1ff, and of W. H. S. Jones, 1923–31, II p. 138. Grensemann square-brackets the first sentence I have translated: but even if this is a gloss, the idea it expresses is genuine enough, being repeated in a slightly different form at the beginning of ch. 2 (para. 2 (G) = ch. 5 (J)) and cf. ch. 18 (para. 1 (G) = ch. 21 (J)).

[35] Ch. 1 paras. 10–12 (G) (cf. L VI 354.12ff).

[36] The identity of these opponents cannot be determined precisely, but see further below, pp. 37f.

[37] E.g. ch. 1 para. 28 (G) (cf. L VI 358.16ff): 'And yet I believe that all these professions – as they think – of piety are really more like impiety and a denial of the existence of the gods, and all their piousness and talk of the divine is impious and unholy, as I shall demonstrate.' Cf. also ch. 1 paras. 30, 39 and 44 (G) (L VI 360.3ff, 362.6ff, 16ff).

and the *Decline of Magic*.[38] It is essential, then, both to analyse the precise nature of the attack on the 'mages' in *On the Sacred Disease* – in particular to identify where that attack departs from patterns that may easily be paralleled elsewhere – and then also to assess what the Hippocratic author offers in place of the doctrines he rebuts.

A comparison may be suggested, first, in respect of certain accusations of dishonesty and fraudulence. Discussing attitudes to witch-doctorhood among the Azande, Evans-Pritchard wrote:

I was surprised to find a considerable body of sceptical opinion in many departments of Zande culture, and especially in regard to their witch-doctors. Some men are less credulous than others and more critical in their acceptance of statements made by witch-doctors...Many people say that the great majority of witch-doctors are liars whose sole concern is to acquire wealth. I found that it was quite a normal belief among Azande that many of the practitioners are charlatans who make up any reply which they think will please their questioner, and whose sole inspiration is love of gain.[39]

Similarly the author of *On the Sacred Disease* both explicitly accuses his opponents of ignorance,[40] and suspects that their motive is love of gain:

But perhaps these claims are not true and it is men in search of a living (βίου δεόμενοι) who invent all these fancy tales about this particular disease and all the others too – attaching the responsibility for each of the different forms of the complaint to a god, for they hold not just one, but several gods responsible for these.[41]

Next there are charges of special pleading, or of recourse to what we may call secondary elaborations. Analysing the factors that contributed to the reputation enjoyed by the 'cunning men' in Tudor and Stuart England, and in particular the defences available to them when they failed actually to produce a cure, Thomas wrote:

When failure was unavoidable the belief in witchcraft provided a ready excuse. By informing their clients that they had been 'overlooked' or 'forspoken', the cunning men could imply that if only the disease had been natural they would have been able to cure it. Even the Catholic who held charming sessions at St James's in

---

38 Thomas 1971.
39 Evans-Pritchard 1937, p. 183.
40 E.g. ch. 1 para. 11 (G) (L vi 354.15ff) quoted above, p. 16.
41 Ch. 1 para. 32 (G) (L vi 360.9ff). Among many other passages in Greek literature, one may compare Oedipus accusing Teiresias of prophesying for gain (*OT* 387ff: he calls Teiresias μάγον and ἀγύρτην) though he does not deny the art of prophecy as a whole. Similarly accusations of greed and fraudulence are particularly common in the many scenes in which Aristophanes satirises both named prophets and soothsayers and their kinds in general, e.g. *Pax* 1045–1126, *Av.* 958–91, *Eq.* 115ff, 1002ff, cf. Plato, *Lg.* 909ab. Homer already provides examples of attacks on particular prophets or prophecies, e.g. *Il.* 1 106ff, *Od.* 11 178ff, and in a famous speech at *Il.* xii 231ff Hector, dismissing Polydamas' interpretation of an omen, says that he does not care whether birds fly to the right or to the left: there is one best omen, to fight for the fatherland.

1664 was prepared to fall back on this. In this way the wizard's procedure could be virtually foolproof. For if the patient recovered it was a tribute to the cunning man's perception, and if he died then the witch was to blame.[42]

Although it is not witches or other magicians, but the gods, whom the charlatans attacked in *On the Sacred Disease* invoke, the way they are said to excuse themselves is strikingly similar. 'They also employ', the Hippocratic writer reports, 'other pretexts so that, if the patient be cured, their reputation for cleverness is enhanced, while, if he dies, they can excuse themselves by explaining that the gods are to blame while they themselves did nothing wrong.'[43]

Yet whilst references to anthropological and other sources shows that there are certain similarities between points that *On the Sacred Disease* makes in its attacks on the purifiers and what can readily be found elsewhere, the criticisms in our Hippocratic text do exhibit certain exceptional features. Evans-Pritchard, for instance, emphasised that although many Azande suspect individual witchdoctors of being frauds, there is no scepticism about witchdoctorhood in general: 'I particularly do not wish to give the impression that there is any one who disbelieves in witch-doctorhood. Most of my acquaintances believed that there are a few entirely reliable practitioners, but that the majority are quacks.'[44] He observed that 'faith and scepticism are alike traditional. Scepticism explains failures of witch-doctors, and being directed towards particular witch-doctors even tends to support faith in others.'[45] Similarly, although there is this major difference in the material that Thomas dealt with, that general scepticism about witchcraft *was* occasionally expressed in sixteenth- and seventeenth-century England,[46] Thomas too drew

---

[42] Thomas 1971, p. 247, and cf. p. 401 on astrology.

[43] Ch. 1 para. 20 (G) (L VI 356.9ff).

[44] Evans-Pritchard 1937, p. 185.

[45] Evans-Pritchard 1937, p. 193.

[46] For example by Reginald Scot, in his *Discoverie of Witchcraft* (1584) 1964, on which see Thomas 1971, pp. 684f especially. Although Scot has a four-fold classification of witches, and admitted they existed in the sense that he admitted the reality of 'impostors, poisoners, scolds and deluded persons', the key point is that he denied that any of them had any supernatural power. Although Scot had some followers, Thomas went on to note (p. 685) that 'most members of the educated classes remained slow to accept the full implications of his thesis...Scot's position remained that of a self-conscious minority.' One may also compare J. Needham's account (1954–, II Section 14, pp. 346ff) of the sceptical tradition in Chinese thought. There are some admittedly rather limited signs of critical and rationalistic attitudes towards divination in two third-century B.C. writers, Hsün Chhing (see Dubs 1927, pp. 68ff, and 1928, pp. 179ff) and Han Fei (see Liao 1939, e.g. pp. 156ff, and 1959, e.g. p. 308), and a more general attack in Huan T'an (43 B.C.–A.D. 28). In Fragment 210 of Huan T'an (Pokora 1975, p. 239) we read: 'Today all the artful and foxy, magicians of small talent, as well as the soothsayers, disseminate and reproduce diagrams and documents, falsely praising the records of prognostication. By deception and misinformation, by greed and dishonesty,

attention to the way in which failures in the predictions of individual astrologers did nothing to undermine, and even confirmed, belief in astrology as a whole.

Everyone knew that some practitioners were better than others and that the profession was infested by charlatans and quacks...The paradox was that the mistakes of any one astrologer only served to buttress the status of the system as a whole, since the client's reaction was to turn to another practitioner to get better advice, while the astrologer himself went back over his calculations to see where he had slipped up.[47]

What is important in the attack expressed by the author of *On the Sacred Disease* is that it is directed against *all* the purifiers, and against *any* idea that the sacred disease or any other disease is the result of divine intervention, indeed against any idea that ritual purifications can influence natural phenomena in any way. He writes:

If these people claim to know how to draw down the moon, cause an eclipse of the sun, make storms and fine weather, rain and drought, to make the sea too rough for sailing or the land infertile, and all the rest of their nonsense, then, whether they claim to be able to do it by rites or by some other knowledge or practice, they seem to be impious rogues.[48]

The Hippocratic author here and elsewhere clearly has in view not just this or that practitioner, but such practitioners *as a whole*, not just this or that instance of the belief in divine intervention causing diseases or in the ability to influence natural phenomena by ritual practices, but, again, such beliefs *in general*.

The author of *On the Sacred Disease* is evidently confident enough to attack his opponents' underlying assumptions as such, and this

they lead the ruler astray. How can we fail to suppress and banish such things?' (cf. also Frr. 40, 58, 68, 157, Pokora 1975, pp. 31, 50f, 65, 156f). The position of Wang Chhung (A.D. 27–97) is particularly interesting: as both J. Needham 1954–, II pp. 368ff, and Forke 1907, pp. 16ff, point out, many of his criticisms of teleology, of superstitions and of imaginary causal connections between things are strikingly similar to those that can be cited from Greco-Roman sources, especially Lucretius. While Wang Chhung did not reject omens and portents completely, he attacked not just particular groups of diviners, but also the general assumptions on which common methods of divination were based, as for example those using milfoil and tortoise shells in ch. 71 of his work *Lun Hêng* (Forke 1907, ch. 14, pp. 182–90). 'As a matter of fact, diviners do not ask Heaven and Earth, nor have weeds or tortoise shells spiritual qualities.' Nevertheless 'when a lucky man cuts up a tortoise, he finds auspicious omens, whereas an unlucky one, grasping the milfoil, obtains contrary signs' – even though this is not Heaven replying to the diviner, but a matter of chance (cf. also Forke 1907, pp. 173ff). Similarly he rejects the idea that dead men become ghosts, but not that there are ghosts or phantoms – which he explains as being formed by the Yang fluid. 'Thus we hold that the dead do not become ghosts, are not conscious and cannot hurt people. Consequently, it is evident that the ghosts, which are seen, are not the vital force of dead men, and that, when men have been hurt, it cannot have been done through this vital force' (Forke 1907, p. 201, cf. pp. 239ff).

[47] Thomas 1971, p. 401. The argument that lack of skill was to blame for failures was common in antiquity, e.g. Cicero, *Div.* I 52.118.

[48] Ch. 1 paras. 29f (G) (cf. L VI 358.19ff) and cf. ch. 1 para. 31 (G) (L VI 360.6ff).

immediately raises the question of what his own explanation of the sacred disease was. His account, which brings epilepsy under a general theory of diseases, is explicit, detailed and, in parts, surprising. 'The brain is responsible for this disease', he says, 'as it is for the other very severe diseases. I shall explain clearly the manner in which it comes about and the reason (πρόφασις) for it.'⁴⁹ There are 'veins'⁵⁰ leading up to the brain from all over the body, and he proceeds to give a quite complex account of these to which I shall return. These 'veins', he believes, normally carry air, air being responsible, in his view, for, among other things, sensation and consciousness.⁵¹ But if the air in the 'veins' 'remains still and is left behind in some part of the body, then that part becomes powerless'.⁵² He goes on to describe a variety of other conditions that may arise when the air is obstructed by discharges, especially by phlegm, and then applies this general theory to epilepsy which he describes as follows: 'Should these routes for the passage of phlegm from the brain be blocked, the discharge enters the veins which I have described. This causes loss of voice, choking, foaming at the mouth, clenching of the teeth and convulsive movements of the hands; the eyes roll, the patient becomes unconscious and, in some cases, passes a stool.'⁵³ He then promises, and gives, an explanation of each of these symptoms in turn. Thus 'loss of voice', he says,

occurs when the phlegm suddenly descends in the veins and blocks them so that air can pass neither to the brain nor to the hollow veins nor to the body cavities, and thereby inhibits respiration...Therefore, when the veins are shut off from this supply of air by the accumulation of phlegm and thus cannot afford it passage, the patient loses his voice and his wits.⁵⁴

This account is supported by remarks concerning the observed or supposed differences in the incidence of the sacred disease among different sections of the population. He suggests that the disease attacks the phlegmatic, but not the bilious.⁵⁵ He notes that older people are not killed by an attack of the disease,⁵⁶ but that the young

---

⁴⁹ Ch. 3 para. 1 (G) (L vi 366.5ff).
⁵⁰ I use the conventional translation for φλέβες, though it should be understood that the vessels in question are imagined as carrying air and phlegm, for example, as well as blood.
⁵¹ The chief proponent of the view that air is responsible for intelligence was Diogenes of Apollonia (Frr. 4 and 5): cf. also Anaximenes Fr. 2.
⁵² Ch. 4 para. 2 (G) (L vi 368.5f).
⁵³ Ch. 7 para. 1 (G) (L vi 372.4ff). This account was considered accurate enough to be paraphrased by Osler 1947, p. 1364, in his own description of Grand Mal, or major epilepsy.
⁵⁴ Ch. 7 paras. 3 and 7 (G) (L vi 372.10ff, 22f).
⁵⁵ E.g. ch. 5 para. 1 (G) (L vi 368.10f).
⁵⁶ E.g. ch. 9 para. 1 (G) (L vi 376.17f).

are particularly prone to it.[57] He maintains that 'the discharge of phlegm takes place more often on the right side of the body than on the left because the veins on that side are more numerous and of greater calibre than on the left',[58] and he states that 'attacks are most likely to occur when the wind is southerly; less when it is northerly, less still when it is in any other quarter'[59] arguing that the winds have a direct effect on the body, especially the brain.

Finally at the end of the work he puts forward a general aetiology of diseases:

This so-called 'sacred' disease is due to the same causes (προφάσιες) as all other diseases, to the things we see come and go [i.e. to and from the body], the cold and the sun too, the changing and inconstant winds...Each [disease] has its own nature (φύσις) and power (δύναμις) and there is nothing in any disease which is unintelligible or which is insusceptible to treatment. The majority of maladies may be cured by the same things as caused them...A man with the knowledge of how to produce by means of a regimen dryness and moisture, cold and heat in the human body, could cure this disease too provided that he could distinguish the right moment for the application of the remedies. He would not need to resort to purifications (καθαρμοί) and magic (μαγίη)[60] and all that kind of charlatanism.[61]

As these quotations indicate, the writer exhibits an extraordinary self-assurance in the theories and explanations he advances not only about the causes and cures of epilepsy and other diseases, but also about the internal structures and functioning of the body. Yet many of those theories and explanations are quite fanciful. His account of respiration is that 'when a man draws in breath through the mouth and nose, the air passes first to the brain and then the greater part goes to the stomach, but some flows into the lungs and some to the veins. From these places it is dispensed throughout the rest of the body by means of the veins.'[62]

His descriptions of the 'veins' themselves too is very largely imaginary. Like many other early Greek anatomists,[63] he speaks of two particularly important vessels, one connected with the liver and the other with the spleen, and some of what he says may be thought to reflect some knowledge of the main trunks of the inferior vena cava and the abdominal aorta. Describing the vein connected with the

---

[57] E.g. ch. 8 paras. 1f and ch. 10 para. 2 (G) (L vi 374.21ff and 378.12ff).
[58] Ch. 10 para. 1 (G) (L vi 378.10f).
[59] Ch. 13 para. 1 (G) (L vi 384.4ff).
[60] Littré reads μαγευμάτων, Jones μαγείης, for μαγίης (Grensemann).
[61] Ch. 18 paras. 1ff (G) (L vi 394.9–396.9).
[62] Ch. 7 para. 4 (G) (L vi 372.14ff).
[63] The notion of two vessels, one connecting the liver with the right arm, the other the spleen with the left, occurs in Diogenes of Apollonia Fr. 6 (Aristotle, *HA* 512a4ff, 9ff, 29ff) and Polybus (Aristotle, *HA* 512b32ff = *Nat. Hom.* ch. 11, L vi 60.1ff) and reappears in a modified form in Aristotle himself (*HA* 514a32ff, b3ff).

liver, he says:[64] 'one half runs down on the right side in relation with the kidney and the lumbar muscles, to reach the inside of the thigh and then continues to the foot. It is called the "hollow vein".'[65] But then he goes on:

The other half courses upwards through the right side of the diaphragm and the right lung; branches split off to the heart and to the right arm while the remainder passes up behind the clavicle on the right side of the neck and there lies subcutaneously so as to be visible. It disappears close to the ear and then divides; the thickest and largest and most capacious part finishes in the brain while smaller branches go separately to the right ear, the right eye and to the nostril.[66]

Although the account of the lower part of the liver-vein may be thought to correspond, very roughly, to the inferior vena cava, this identification breaks down when we find the liver- and spleen-veins correlated with the right and left sides of the body respectively.[67] His picture of the vascular system – like that of many other Greek writers[68] – is strongly coloured by his expectations of general bilateral symmetry and by a firm conviction in the superiority of the right-hand side.[69] Thus on the spleen-vein he simply notes: 'It is similar to that coming from the liver, but is thinner and weaker.'[70]

The boldness of his general pathology and therapeutics is equally striking. The idea that certain diseases are cured by what causes them, or by their opposites, is a common one in Greek medical writings. Here we find the principle generalised: 'The majority of maladies may be cured by the same things as caused them.'[71] It is particularly remarkable that he should claim that there is no disease – not even epilepsy – that is not susceptible to treatment, and indeed by fairly simple means, to judge from his reference to the control of dryness, moisture, cold and heat by diet.[72]

Although the description the writer gives of an epileptic attack is accurate enough as far as it goes, and so too are some of his remarks concerning the incidence of the disease,[73] most of the pathological,

---

[64] Ch. 3 para. 4 (G) (L vi 366.12ff).
[65] κοίλη φλέψ, the regular term, in Greek anatomists, for the vena cava.
[66] Ch. 3 paras. 5–7 (G) (L vi 366.15ff).
[67] Note especially the reference to the right ear and the right eye, as well as the right arm, in the account of the connections of the liver-vein.
[68] See further below, pp. 157f.
[69] Cf. Lloyd 1966, pp. 48ff, and 1973.
[70] See ch. 3 para. 8 (G) (L vi 366.23ff).
[71] Ch. 18 para. 3 (G) (L vi 394.15f), quoted above, p. 21.
[72] Ch. 18 paras. 2 and 6 (G) (L vi 394.14f, 396.5ff), quoted above, p. 21. He notes, however, that epilepsy may not be curable if it is firmly established (ch. 2 para. 3 (G) (L vi 364.12ff)).
[73] For example that the young are more prone to the disease than older people (see above, pp. 20f). Cf. Osler 1947, p. 1363: 'In a large proportion the disease begins shortly

anatomical and physiological theories are highly speculative and schematic, and this prompts one to ask how far he attempted to support his ideas by observation and research. Among the – fairly rare – occasions on which we find attempts made to collect and use empirical evidence, two are worth considering especially. First, when he speaks about the role of the winds in the disease, he suggests that the effects of the south wind in particular on the fluids in the body can be inferred from the changes it brings about on things outside the body. 'Jars in the house or in the cellars which contain wine or any other liquid are influenced by the south wind and change their appearance.'[74] Although it is not clear precisely what change the writer had remarked or had in mind,[75] he was evidently *attempting* to point to observable data outside the body in order to establish or support conclusions about what happens inside it.[76]

The second passage is more striking. In this the writer sets out to justify his suggestion that the sacred disease is due to the brain being flooded with phlegm especially when the wind is southerly. It is particularly hard to cure then since 'the brain has become more moist than normal and is flooded with phlegm. This renders discharges more frequent. The phlegm can no longer be completely separated out; neither can the brain, which remains wet and soaked, be dried up.'[77] But then the writer goes on:

This observation results especially from a study of animals, particularly of goats which are liable to this disease. Indeed, they are peculiarly susceptible to it. *If you cut open the head* to look at it, you will find that the brain is wet, full of fluid and foul-smelling, convincing proof that disease and not the deity is harming the body.[78]

It is clear from this passage that the idea of carrying out a post-mortem examination on an animal had occurred to this writer, and this is quite exceptional not only for the period at which the treatise was composed, but for any period in antiquity, since post-mortem investigation to establish the cause of death or to throw light on the aetiology of diseases never became a *regular* procedure in the ancient

before puberty. It is well always to be suspicious of "epilepsy" beginning in adult life, for in a majority of such cases the disease is not epilepsy.'
[74] Ch. 13 para. 8 (G) (L vi 384.22ff).
[75] The writer seems to have in mind not so much a change in the shape of the jars (as some translations imply) as in their appearance or – more plausibly – in that of the liquids they contain.
[76] Cf. Anaxagoras' dictum ὄψις τῶν ἀδήλων τὰ φαινόμενα, 'things that are apparent are the vision of things that are unclear' (Fr. 21a), on which see below, p. 134.
[77] Ch. 11 para. 2 (G) (L vi 382.2ff).
[78] Ch. 11 paras. 3–5 (G) (L vi 382.6ff).

world.[79] It is, to be sure, not certain that the writer of *On the Sacred Disease* actually carried out the inspection he suggests: if he did not, that would not be the first nor the last time that a test that could be conducted in practice was treated by an ancient writer as a hypothetical exercise – a thought experiment. But if we assume, as perhaps we may, that he did do the test he describes, the result is as interesting for what is omitted as for what is included. The statement that 'the brain is wet, full of fluid and foul-smelling' does indeed help to achieve what the writer wanted, namely to establish that the 'sacred disease' is the result of natural causes: 'disease, and not the deity, is harming the body'. At the same time we may remark that it apparently did *not* occur to the writer to check the description of the veins leading to the brain which he had set out in explaining the origin of the disease.[80] Yet much of what he presents by way of what we should call anatomical theories could have been verified by observation. Although the possibility of direct inspection, using dissection, is mentioned in this one context, at least, *in fact* the writer evidently tested very few, if any, of his general anatomical doctrines by this method.

These texts certainly show that this writer occasionally thought to support his theories by appealing not just to what could easily be observed, but to the results of deliberate research. But they also illustrate just how limited the research in question was. Many of his doctrines are not so supported at all. Furthermore many could have been disproved, or at least seriously undermined, by the use of quite simple techniques of investigation, including techniques (such as post-mortem dissection) that the writer himself refers to.

But while his attempts to provide empirical backing for his own ideas are often feeble and abortive, the deploying of critical and destructive arguments to defeat his opponents is clearly one of his strengths. As we have remarked, he uses a wide variety of arguments

---

[79] Herodotus (IV 58) says that the fact that the grass in Scythia is very 'bilious' may be judged by opening the bodies of the cattle (though he does not describe this further). Otherwise our evidence is late. The nearest ancient parallel to the text in *Morb. Sacr.* is, perhaps, the story in Plutarch (which may well be apocryphal) that Anaxagoras had the head of a one-horned ram opened in order to demonstrate that its deformity was due to natural causes (*Pericles* ch. 6). As regards post-mortem dissection of men, this seems to be implied by Pliny (*Nat.* XIX 26.86) when, in mentioning that radish juice is a specific for certain diseases of the internal organs, he says that the kings in Egypt had the bodies of the dead dissected (he does not specify whether men or animals, but the former seems more likely in the context). Yet if carried out at all, such a procedure was clearly not a regular one. On the early history of dissection, see further below, pp. 156ff.

[80] See above, pp. 21f, on ch. 3 paras. 3–8 (G) (L VI 366.10ff).

against the 'mages' and 'purifiers', and some of these are particularly interesting when considered as techniques of refutation. At one point, for instance, he mentions that the purifiers prohibit the eating of goat meat, the wearing of goat skins and the use of goat skin blankets. 'I suppose', he says, 'that none of the inhabitants of the interior of Libya can possibly be healthy seeing that they use goat skins and eat goat meat. In fact, they possess neither blanket, garment nor shoe that is not made of goat skin, because goats are the only animals they keep.'[81] If we supply what the writer merely leaves implicit here, we have an argument of the form that later came to be known as Modus Tollens[82] ('If *A*, then *B*; but not *B*; therefore not *A*'). If goat skins are responsible, then the Libyans would be expected to suffer especially from the disease; but that is not the case; so goat skins cannot be held to be responsible.

A second instance of a similar type of argument occurs when he adopts as one of his premisses the supposed distinction in the incidence of the disease among the phlegmatic and the bilious. 'Another important proof that this disease is no more divine than any other lies in the fact that the phlegmatic are constitutionally liable to it while the bilious escape. Yet if its origin were divine, all types would be affected alike without this particular distinction.'[83] Again the implied argument is a Modus Tollens: if the disease is divine, it should attack all equally; but it does not do so; so the disease is no more divine than any other.

Although Modus Tollens as such is not stated in general terms until Aristotle,[84] and not formally analysed until the Stoics in the early Hellenistic period,[85] we find plenty of examples of the use of arguments of that general type in the philosophers and medical writers – and indeed in other authors – before Plato. Here then is one powerful technique of refutation, the development of which we shall consider in detail later.[86] We may observe here, however, that in both the examples we have taken from *On the Sacred Disease* the writer presupposes what is at issue between him and his opponents,

---

[81] Ch. 1 para. 22 (G) (L vi 356.15ff). The writer continues (para. 23) with a second argument based on his opponents' premisses: see below, p. 55.

[82] Now more often called Denying the Consequent.

[83] Ch. 2 paras. 6–7 (G) (L vi 364.20ff).

[84] Thus in the context of showing that it is not possible to draw false conclusions validly from true premisses, Aristotle states that 'If, when *A* is, it is necessary that *B* is, then, when *B* is not, it is necessary for *A* not to be' (*APr.* 53 b 11ff).

[85] The schema of the second of the Stoics' elementary arguments is: 'If the first, then the second; but not the second; and so not the first.' See, for example, Sextus, *M.* viii 227, cf. 225, and for discussion, see Mates 1961, pp. 70ff, Frede 1974, pp. 127ff, 148ff.

[86] See below, ch. 2.

namely the doctrine of the uniformity of nature, the regularity of natural causes and effects. If a factor is to be held to be a cause or contributory agent in bringing about a disease, then the action of that factor must be supposed to be uniform. If wearing goat skins is relevant, then this must be so whenever and wherever that is done. Indeed the gods too (whom his opponents invoke) are assumed by the Hippocratic author to be uniform in their behaviour: he takes it for granted that they would not discriminate between the phlegmatic and the bilious.

The two interrelated concepts of nature, φύσις,[87] and cause, to express which he uses such terms as αἰτίη, αἴτιος and πρόφασις,[88] provide the key to the writer's own position. 'Nature', for him, implies a regularity of cause and effect. Diseases, like everything else that is natural, have determinate causes and this rules out the idea of their being subject to divine ('supernatural') intervention or influence of any sort. Interestingly enough, however, the writer of *On the Sacred Disease* does not exclude the use of the notion of the 'divine' altogether. Indeed his view is not that no disease is divine, but that all are: all are divine and all natural.[89] For him, the whole of nature is divine,[90] but that idea does not imply or allow any exceptions to the rule that natural effects are the result of natural causes.

This suggests that what we are dealing with has some of the features of a paradigm switch: the author and his opponents disagree fundamentally on what sort of account to give of the 'sacred disease', that is on what would count as an 'explanation' or 'cause' of this and other phenomena. Unlike the Zande sceptics described by Evans-Pritchard, the Hippocratic writer rejects the notion of supernatural

---

[87] Ch. 1 para. 2, ch. 2 paras. 1, 2, 6, ch. 11 para. 2, ch. 13 paras. 9, 10, ch. 14 paras. 5, 6, ch. 17 para. 4, ch. 18 para. 2 (G) (L vi 352.2f, 364.10f, 366.1, 382.3, 386.4, 388.4–7, 392.11f, 394.14). Cf. Holwerda 1955.

[88] αἰτίη, αἴτιος ch. 1 paras. 20, 21, 23, 25, 32, 33, 34, 37, 43, ch. 3 para. 1, ch. 17 paras. 5, 6, 8 (G) (L vi 356.13, 15, 358.3, 10, 360.12, 15, 16, 362.3, 16, 366.5, 392.13, 17, 394.2). πρόφασις ch. 1 paras. 2, 7, 20, ch. 2 para. 2, ch. 3 para. 1, ch. 10 paras. 4, 7, ch. 15 para. 2, ch. 18 para. 1 (G) (L vi 352.4, 354.5, 356.10, 13, 364.10f, 366.7, 378.18, 380.8, 388.16f, 394.9f). See especially the studies of Deichgräber 1933c, Weidauer 1954, pp. 8ff, 32ff, Nörenberg 1968, pp. 49ff, 61ff, Rawlings 1975, pp. 36–55, and cf. further below, p. 54 n. 231.

[89] As he puts it in the final chapter, for example: 'This so-called "sacred" disease is due to the same causes as all other diseases, to the things we see come and go, the cold and the sun too, the changing and inconstant winds. These things are divine so that there is no need to regard this disease as more divine than any other; all are alike divine and all human. Each has its own nature and power and there is nothing in any disease which is unintelligible or which is insusceptible to treatment' (ch. 18 paras. 1–2 (G) (L vi 394.9ff). Cf. H. W. Miller 1953, Kudlien 1967, p. 58, Nörenberg 1968, pp. 68ff, Ducatillon 1977, pp. 159ff.

[90] One may compare the evidence, noted above, p. 11 n. 9, that some philosophers too held that that from which the world originates is divine.

intervention in natural phenomena *as a whole*, as what might even be called a category mistake. Even when we have to deal with the divine, the divine is in no sense *super*natural. We have, however, seen that, although appeals to observation and research are made, the empirical support for his own theories and explanations is often weak, and indeed many of his ideas could have been undermined by quite simple tests. Again, although he deploys a range of techniques of refutation to good effect, the key notion of the uniformity of nature is an assumption, not a proposition for which he explicitly argues.

On the Sacred Disease provides a full and in general clear statement of a controversy concerning the origin and treatment of the sacred disease as seen from the Hippocratic writer's side. But we must now place this work in the wider context of debate in which it was composed. First there are other texts that afford further illustrations of the criticism of the belief in the supernatural intervention in diseases. At the same time that belief continued to be maintained in different forms by a variety of writers in the fifth and fourth, not to mention subsequent, centuries. The development of the notions of nature and of cause, and the survival of certain traditional beliefs, present, as we shall see, a complex set of interrelated issues. Our task now is to set out the chief evidence from both philosophy and medicine that will help to define the interaction of criticism and popular assumptions.

The closest parallel to what we find in *On the Sacred Disease* comes in the treatise *On Airs Waters Places*, another work of the late fifth or early fourth century,[91] which expresses such similar views to those in *On the Sacred Disease* on certain topics that it has sometimes been thought to have been by the same author.[92] In

---

[91] No precise date can be assigned to *Aër.* (which may, in any case, not be a unity, see below, n. 92) any more than to *Morb. Sacr.* There are possible echoes of views of Diogenes of Apollonia in the account of evaporation in ch. 8 (cf. DK 64 A 17), and it has been thought that ch. 22 echoes Euripides, *Hippolytus* 7f (which would give a date for that chapter after 428) although the sentiment expressed – that the gods are pleased by the honours they receive from men – is a commonplace. There are many similarities between *Aër.* and *Morb. Sacr.*, although there is no agreement as to which treatise was written first (for *Aër.* being the earlier, see, for example, Heinimann 1945, p. 209: for *Morb. Sacr.* being the earlier, see, for example, H. Diller 1934, p. 100, Pohlenz 1938, p. 35). It seems reasonable to suppose, however, that both were composed within about 20 years of the turn of the fifth and the fourth centuries.

[92] See, for example, Wilamowitz 1901, pp. 16ff, H. Diller 1934, pp. 94ff (for identity of authorship of *Morb. Sacr.* and *Aër.* chh. 1–11), and cf. Grensemann 1968, pp. 7–18. But contrast W. H. S. Jones 1923–31, II pp. 131f, Edelstein 1931, p. 181 n. 1, Heinimann 1945, pp. 181ff. Yet whether *Aër.* as a whole, as we have it, was composed by the same man is itself not certain. That the treatise falls into two main halves (chh. 1–11 and chh. 12–24) has been generally recognised at least since Fredrich 1899, p. 32 n. 2. Although Deichgräber 1933a, pp. 112ff, Pohlenz 1938, pp. 3ff, 31ff, and Heinimann

ch. 22[93] the writer discusses the impotence that affects certain Scythians, the so-called Anarieis. 'The Scythians themselves', he says, 'attribute this to a divine visitation and hold such men in awe and reverence, because they fear for themselves.' His own view on the general issue is identical with that put forward in *On the Sacred Disease*: he believes that all diseases are divine, but equally all are natural. As he puts it: 'Each disease has a natural cause (φύσις) and nothing happens without a natural cause.' He goes on to offer his own view of the cause of the Anarieis' condition. Horse-riding, he suggests, leads to varicose veins, which the Scythians then treat by cutting the vein that runs behind each ear. It is this treatment, he claims, that causes impotence: 'My own opinion is that such treatment destroys the semen owing to the existence of veins behind the ears which, if cut, cause impotence and it seems to me that these are the veins they divide.' As with *On the Sacred Disease*, we may remark the quite speculative nature of the anatomical theory implied (the idea of a vein linking the ears and the seminal vessels). And as in that treatise, so too this writer refutes the idea of divine intervention by an implied Modus Tollens argument. He states that the rich Scythians suffer more from the condition than the poor – since the poor ride less than the rich – and he proceeds: 'Yet, surely, if this disease is more to be considered a divine visitation than any other, it ought to affect not the most noble and richest of the Scythians only, but everyone equally.'[94]

A third Hippocratic treatise that adopts a similarly naturalistic attitude towards particularly frightening conditions is *On the Diseases of Young Girls*.[95] This provides a brief account of the sacred disease, of apoplexies and of 'terrors' in which patients believe they see evil δαίμονες. Young women who do not marry when of the age to do so are, the writer says, particularly liable to such complaints, which he explains as due to a retention of blood. He remarks that when they

---

1945, pp. 170ff, have argued that the two main parts are by the same man, that view has been contested: see, for instance, Edelstein 1931, pp. 57ff, and H. Diller 1934, pp. 89ff (but cf. H. Diller 1942, pp. 65ff).

93 *CMG* i, 1 74.10–75.25. My translations are again based on those of Chadwick and Mann 1978.

94 *CMG* i, 1 75.5ff. The writer goes on, however, to consider the possibility that the gods may *not* behave uniformly in respect of the rich and the poor. If there is any truth in the belief that the gods take pleasure in sacrifices, one would expect the poor to be more liable to this condition, not less (as the writer claims is in fact the case because the poor do not ride). 'Surely it is the poor rather than the rich who should be punished.' But he then proceeds: 'Really, of course, this disease is no more of "divine" origin than any other. All diseases have a natural origin and this peculiar malady of the Scythians is no exception' (*CMG* i, 1 75.13–17).

95 L viii 466–470.

recover, women are often deceived by diviners (μάντιες) into dedicating costly garments to Artemis, although their recovery is to be attributed – he claims – merely to the evacuation of blood, and his own recommendation for treatment in such cases is that the girls should marry as soon as possible.[96]

THE PERSISTENCE OF TRADITIONAL BELIEFS: HERODOTUS

Yet whilst in certain medical circles, at least,[97] the belief in the possibility of supernatural intervention in diseases and in the efficacy of spells and purifications was vigorously attacked, such beliefs not only persisted widely among ordinary people in the fifth and fourth centuries,[98] but can be found in leading writers some of whom are generally claimed as representatives, if not of the 'enlightenment', at least of the more advanced thought of their period. The evidence in Herodotus is particularly suggestive. On the one hand his work includes not only much natural history (topography, descriptions of flora and fauna), but also attempted explanations of such problematic phenomena as the flooding of the Nile (II 20ff), explanations that are directly comparable with those attributed to the Presocratic philo-

---

[96] L VIII 468.17ff.

[97] But not in all: cf. below, pp. 40ff.

[98] Such beliefs can be attested from Homer and Hesiod (e.g. *Il.* 1 43–52, *Od.* v 395f, IX 411, XIX 455ff, Hesiod, *Op.* 240–5, cf. 102ff) to late antiquity (as we can see from, for example, Plutarch, *De Superstitione* 168 bc, Galen, *CMG* v, 9, 2 205.28ff = K XVIII B 17.9ff, Plotinus, *Enneads* II 9.14, Porphyry, *De Abstinentia* II 40, as well as from a mass of magical papyri). In the period that particularly concerns us, the fifth and fourth centuries B.C., such texts as Pindar, *P.* III 51ff, Aeschylus, *A.* 1019ff, *Eu.* 649f, Sophocles, *Aj.* 581f, *Tr.* 1235f, Aristotle, *HA* 605a4ff, are evidence of popular beliefs in supernatural interventions in diseases and in the power of spells, whilst pseudo-Demosthenes, *Against Aristogeiton* XXV 79–80 (with Plutarch, *Demosthenes* ch. 14) implies that the practice of magic could be the subject of legal action. Plato took those who claimed to have special magical powers and to be able to control the gods by sacrifices and spells sufficiently seriously to issue a warning against their evil influences in the *Republic* 364b ff and to legislate against them in the *Laws* 909a–d, 933a ff (the latter passage notes how difficult it is to get to the truth of the matter in such cases). At *Phdr.* 244d–245a Socrates, referring to the second kind of 'divine madness', speaks of maladies that afflict certain families because of ancient sins, and says that relief may be procured from these by means of worship involving rites and purifications (cf. also *Chrm.* 155e ff, *Smp.* 202e–203a, *R.* 426b, *Tht.* 149cd and *Plt.* 280e among other Platonic texts). To this literary evidence may be added the mainly epigraphical data concerning the continued belief in god- or hero-healers, Apollo, Paean, Hygieia, and a variety of local heroes (see, for example, Kutsch 1913), whilst the cult of Asclepius himself grew in importance and spread during the latter part of the fifth, and in the fourth, century (see, for example, Herzog 1931, Edelstein and Edelstein 1945 and cf. further below, pp. 40f). The whole topic of such popular beliefs has been extensively discussed and documented: see especially Heim 1893, Tambornino 1909, Weinreich 1909, Wächter 1910, Deubner 1910, Stemplinger 1922 and 1925, Halliday 1936, Edelstein (1937) 1967, pp. 205ff, Dodds 1951, Moulinier 1952, Lanata 1967, Kudlien 1968.

sophers.⁹⁹ In his descriptions of the habits of the crocodile (II 68) and of the form of the hippopotamus (II 71) Herodotus employs the term φύσις – 'nature', 'character' or 'growth' – much as it is used in connection with the philosophers' 'inquiry into nature' (περὶ φύσεως ἱστορία) or in the Hippocratic Corpus.¹⁰⁰ Moreover in reporting beliefs and stories that invoke the marvellous or the supernatural he often records his own doubts or frank disbelief.¹⁰¹

On the other hand there are other passages where he voices no such doubts,¹⁰² and on several occasions he himself endorses the idea that misfortunes of many kinds, including diseases, may be the result of divine displeasure. Thus in discussing Cleomenes' madness and suicide he first recounts three views all of which associated Cleomenes' fate with some offence against the gods (VI 75). Most Greeks said that his misfortunes occurred because he suborned the Pythian priestess to give judgement that Demaratus was not the son of Ariston; the Athenians, however, said it was because he invaded the precinct of the gods at Eleusis, whilst the Argives held that it was because he desecrated the temple of Argus. He later notes (VI 84) that the Spartans said that 'heaven had no hand in Cleomenes' madness' – ἐκ δαιμονίου μὲν οὐδενὸς μανῆναι Κλεομένεα – which came about rather because he had consorted with Scythians and become a drinker of neat wine – but Herodotus concludes his account by endorsing what he had represented as the general view, namely that Cleomenes paid the penalty for what he had done to Demaratus.¹⁰³ Again after describing the death of Pheretime following a disease in

---

⁹⁹ See Aetius IV 1.1ff and the other testimonies collected at DK 11A1 (37) (Thales), 35A1 (Thrasyalkes), 41A11 (Oenopides), 59A91 (Anaxagoras) and 64A18 (Diogenes of Apollonia).

¹⁰⁰ See Holwerda 1955, pp. 18 and 64, and cf. Heidel 1909–10, Deichgräber 1939, Heinimann 1945.

¹⁰¹ Thus he reserves judgement, for example, about the story of Salmoxis (IV 94–6), about whether the Athenians were right to claim that it was in response to their prayers that the North Wind struck the Persian fleet (VII 189), and about whether the Magi were responsible for the wind's abating (VII 191); he rejects, for instance, Egyptian fables about the phoenix (II 73), stories about men with goat's feet and men who sleep six months of the year (IV 25) and Scythian tales about were-wolves (IV 105).

¹⁰² Thus at I 167 he records that men and animals from Agylla became crippled and palsied when they passed the place where the Agyllaeans had stoned certain Phocaeans to death; at VI 98 he says that an earthquake on Delos was sent by god as a portent of the evils to come and at VII 129 he endorses, but rationalises, the Thessalian story that the vale of Tempe was caused by Poseidon, a reasonable belief because Poseidon is the earthshaker and it was an earthquake that caused the rift in the mountains. Cf. also I 19ff, 138, 174, II 111, VI 27, VII 133 and IX 100.

¹⁰³ Cf. also III 33, where he says that Cambyses became mad either because of the Egyptian god Apis (whose sacred calf Cambyses had killed) or because Cambyses suffered from the sacred disease. It is clear that Herodotus here treats the sacred disease primarily as a condition of the body, though one that can affect the mind also.

which her body became infested with worms, Herodotus comments: 'thus, it would seem, over-violent human vengeance is hated by the gods'.[104] Finally a text in which he mentions the Scythian Enareis (no doubt the same group as that called Anarieis in *On Airs Waters Places* ch. 22) enables a direct comparison to be made between him and the Hippocratic author. Whereas *On Airs Waters Places* directly refutes the idea that the impotence of the Anarieis is caused by a god,[105] Herodotus reports that it was the men who pillaged the temple of Heavenly Aphrodite at Ascalon – they and their descendants – who were afflicted by the goddess with the 'female sickness'. He makes it clear that he had this story from the Scythians themselves, but there is no hint of his doubting or rejecting it (1 105).

The evidence in Herodotus shows that it was perfectly possible to *combine* engaging in inquiries concerning the 'nature' of various phenomena with adherence to such beliefs as that diseases could be brought about by the gods. Such a belief was not threatened by an interest in – even by quite sustained research into – the character of *particular* phenomena, only by the generalistion that *all* such phenomena have natural causes. What counted was not just any notion of the nature or character of particular things – the term φύσις itself was already used, after all, in a passage in the *Odyssey* where Hermes indicates the 'nature' of a plant to Odysseus[106] – but rather the application of that notion in the form of a *universal* rule, that every physical object has a nature, that is, it manifests, or conforms to, certain regularities and has a determinate physical cause or causes. Nature may be thought of as itself divine, as in *On the Sacred Disease*.[107] But once it was believed that natural phenomena form a set every member of which has determinate physical causes, then it was no longer enough to cite a god or supernatural being as responsible for events (either for a specific occurrence of a phenomenon, or even for a group of phenomena such as a type of disease). The notion of divine intervention had, then, either to be abandoned or to be redefined: if maintained, it had now to be seen either as the suspension cf nature

---

[104] ὡς ἄρα ἀνθρώποισι αἱ λίην ἰσχυραὶ τιμωρίαι πρὸς θεῶν ἐπίφθονοι γίνονται (IV 205). The excessive revenge that Pheretime had exacted on the people of Barce is described at IV 202. [105] Cf. above, p. 28.

[106] *Od.* x 302ff: Odysseus says that Hermes offered him a 'drug' (φάρμακον) 'pulling it from the earth, and he showed me its nature (καί μοι φύσιν αὐτοῦ ἔδειξε): it had a black root, but a flower like milk; the gods call it "moly", but it is difficult for mortal men, at least, to dig up'. φύσις, interpreted by Holwerda 1955, p. 63, as 'appearance' here, may also have some of the other primary sense of 'growth', the natural form being thought of as the result of growth.

[107] Cf. above, p. 26 on *Morb. Sacr.* ch. 18, p. 28 on *Aër.* ch. 22 and p. 11 n. 9 on the evidence for the Milesian philosophers.

(that is, in later terminology, a miracle) or as in addition to it (when the event would be 'doubly determined', brought about both by gods and by natural causes, the former working through the latter).[108]

THE PHILOSOPHICAL BACKGROUND

Now the origins of the idea that *all* natural phenomena are law-like are, fairly evidently, to be sought not in the medical writers themselves, so much as in the Presocratic philosophers, particularly in the group whom Aristotle calls the φυσιολόγοι, 'the inquirers into nature'. That some such general principle had been explicitly formulated by the time we come to the end of the Presocratic period can be affirmed on the basis of Leucippus Fr. 2, which states that 'Nothing comes to be at random, but everything for a reason and by necessity.'[109] The question is, rather, how much earlier a similar principle was expressed or at least used, and here the lack of original texts for most of the earlier Presocratics proves a serious handicap. As we noted at the outset (p. 11) our secondary sources ascribe to Thales, Anaximander and Anaximenes a number of theories and explanations concerning a variety of what we should call natural phenomena. What our sources report generally takes the form of a naturalistic account,[110] one that refers the phenomenon to be explained to a determinate physical cause, and one in which personal deities play no role. Moreover a high proportion of the theories and explanations recorded relate to phenomena such as lightning and thunder, earthquakes or eclipses, that were either terrifying or rare or both and that had often, in mythology, been associated with gods. We cannot know how far that predominance reflects the particular interests of our doxographic sources,[111] rather than those of the

---

[108] There is, to be sure, an element of 'double determination' (the combination of a 'natural' and a divine cause) in the account of Pheretime's death in Herodotus IV 205, though it is absent, for instance, from the story about the Scythian Enareis, where divine displeasure alone is mentioned (I 105). What must remain in some doubt is the extent to which Herodotus saw nature as a *universal* principle, and *all* natural phenomena as law-like.

[109] οὐδὲν χρῆμα μάτην γίνεται, ἀλλὰ πάντα ἐκ λόγου τε καὶ ὑπ' ἀνάγκης. Our source for this, Aetius, is, admittedly, late: nor can we say with confidence just how strictly Leucippus intended the principle to be applied, although the double formulation, both negative and positive ('nothing...' 'everything...') may, if original, suggest at least an attempt at emphasis.

[110] E.g. the theory of lightning and thunder ascribed to Anaximander by Aetius (III 3.1, DK 12A23), namely that these phenomena happen when wind, enclosed in a dense cloud, bursts out violently. Even the speculative cosmogony attributed to Anaximander in pseudo-Plutarch, *Strom.* 2 (A 10) takes a similar, naturalistic, form.

[111] There is a whole literature devoted to problematic or marvellous phenomena stretching from the fourth (if not the fifth) century B.C. to late antiquity. Already

Milesians themselves, but at least we may presume that they paid considerable attention to marvellous phenomena. Furthermore our sole surviving fragment of Anaximander is generally and surely rightly interpreted as conveying an idea of the world-order through the legal metaphors of justice and reparation for wrong-doing,[112] and if that is correct, then it may be that he had some conception of natural phenomena as a totality as subject to determinate physical causes.[113] Nevertheless we must recognise that this is far from certain. What would help to remove doubt would be an explicit statement either like that of Leucippus Fr. 2 or – clearer still – like some Hippocratic formulations, as when the writer of *On Airs Waters Places* puts it, in connection with diseases, that 'each has a nature and nothing happens without a natural cause',[114] or the author of *On the Art* writes: 'indeed, upon examination, the reality of the spontaneous (τὸ αὐτόματον) disappears. Everything that happens will be found to have some cause, and if it has a cause, the spontaneous can be no more than an empty name.'[115] But no such assertion is to be found in our extant evidence for the Milesians.[116]

Moreover when we turn to the work of some of the later Presocratics for whom our information is both fuller and more reliable, we find further evidence[117] of the dangers of assuming that engagement in the inquiry into nature was necessarily accompanied by a sceptical attitude towards traditional beliefs in, for example, the possibility of wonder-working. Empedocles[118] illustrates the point

Herodotus pays particular attention to striking natural phenomena, and Aristotle devoted a treatise to problematic phenomena of many different kinds (though the *Problemata* that passes by his name is not authentic).

[112] διδόναι γὰρ αὐτὰ δίκην καὶ τίσιν ἀλλήλοις τῆς ἀδικίας κατὰ τὴν τοῦ χρόνου τάξιν (DK 12 B 1) 'For they pay the penalty and recompense to one another for their injustice according to the assessment of time.' On the differing interpretations of this fragment, see, for example, Kahn 1960, pp. 166ff, Guthrie 1962, pp. 76–83, Classen 1970, col. 56ff.

[113] Thereafter 'necessity' and 'justice' are used to express the law-like behaviour of the cosmos in, for example, Heraclitus (Fr. 94: though for him 'justice' is 'strife', Fr. 80) and Parmenides' *Way of Seeming* (Fr. 10.6f). The importance of the notion of 'necessity' in particular in conveying the orderliness of nature was especially stressed by Cornford 1912, chh. 1 and 2, who saw the idea as having pre-philosophical origins. It should, however, be noted that general references to a principle of necessity are not equivalent to a statement of a universal rule to the effect that *all* phenomena have natural causes.

[114] Ch. 22, *CMG* I, 1 74.17, cf. also 75.16.

[115] Ch. 6, *CMG* I, 1 13.1–4.

[116] Neither in the meagre citations, nor indeed in the secondary comments of our ancient sources.

[117] In addition to that from Herodotus, considered above, pp. 29ff.

[118] Admittedly Empedocles belongs to the West Greek philosophical tradition and the influences both of Pythagoreanism and of the doctrines of Parmenides are clear from his fragments. But though there are obvious broad distinctions between this and the Ionian tradition represented by the Milesians, Anaxagoras and the atomists, for example, the question at issue here is on a point where Empedocles shares an interest

dramatically. His place in the history of physical theory is assured. After Parmenides had denied the possibility of change and rejected the senses as unreliable, Empedocles reinstated sense-perception and interpreted coming-to-be in terms of the mixing and separating of the four 'roots', earth, water, air and fire. With this doctrine of 'roots' Empedocles was responsible for the first clear statement of the idea of an element in the sense of the simple substances into which other things can be analysed, and the particular four-element theory he put forward was to prove, in one version or another, the most influential physical theory not only in antiquity but through the Middle Ages and right down to the seventeenth century. Yet apart from the work *On Nature*[119] Empedocles wrote another poem called the *Purifications*, Καθαρμοί, which was concerned with the downfall, wanderings and eventual redemption of the δαίμων. In Fr. 112 (which is reported to have come at the beginning of the poem) he speaks of himself as coming to the people of Acragas as 'an immortal god, no longer mortal', and he describes how they throng to him 'asking where the way towards gain lies, some desiring oracles, others seeking to hear the word of healing for every kind of disease'. Whether this 'word of healing' consisted of the sort of advice we find in such Hippocratic works as *On Regimen* and *On Affections*, or whether it was a matter simply of spells or charms – ἐπῳδαί – is not clear from the text, but the fact that the term for 'word' is βάξις – used of the pronouncements of oracles in particular – suggests that the latter is more likely. Nor, it seems, is it only in the *Purifications* that such claims are made. In another fragment (111) which appears to belong to the poem *On Nature*[120] he promises to teach φάρμακα ('drugs', or perhaps more generally 'remedies'[121]) that are a defence for ills and old age, and he states that his listener will be able to control the winds and rain and drought, and even will bring the dead back to life.

The relationship between the poem *On Nature* and the *Purifications* –

with the Ionians and a direct comparison is possible between him and them, namely on how the 'inquiry concerning nature' was viewed.

[119] Περὶ φύσεως. This title was attached rather indiscriminately (as Kirk and Raven put it) to works by early philosophers (including Anaximander, Xenophanes and Heraclitus), but we have no good grounds to doubt its applicability to Empedocles' physical poem. A text in *VM* ch. 20, *CMG* I, 1 51.10f, already implies, if genuine (though cf. Dihle 1963, pp. 145ff), that Empedocles wrote περὶ φύσεως (whether or not that was the actual title of his work) and his physical poem is referred to as τὰ φυσικά by both Aristotle (*Mete.* 382a1) and Simplicius (*In Ph.* 157.27, 300.20, 381.29: he speaks of the work in two books).

[120] *On Nature* is addressed to Pausanias (Fr. 1), the *Purifications* to the Acragantines (Fr. 112). Since the addressee of Fr. 111 is singular, there is at least a prima facie presumption that that fragment belongs to the work *On Nature*.

[121] On the range of meaning of the term, see below, p. 44.

and more generally that between 'science' and 'religion' in the thought of Empedocles – are among the most controversial topics in the interpretation of Presocratic philosophy.[122] But in any case no simple hypothesis – for example that he had abandoned the views and interests of the one work when he came to compose the other – will meet the point that he appears to make claims as a wonder-worker in both poems. As to how Empedocles himself saw the relationship between those claims and his investigations into natural phenomena, we have no direct evidence, and in particular the exact status of the marvellous effects he refers to is not clear. It is certain that they are not thought of as produced at the whim of personal divine agencies like the Olympian gods. Rather they are brought about by the man with special knowledge. But the question that remains unresolved is whether Empedocles held that the wise man's knowledge enables him to *suspend* natural laws (to perform miracles), or whether the wise man merely exploits the *hidden* powers of nature to produce effects that are contrary to nature not in the sense of the supernatural, but only in the sense of the extra-ordinary.[123] Considerations might be suggested in favour of each of these views, and in the final analysis it may be that – whether deliberately or not[124] – Empedocles himself was ambivalent on the issue. On the one hand the poem *On Nature* was clearly largely devoted to how things are and how they come to be:[125] it included accounts of the material constitutions of compound substances and went into such problems as the processes of vision and respiration in some detail.[126] On the other hand the extravagant character of the claims he made in Frr. 111 and 112 – and the language he made them in – immediately tend to align Empedocles with other wonder-workers.[127]

If the Milesians may be said to have initiated the inquiry into natural phenomena as a more or less systematic investigation, the

---

122 For a survey of the views that have been put forward on this topic, see, for example, Guthrie 1965, pp. 122ff, 132ff.

123 In the former case he would, in the latter he would not, have denied the principle that *all* phenomena are law-like.

124 It may be that the question had not occurred to Empedocles: but it is also possible that it had, and that he was deliberately hedging on the issue, even deliberately allowing some of his audience (at least) to be misled by the language of Frr. 111 and 112 (cf. the discussion of ἀπάτη in Greek thought in Detienne 1967, especially ch. 6, and Detienne and Vernant 1978).

125 Although he denies that there is any absolute coming-to-be, i.e. from nothing: e.g. Fr. 8, where the term φύσις is now generally interpreted as 'birth'.

126 Frr. 96 and 98 deal with compound substances, Frr. 84 and 100 with vision and respiration.

127 Note particularly that Empedocles suggests that the person whom he addresses will be able to control the winds 'at will', Fr. 111.5.

aims and presuppositions with which that inquiry was undertaken varied greatly from one Presocratic philosopher to another. It could be, and often was, conducted by men who did not make use of, and may have intended directly to supplant,[128] traditional beliefs in divine interventions in natural phenomena, who sought determinate physical causes of whatever appeared striking or exceptional, and who held that every physical phenomenon could be so explained. At the same time it was sometimes assumed that the knowledge gained from the investigation could be used to bring about effects that – at the least – run counter to the regularities of nature herself. When Aristotle records the views of the 'physiologists', the emphasis is very much on their accounts of the material causes of things, of change and coming-to-be, and on their attempts to provide explanations of particular natural phenomena.[129] Again Plato, in some of his comments on those who investigated nature,[130] particularly attacks those[131] whom he represents as atheists because they saw the world as a whole as the product of 'nature' and 'chance' as opposed to 'reason' 'god' and 'art', where 'nature' stands primarily for the interplay of mechanical causes and effects,[132] and where the chief thrust of Plato's polemic is that these theorists denied or neglected the role of a benevolent and divine creative intelligence.[133] Yet on the other side Empedocles can be taken as the prime[134] representative

---

[128] This may be thought likely in the case of Democritus, in particular, if he saw belief in the gods as in part a mistaken inference from terrifying natural phenomena (Sextus, *M.* IX 24, cf. above, p. 14). Cf. also his reported enthusiasm for αἰτιολογίαι (Fr. 118, together with the titles of a series of works in the list in Diogenes Laertius, IX 47).

[129] To Aristotle (as also to Plato, see below, n. 132) some of the natural philosophers, and especially the atomists, appeared as determinists, that is as having explained everything in terms of necessity, but this is chiefly because they denied teleology. He himself reinstates 'chance', τύχη, as well as 'the spontaneous', τὸ αὐτόματον, against those who denied that it existed at all (*Ph.* 195b36ff), but for him 'chance' events are themselves capable of explanation in other terms (*Ph.* II chh. 4–6 especially). Nature is a matter of what happens 'always or for the most part': but what happens παρὰ φύσιν, contrary to nature, is what is unusual, irregular, not 'supernatural'. Cf. e.g. Wieland 1962, pp. 256ff.   [130] Especially *Lg.* x 888e ff.

[131] Again it is likely that he had the atomists particularly in mind. Two prominent natural philosophers had, in fact, attempted cosmologies in which reason, νοῦς, plays an important role, namely Anaxagoras (Fr. 12, especially) and Diogenes of Apollonia (Frr. 3 and 5). But Plato makes Socrates complain that Anaxagoras failed to put his principle to adequate use (*Phd.* 97b ff).

[132] As is clear from the example of the interactions of hot and cold, dry and wet, soft and hard things, at *Lg.* 889bc.

[133] Cf. Vlastos' comment, 1975, p. 97 (cf. also p. 66), on the role of the Craftsman in Plato's own cosmology: 'If you cannot expunge the supernatural, you can rationalize it, turning it paradoxically into the very source of the natural order, restricting its operation to a single primordial creative act which insures that the physical world would be not chaos but cosmos forever after.'

[134] But it may well be not the only one: see below, p. 37 and n. 135 on the evidence for the Pythagoreans.

of a very different view, according to which the knowledge of nature might be used in some sense to transcend nature herself.

HEALING AND HEALERS IN THE CLASSICAL PERIOD

If we now turn back to *On the Sacred Disease*, we can see that the relationship between that treatise and the work of those whom we conventionally group together as the Presocratic philosophers is an intricate one. On the one hand the insistence that all diseases have natural causes may be compared with similar assumptions underlying the philosophers' more general physical investigations and with Leucippus' statement of the principle that everything happens for a reason and by necessity. On the other, Empedocles has, from some points of view, more in common with the opponents of the Hippocratic author than with the Hippocratic author himself. Where Empedocles Fr. 111 talks of raising and quelling the winds, and of bringing rain or drought,[135] *On the Sacred Disease* attacks those who 'claim to know how to...make storms and fine weather, rain and drought...and all the rest of their nonsense', calling them all 'impious rogues'.[136] Moreover among the prescriptions he attributes to his opponents are some that can be paralleled in our admittedly late evidence for Pythagorean beliefs.[137] Thus he says that the quacks recommend not eating certain fish, including the mullet and the blacktail,[138] and we find similar prohibitions in our sources for Pythagoreanism.[139] Again the quacks are said to recommend avoiding black clothing,[140] and Diogenes Laertius, for example, attributes to Pythagoras an association of black with evil.[141]

Now despite what has sometimes been suggested,[142] the conclusion

---

[135] Our secondary literature for Empedocles contains a variety of stories – most, if not all, no doubt apocryphal – relating to his wonder-working, see, e.g., D.L. VIII 59–61. Pythagoras, too, was frequently represented as a wonder-worker, perhaps, indeed, already by Empedocles (Fr. 129): see also Heraclides Ponticus in D.L. VIII 4, Timon in D.L. VIII 36, as well as D.L. VIII 11, 14, 21, 38, Iamblichus, *VP* 6off, 134ff, 14off (cf. Porphyry, *VP* 23ff, 27ff), and cf. Burkert 1972a, pp. 136ff.

[136] *Morb. Sacr.* ch. 1 paras. 29f and 31 (G) (L VI 358.19ff), see above, p. 19.

[137] Cf. especially Burkert 1972a, pp. 176ff, who mentions other evidence relating, for example, to initiation rites and to the mystery religions.

[138] *Morb. Sacr.* ch. 1 para. 13 (G) (L VI 356.1).

[139] E.g. Diogenes Laertius VIII 19 and 33, Porphyry, *VP* 45, Iamblichus, *Protr.* 21 (5). With the prohibition on eating certain birds, including the cock, mentioned at *Morb. Sacr.* ch. 1 para. 15 (G) (L VI 356.4), one may compare the Pythagorean prohibition on eating or sacrificing a white cock (see D.L. VIII 34, Iamblichus, *VP* 84 and cf. *Protr.* 21 (17)).

[140] *Morb. Sacr.* ch. 1 para. 17 (G) (L VI 356.6f).

[141] D.L. VIII 34.

[142] See, for example, Wellmann 1901, p. 29 n. 1, Burnet (1892) 1948, p. 202, Jouanna 1961, pp. 46off, for a connection with followers of Empedocles. For one with Pytha-

we should draw from all this is not that the opponents of *On the Sacred Disease* are to be identified as Pythagoreans or as followers of Empedocles. On the contrary, there are good grounds for resisting any such hypothesis. First, some of the similarities in question merely reflect popular Greek beliefs,[143] such as the association of black with misfortune. Secondly, whereas the Hippocratic writer's opponents are suggesting remedies *for a particular illness*, the Pythagorean rules are rules *for general behaviour*.[144] Thirdly, the idea that sufferers from the sacred disease may be purified with blood[145] is one that Empedocles himself, at least, with his horror of blood-shedding, would certainly have repudiated.[146] Yet if any such simple identifications should be ruled out, the comparison between these texts certainly illustrates the survival and systematisation of certain popular or traditional beliefs in parts of Presocratic philosophy and shows that on certain issues the Hippocratic author not only did not endorse, but was concerned to expose, a view that can be exemplified in an important natural philosopher.

We have seen in considering Empedocles how complex and ambivalent the assumptions underlying the Presocratic 'inquiry concerning nature' could be. The writer of *On the Sacred Disease*, for his part, exemplifies only one of the many different strands that go to make up Greek medicine in the fifth and fourth centuries B.C. Apart from the various kinds of doctors represented in the Hippocratic Corpus,[147] many others laid some claim to be able to alleviate diseases. They included people who would be known not as ἰατροί, but as herb-collectors or 'root-cutters' (ῥιζοτόμοι), 'drug-sellers' (φαρμακοπῶλαι), midwives and gymnastic trainers,[148] as well as priests and attendants who practised 'temple-medicine' at the shrines of healing gods and heroes,[149] and the dividing lines between some of these broad categories were far from sharply defined. There was, in

goreanism, see Delatte 1922, p. 232, Boyancé 1937, pp. 106f, Burkert 1972a, p. 177 n. 87, but cf. the more cautious assessment in Moulinier 1952, pp. 134ff.

[143] This emerges clearly from the analysis of Greek popular assumptions concerning the pure and the impure in R. C. T. Parker 1977.

[144] As was noted by Boyancé 1937, p. 106.

[145] *Morb. Sacr.* ch. 1 para. 40 (G) (L vi 362.8ff).

[146] See Empedocles Frr. 128, 136 and 137 especially, and cf. also Heraclitus Fr. 5, quoted above, p. 12. Contrast, e.g., A. *Eu.* 28off.

[147] The Corpus includes some treatises, such as *de Arte* and *Flat.*, that are sophistic displays and are probably not the work of men who actually practised as doctors (see further below, ch. 2, pp. 88f). Moreover the doctrinal positions of the authors who did so practise varied enormously, see, for example, Lloyd 1975b, pp. 183ff.

[148] Surgeon-barbers would be a later addition to this list.

[149] The priests and attendants gave advice and suggested 'treatment' usually on the basis of the interpretation of the dreams and signs that supposedly came from the god

the ancient world, no equivalent to the modern, legally recognised, professional medical qualification. It was undoubtedly an advantage to an ancient doctor – when dealing with certain types of client or employer – to have been associated with one of the centres of medical training, such as Cos or Cnidus.¹⁵⁰ Yet even if he could claim such an association, a doctor's title to practise might always be called in question. An accusation of charlatanry (ἀλαζονεία) was easy to make and hard to rebut,¹⁵¹ and, understandably, many Hippocratic authors were evidently much concerned to establish that medicine, as they practised it, is a true art, and to insist on the distinctions between doctors and laymen on the one hand and between true doctors and quacks on the other.¹⁵²

In some cases there were, to be sure, certain fairly well-marked differences in the doctrines and procedures of some of the medical writers and those of some of the groups from which they were keen to be dissociated. Yet there was also, in practice, a considerable

to the faithful (see further below, pp. 40f). They were, however, generally much more closely integrated into the state religion than the purifiers attacked as 'vagabonds' in *Morb. Sacr.* (that the latter did *not* take their patients to the temples seems to be implied at *Morb. Sacr.* ch. 1, paras. 41ff (G) (L vi 362.10ff), see below, p. 48 n. 209). We should, in fact, recognise differences and gradations within 'religious', as much as within 'rationalistic', medicine. (I am grateful to Professor Vernant for first stressing this point to me.)

¹⁵⁰ This is clear from the high proportion of doctors from Cos who – at least from the third century on – were given appointments as 'public physicians': see Cohn-Haft 1956.

¹⁵¹ In some of the (generally rather late) Hippocratic works that deal with medical etiquette there are some interesting, and conflicting, evidences on the question of the sanctions exercised against the medical profession. Thus the treatise *Lex* complains that the only sanction used against bad medical practice is that of dishonour (ch. 1, *CMG* i, 1 7.5ff) and a similar view seems to be implied in *Praec.* ch. 1 (*CMG* i, 1 30.18ff). Yet in *Decent.* ch. 2 (*CMG* i, 1 25.14f) reference is made to the banishment of corrupt practitioners from certain states. Antiphon iv 3.5 is one classical text that shows that the law absolved the physician of blame if his patient died.

¹⁵² Apart from the frequent references to these themes in the treatises dealing with medical etiquette, the work *de Arte* is devoted to showing that medicine is a veritable art (see, e.g., ch. 8, *CMG* i, 1 14.23ff on the difference between true physicians and those who are doctors only in name). The contrast between what is brought about by the art and what is due merely to chance recurs, e.g., in *Morb.* i chh. 7 and 8, L vi 152.9ff, 154.5ff, *Aff.* ch. 45, L vi 254.9ff, and, especially, *Loc. Hom.* ch. 46, L vi 342.4ff. For the distinction between the doctor and the layman, see, e.g., *Acut.* ch. 1, L ii 224.3ff, ch. 2, 234.2ff, ch. 11, 316.13ff, *VM* ch. 2, *CMG* i, 1 37.7ff and 17ff, ch. 9, 42.6ff, ch. 21, 52.17ff: for that between the doctor and the quack, see, e.g., *Acut.* ch. 2, L ii 236.4ff, *VM* ch. 9, *CMG* i, 1 41.25ff, *Art.* ch. 42, L iv 182.15ff, ch. 46, 198.5ff, *Fract.* ch. 1, L iii 414.1ff. References to bad practice are especially frequent in the surgical treatises, see also *Art.* ch. 1, L iv 78.5ff, ch. 11, 104.20ff, ch. 14, 120.7ff, *Fract.* ch. 2, L iii 418.1ff, ch. 3, 422.12ff, ch. 25, 496.11ff, ch. 30, 518.1ff, ch. 31, 524.17ff, and cf. further below, pp. 89ff and 91 n. 174. Interestingly enough the writer of *VM* suggests that medicine originated from dietetics (ch. 4, *CMG* i, 1 38. 27ff) and he compares the doctor with the gymnastic trainer to make the point that both arts are being continually improved (ch. 4, 39.2ff).

overlap both in ideas concerning the nature of some diseases[153] and in techniques of treatment. Once again *On the Sacred Disease* provides evidence on the point. The author describes his opponents as not merely using charms or spells (ἐπαοιδαί) and purifications (καθαρμοί) as remedies for the sacred disease, but also making certain dietary and other recommendations, although these were of a negative sort, about what was to be avoided, rather than about what was to be taken.[154] Moreover when reporting some of their dietary rules, the Hippocratic writer sometimes adds his own glosses to the effect that the foods in question are indeed harmful to the sick,[155] thereby indicating that he saw some point in their recommendations in these instances, even though he would probably have given rather different reasons as their justification.

A further aspect of this overlap can be illustrated by referring to the inscriptions relating to the cult of Asclepius at Epidaurus.[156] These show that apart from cases where the treatment involved the god touching a patient's body with a ring, for example,[157] the god was sometimes represented as employing foods or drugs, for instance in one case an emetic, to heal the sick.[158] Indeed on several occasions the god appears in a vision or a dream in the role of a surgeon, using the knife to effect spectacular, in some cases quite fantastical, cures.[159] Clearly the faithful who attended the shrines of Asclepius were used to the god behaving – and they expected the god to behave – in

---

[153] As Kudlien has suggested in relation to some of the diseases discussed in the pathological treatise *Morb.* II especially, for example the 'bad-sorrow' disease of ch. 72 (L VII 108.25ff) and the 'murder' fever of ch. 67 (102.4ff), see Kudlien 1968, pp. 326ff, 330f.

[154] E.g. the recommendation to abstain from baths, ch. 1 para. 12 (G) (L VI 354. 20).

[155] See ch. 1 para. 13 (G) οὗτοι γὰρ ἐπικηρότατοί εἰσι ('for these are most dangerous', cf. L VI 356.2) and para. 14 (G) (L VI 356.3f) ταῦτα γὰρ κρεῶν ταρακτικώτατά ἐστι τῆς κοιλίης ('for of meats these most disturb the digestive organs'). The present indicatives indicate that these statements contain the writer's own views. Contrast the infinitive in para. 19 (G) (L VI 356.9) πάντα γὰρ ταῦτα κωλύματα εἶναι ('for all these are impediments') where he is reporting his opponents' beliefs in oratio obliqua.

[156] *IG* IV 951–953, *IG* 4² 1, 121–4. The inscriptions belong to the latter part of the fourth century B.C. They have subsequently been edited by Herzog 1931, and cf. also Edelstein and Edelstein 1945, 1 pp. 221ff.

[157] As in case 62, where an epileptic patient is cured after seeing the god touching parts of his body with a ring in a dream: see Herzog 1931, pp. 32 and 109ff.

[158] As in case 41 (Herzog 1931, p. 24). Other cases where the god is represented in visions or dreams as using drugs are case 9 (to cure an eye complaint, Herzog 1931, p. 12) and case 19 (to cure baldness, Herzog 1931, p. 16). While that does not prove that the temple treatment involved the actual use of drugs in those cases, it is likely enough, to judge from the later evidence in such writers as Aelius Aristides, that it sometimes did so.

[159] As in cases 13, 21, 23, 25 and 27 (Herzog 1931, pp. 14–18 and cf. pp. 75ff).

visions in ways which were in certain respects very similar to those of the doctors represented in our extant Hippocratic treatises.[160]

What we know of the practice of religious medicine in later periods confirms this picture. Thus the instructions that Aelius Aristides claimed to have had from the god (usually through dreams) include not only, for example, a command to take a ritual mud bath and run three times round the temples at Pergamum (*Or.* XLVIII 74f) but also prescriptions concerning foods (e.g. XLVII 45, XLIX 6, 24, 34, 35, 37), and drugs (XLVIII 13, where the sign from the god is interpreted as referring to hellebore), the use of poultices (e.g. XLIX 25) and blood-letting (e.g. XLVIII 47). But if Asclepius' treatment is often strongly reminiscent of that of contemporary medical men, there is this difference, that his diagnoses and cures are deemed to be infallible. Aristides is in no doubt as to whose advice to follow when, as frequently occurs, merely mortal physicians, and the true, immortal healer are in disagreement.[161]

Conversely it was not merely in a spirit of conventional piety that some of the medical writers of the classical period invoke divine patronage for their art. Apollo the healer, Asclepius, Hygieia (Health) and Panacea ('All-Heal') are called as witnesses at the beginning of the Hippocratic *Oath*;[162] the *Law* borrows the language of the mystery religions when talking of the secrets of the art;[163] and *On Ancient Medicine* says that the art is rightly dedicated to a god.[164]

---

[160] Edelstein and Edelstein 1945, II p. 112 n. 4 ('it is interesting to observe again and again how closely the concept of the god resembles that of the medical practitioner').

[161] See Behr 1968, pp. 168f, and cf. Ilberg 1931, p. 32, commenting on a fragment of Rufus preserved in Oribasius XLV 30 (*CMG* VI, 2, 1 191.1ff, Raeder, IV 83.1ff, Bussemaker and Daremberg): 'Der Gott hat offenbar Medizin studiert, man sieht den Einfluss der Wissenschaft auf die Tempelpraxis um 100 nach Chr.'

[162] *Jusj.* 1, *CMG* I, 1 4.2ff. Although many of its ideals were widely shared, the *Oath* as such probably belongs to a group of practitioners, not to Greek doctors as a whole: certainly some of the specific injunctions it contains, for example not to operate 'even for the stone', run counter to common Greek medical practices of the fifth and fourth centuries B.C. Cf. e.g. Edelstein (1943) 1967.

[163] *Lex* ch. 5, *CMG* I, 1 8.15ff τὰ δὲ ἱερὰ ἐόντα πρήγματα ἱεροῖσιν ἀνθρώποισι δείκνυται, βεβήλοισι δὲ οὐ θέμις, πρὶν ἢ τελεσθῶσιν ὀργίοισιν ἐπιστήμης. 'Holy things are revealed only to holy men. Such things must not be made known to the profane until they are initiated into the mysteries of knowledge.'

[164] *VM* ch. 14, *CMG* I, 1 45.17f. Cf. *Vict.* I ch. 11 (L VI 486.14f) which implies that men learnt the arts from the gods, and *Vict.* IV ch. 93 (662.8f) where the writer says that his discoveries in regimen have been made with the help of the gods. To these passages may be added others whose interpretation is more obscure. In *Decent.* the writer, having just spoken of medicine as wisdom and of the physician as 'having most things', says that knowledge of the gods is entwined with medicine in the mind (ch. 6, *CMG* I, 1 27.13, reading αὐτῇ, as opposed to Littré's αὐτή). Nor is it clear precisely what the author of *Prog.* had in mind when he wrote that one of the tasks of the doctor is to learn whether there is anything divine in diseases (εἴ τι θεῖον ἔνεστι ἐν τῇσι νούσοισι, ch. 1, L II 112.5f, cf. also *Nat. Mul.* ch. 1, L VII 312.1ff and 9). The interpretation of

Although many popular remedies were implicitly[165] or explicitly rejected by certain of the medical writers, such questions as the efficacy of amulets (περίαπτα), of spells and prayers, and of music continued to be much debated. Thus amulets[166] were counted among the 'natural remedies' by Rufus (Fr. 90), and even Soranus, who rejects them, suggests that they should not be forbidden since they may perhaps make patients more cheerful.[167] Galen, who is, in general, critical,[168] offers a naturalistic explanation of one amulet that he claims to have tested and found to be effective: either parts of the root used as the amulet came off as effluences and were inhaled, or the air round the root was itself modified in some way.[169] Although incantations are firmly rejected by *On the Sacred Disease*,[170] the writer of *On Regimen* IV first criticises those who rely on prayer alone on the grounds that, while prayer is good, men should also help themselves at the same time as they call on the gods,[171] but then goes on to give some specific instructions about which gods to pray to when the signs seen in dreams are favourable or unfavourable.[172] Stories about healing by music were common,[173] but although, of the later medical writers, Soranus was critical of the use of music as a remedy,[174] that was not the only view expressed. Galen, who wrote

that text (which some modern editors, such as Kühlewein and Jones, have treated as an interpolation) was already the subject of dispute among the ancient commentators, as we learn from Galen, who believed that 'divine' here must be taken to refer to atmospheric influences (*CMG* v, 9, 2 205.28ff, K xviii b 17.9ff) (see most recently Kudlien 1966, pp. 38f, Thivel 1975 and Laín Entralgo 1975, pp. 315ff).

[165] Thus the final aphorism (*Aph.* vii 87, L iv 608.1ff) gives as possible types of treatment drugs, the knife and cautery (though the term for 'drugs', φάρμακα, is capable of a wide extension, see below, p. 44).

[166] Theophrastus is one non-medical writer who is critical of the use of amulets, claiming that most of what is said about them is the work of men 'who wish to magnify their own arts' (*HP* ix 19.2–3). On the whole subject see Stemplinger 1919, pp. 82ff.

[167] *Gyn.* iii 10.42, *CMG* iv 121.26ff, cf. i 19.63, *CMG* iv 47.16ff.

[168] E.g. K xi 792.14ff.

[169] A boy never had epileptic fits when he wore the amulet in question, but did when it was removed, only again to cease to have fits when he wore it once more: K xi 859.12ff, cf. also xii 573.5ff.

[170] The uselessness of incantations and purifications in the treatment of epilepsy, insisted on in *Morb. Sacr.*, can be paralleled, outside medical literature, by Thucydides' remarking, in his account of the plague at Athens, that supplications and oracles were useless (though so indeed were all the other remedies tried, ii 47) and cf. Democritus Fr. 234 (men seek health from the gods with prayers, but they do not realise that they have power over it in themselves).

[171] *Vict.* iv ch. 87, L vi 642.6ff.

[172] *Vict.* iv ch. 89, L vi 652.17ff and ch. 90, 656.22–658.1.

[173] See, for example, Plutarch, *De Musica* 1146bc. Aulus Gellius (iv 13) quotes Theophrastus as saying that 'many men believe' that flute-playing is good for pain in the hip, and Democritus to the effect that flute-playing cures snake-bites and is good for many other sicknesses (cf. Athenaeus, xiv 624ab). Iamblichus, *VP* 64, 110–11, 164 and Porphyry, *VP* 33, speak of a Pythagorean belief that music contributes to health.

[174] According to Caelius Aurelianus, *Morb. Chron.* v 23, cf. i 175f and 178.

at length on the effects of psychic disturbances on the body as also on those of bodily temperament on the soul, attempted to explain the benefits obtained from music in naturalistic terms.[175]

Although it was generally recognised that dreams could be misleading, it was not only those who advocated the practice of incubation in the temples[176] who saw dreams as indicators – whether of the disease troubling the patient or of its cure. The belief that dreams may be useful guides to diagnosis can be traced in a whole series of medical writers. In the Hippocratic collection the work *On Regimen* IV is devoted to setting out a comprehensive theory of the interpretation of dreams, and other treatises too acknowledge their role in diagnosis.[177] Extraordinarily elaborate theories were developed concerning the different categories of dreams.[178] Of the later medical writers, Herophilus gave a comparatively simple classification,[179] and Galen was prepared to take dreams seriously as signs.[180] Thus at K XI 314.18ff he refers to a therapy suggested to him by a dream, and he sets out some systematic ideas on diagnosis from dreams in his commentary on book I of the *Epidemics*.[181]

Finally, as Artelt and others have long ago shown,[182] there is a deep-seated ambiguity in many of the terms used by the medical

---

[175] *CMG* v, 4, 2 19.24ff., K vi 40.4ff.

[176] The classic study of incubation is that of Deubner 1900: cf. also Hamilton 1906 and Edelstein and Edelstein 1945, II pp. 145ff.

[177] E.g. *Epid.* I 10 (L II 670.8), *Hum.* ch. 4 (L v 480.17), *Hebd.* ch. 45, pp. 66f Roscher (L IX 460.17ff). Aristotle rejects the idea that dreams are sent by the gods, though he says they are δαιμόνια, giving as his grounds for this that nature herself is δαιμονία (*Div. Somn.* 463 b 13ff). He endorses the view he attributes to the more discerning doctors according to which careful attention should be paid to dreams since they may provide information about movements and changes occurring in the body, and he concludes from this that some dreams may be both signs and causes of future events, even though most of what were believed to be prophetic dreams are mere concidences (*Div. Somn.* 463a4 b11).

[178] Our most extensive source on the subject, Artemidorus' *Onirocritica* (second century A.D.), distinguishes two main groups. ἐνύπνια, which include φαντάσματα (visions), indicate what is the case and are not predictive. ὄνειροι, on the other hand, which include ὁράματα and χρηματισμοί (dream-oracles), are signs of what will come to be: they comprise θεωρηματικοί and ἀλληγορικοί ὄνειροι, the former non-allegorical, as when the events themselves seem to be seen in the dream, the latter allegorical or symbolic dreams – and he distinguishes five species of these (I chh. 1–2, pp. 3ff Hercher, 3.9ff Pack). But many other classifications were suggested (see, for example, Behr 1968, ch. 8, pp. 171–95).

[179] One of his three classes of dreams was the 'god-sent': see Aetius v 2.3. Cf. also Rufus, *Quaestiones Medicinales*, *CMG* Suppl. IV 34.13ff Gärtner, 205.3ff Daremberg–Ruelle.

[180] Galen tells us that his father decided that he should take up a medical career after a dream (e.g. K x 609.8ff, XIX 59.9ff).

[181] *CMG* v, 10, 1 108.1ff, K XVII A 214.7ff: the short treatise on the diagnosis from dreams that appears in Kühn's edition, vi 832ff, is thought to be a compilation from this passage.

[182] Artelt 1937, cf. also Wächter 1910, Pfister 1935, and Dodds 1951, e.g. pp. 35ff.

writers and popularly for remedies for 'ills' of one type or another, whether diseases or other kinds of misfortune.[183] The term φάρμακον, which is the regular word for 'drug' and – with or without a qualifying adjective – for 'poison' in medical literature and elsewhere,[184] is also used more generally of any kind of remedy or device.[185] As Moulinier has illustrated in his examination of classical material and as most recently Mary Douglas has emphasised in a more general anthropological context,[186] notions of the 'clean' and the 'dirty' usually reflect fundamental assumptions concerning the natural, and the moral, order, and the Greek terms for purification and cleansing span both spheres and permit no hard and fast distinction between them. Thus καθαρμοί, the term which is used of the purifications criticised in *On the Sacred Disease*,[187] by Empedocles of his religious poem concerning the salvation of the δαίμων,[188] and elsewhere of the rites used to remove pollution, for example after the shedding of blood,[189] is also used of natural evacuations, as, for instance, in Aristotle of the premature discharge of the amniotic fluid in childbirth.[190] The term κάθαρσις covers a similar range. This was the word used by the doctors of natural, or medically induced, evacuations from the body,[191] but it too could refer to ritual purifications after moral pollution.[192]

---

[183] Just as νόσος is used of many other types of ill besides diseases, so conversely ὑγιής is used generally of 'the sound' in many other contexts besides medical ones. In both cases the degree to which these 'extended' uses were understood as metaphors is far from clear.

[184] φάρμακον is generally used in Homer with a qualifying adjective, e.g. ἐσθλά and λυγρά *Od.* IV 230, ἤπια *Il.* IV 218, ὀδυνήφατα *Il.* V 401, οὐλόμενον *Od.* X 394. For φάρμακον used without a qualifying adjective to mean 'poison', see, e.g., Thucydides II 48, Plato, *Phd.* 115 a.

[185] As in Herodotus III 85 (when Oebares says he has a trick to ensure that Darius will become king). Cf. also, e.g., Hesiod, *Op.* 485, Euripides, *Ba.* 283, Plato, *Phdr.* 274 e.

[186] Moulinier 1952, Douglas 1966, and cf. R. C. T. Parker 1977.

[187] *Morb. Sacr.* ch. 1 paras. 4, 12, 23, 25, 39, 42, 46, ch. 18 para. 6 (G) (L VI 352.8, 354.19f, 358.3, 7, 362.6, 13, 364.8, 396.8).   [188] See above, p. 34.

[189] As in Aeschylus, *Ch.* 968, *Eu.* 277, 283, Sophocles, *OT* 99, 1228, cf. Euripides, *Ba.* 77, and the practices referred to by Plato, *R.* 364e f.

[190] *HA* 587b1. Cf. Plato, *Sph.* 226d ff where καθαρμός is a generic term, the genus τὸ καθαρτικὸν εἶδος being divided into two kinds, purgings – καθάρσεις – relating to bodies (which include those brought about by gymnastics and medicine) and those relating to souls.

[191] E.g. of the purging of the menses, *Aër.* ch. 4, *CMG* I, 1 58.31, *Aph.* V 60, L IV 554.7, Aristotle, *HA* 572b29, *GA* 775b5, and of the afterbirth, *Aër.* ch. 7, *CMG* I, 1 60.35, Aristotle, *HA* 574b4. The noun κάθαρσις, like the verb καθαίρω, is regularly applied to the action of purgatives, e.g. *Aph.* II 35 (L IV 480.13), *Acut.* ch. 7 (L II 276.6 and 7), cf. pseudo-Aristotle, *Pr.* 864a34. In *Morb. Sacr.* the term is used in connection with a theory about the origin of phlegmatic constitutions, which arise because of inadequate κάθαρσις of the brain before birth, ch. 5 paras. 1–9 (G) (L VI 368.10ff, e.g. 13).

[192] As in Herodotus I 35, of the purificatory rites used by Lydians and Greeks to remove the pollution of murder, cf. Plato, *Lg.* 872e f. At *Cra.* 405ab Plato expressly links the κάθαρσις and καθαρμοί of doctors and priests.

Two main points that emerge quite clearly from a considerable body of evidence are (1) that the methods of healing used both in what we may call 'rationalistic' and in temple medicine had much in common – the priests had recourse to drugs, prescriptions concerning diet, and phlebotomy,[193] just as some of the rationalistic doctors did not rule out amulets and prayers; and (2) that in describing what they were attempting to bring about the rationalistic doctors might employ some of the very same terms (such as 'purification') that had a wide analogous use in religious contexts. Prognosis, explicitly recognised as an important means of winning over patients to accept treatment (see below, pp. 90f), may well have seemed to some a kind of soothsaying. Indeed it is sometimes referred to by the doctors in terms that are obviously reminiscent of the role of the prophet. Thus the writer of *Prognosis* recommends that the doctor should 'tell in advance' 'the present, the past and the future' in the presence of his patients,[194] and so too does the writer of *Epidemics* I ch. 5.[195]

At the same time, despite these important signs of the overlap between the different strands that go to make up Greek medicine in the fifth and fourth centuries B.C., those strands remain, in certain respects at least, none the less distinct, and indeed the practitioners in question were evidently in direct competition with one another. Some of the common features we have identified appear to reflect a desire not so much to compromise with other approaches, as to outdo them. A theorist such as the author of *On Regimen* IV does not merely accommodate the traditional belief in the predictive value of dreams: he produces a systematic framework for their interpretation as diagnostic signs. Conversely, to be seen to be not just as good as, but far better than, mortal physicians, the god – through his priests or interpreters – saw fit to incorporate many of their techniques, as well as adding some special ones, such as temple incubation, of his

---

193 Thus phlebotomy was practised on the god's command in the time of Aelius Aristides, to judge from XLVIII 47 (cf. above, p. 41).

194 *Prog.* ch. 1, L II 110.2f: προγιγνώσκων...καὶ προλέγων παρὰ τοῖσι νοσέουσι τά τε παρεόντα καὶ τὰ προγεγονότα καὶ τὰ μέλλοντα ἔσεσθαι, cf., e.g., *Il.* I 70 on the prophet Calchas: ὃς ἥδη τά τ' ἐόντα τά τ' ἐσσόμενα πρό τ' ἐόντα and cf. Hesiod, *Th.* 38.

195 *Epid.* I ch. 5, L II 634.6f: λέγειν τὰ προγενόμενα· γιγνώσκειν τὰ παρεόντα· προλέγειν τὰ ἐσόμενα. Cf. such other texts as *Fract.* ch. 35, L III 538.6, *Art.* ch. 9, L IV 100.4 (it is the business of the doctor to foretell, καταμαντεύσασθαι, such things) and ch. 58, 252.14f (which speaks of 'brilliant and competitive – ἀγωνιστικά – forecasts'). On the other hand *Acut.* ch. 3, L II 242.3ff, insists that medicine should not be confused with divination, and *Prorrh.* II chh. 1f, L IX 6.1ff, criticises doctors for 'marvellous' predictions: the author says he will not himself engage in such divinations (ἐγὼ δὲ τοιαῦτα μὲν οὐ μαντεύσομαι, ch. 1, 8.2, cf. προρρηθῆναι ἀνθρωπινωτέρως ch. 2, 8.11), and insists that his own predictions will be based on signs, σημεῖα, e.g. ch. 1, 8.2ff and ch. 3, 10.23ff.

own.[196] We have considered in detail the attack mounted by the author of *On the Sacred Disease* against the 'purifiers': but we also have evidence that the practitioners of temple medicine were critical of ordinary doctors. Thus one of the documents from Epidaurus describes a cure achieved by the god when the first instruction the god gives the patient is to forbid him to follow the treatment (cauterisation) that had been recommended by the doctors.[197]

There is no question of the practitioners of temple medicine *not* claiming to bring about what we can describe as practical results. In the Epidaurus inscriptions this is precisely what is asserted: the god is represented as tackling, and curing, an extraordinary variety of ailments,[198] ranging from headaches and insomnia to cases of stone, worms, gout, dropsy, tumours, consumption, blindness, epilepsy and injuries from wounds of different kinds. Although in some instances the question of what counted as a successful treatment would obviously be highly debatable, in others there was less room for doubt.[199] Of course we cannot now say what – if anything – underlies the cures claimed:[200] we are in no position to assess either the workings of suggestion on the patients,[201] or the elements of wishful thinking – or even plain fraudulence[202] – on the part of the

---

[196] The fact that the Epidaurus inscriptions also record how the god's advice proved efficacious in some non-medical cases as well (as in the consultations about finding hidden treasure, case 46, or a lost child, case 24, or the recovery of a deposit, case 63) suggests another respect in which the priests of the cult of Asclepius would claim superiority to merely mortal medical men.

[197] Case 48, Herzog 1931, p. 28. There may, of course, have been a particular added reason for the god to forbid a treatment that was generally recognised as being drastic (cf. the remarks concerning the hazards and misuse of cauterisation in *Art*. ch. 11, L iv 104.22ff, and Iamblichus' report that the Pythagoreans avoided the use of cautery, *VP* 163, 244). From a later period Aelius Aristides provides many examples where the god overrules the diagnoses or therapies of ordinary physicians, e.g. *Or*. xlvii 61–4, 67–8, cf. 54–7, xlix 7–9.

[198] As well as non-medical problems, see above, n. 196.

[199] In such 'surgical' cases as the extraction of a spear from the jaw (case 12) there could be little doubt about the end-result said to have been achieved. Again in the cases where a barren woman consults the god in order to conceive (e.g. cases 31, 34, 42), whether or not she had a child was fairly easily verifiable.

[200] The various views that have been expressed by modern scholars on the cures claimed at Epidaurus and elsewhere in the ancient world are summarised in Edelstein and Edelstein 1945, ii, ch. 3, especially pp. 142ff.

[201] The need for faith, and the folly of doubting or scoffing at the god, are recurrent motifs in the inscriptions (e.g. cases 3, 4, 9, 10, 35, 37: in case 36 the god punishes a scoffer by crippling him). From a later period we may compare a text in which Galen remarks on the psychological effects of belief in divine healing. At *CMG* v, 10, 2, 2 199.4ff, K xvii b 137.7ff, he observes that the faithful will submit to a course of treatment they would never normally agree to – from ordinary doctors – when they believe that the god recommends it.

[202] We may note, at least, that the question of due recompense to the god is another recurrent theme in the inscriptions (e.g. cases 4, 5, 8, 10, 25: in case 22 a man who was cured for blindness but omitted to make his thank-offering becomes blind again,

priests who had the inscriptions made. But that does not affect the point that the inscriptions *claimed* practical results in a wide variety of cases: they were indeed in all probability set up in large part to *advertise* what the god could do.

The importance of this becomes apparent when we refer back to the anthropologists' debate on the general aims of magical behaviour in traditional societies. As we noted at the outset (pp. 2f), the view that such behaviour should be seen as expressive or affective, rather than as would-be efficacious, has been argued forcefully, and evidently with a good deal of justification, since it provides a clearer understanding of the meaning and function of many magical beliefs and practices. Yet so far as our evidence for Greek medicine of the fifth and fourth centuries B.C. is concerned, the practitioners of temple medicine appear to have accepted a battle on the same grounds as the Hippocratic doctors – in that both sides appeal to, and look to be judged by, the practical results they achieved.

Further confirmation of the point comes from the data provided by our chief Hippocratic text. The symbolic nature of some of the recommendations that are ascribed to the 'purifiers' in *On the Sacred Disease* – for example the prohibition against wearing black or against crossing the hands or legs[203] – is clear enough. At the same time the burden of one of the main charges the Hippocratic writer brings against his opponents is that they neither know what causes the disease nor treat it properly, and he evidently thinks of them as making claims on both scores. He says that they pretend to have superior knowledge, among other things about what causes and cures the disease.[204] Throughout his opening polemic he describes the purifiers as attempting to alleviate epilepsy by the use of charms and the like,[205] even though their ministrations are all useless, and most importantly he says that they take the credit should any of those whom they treat recover, although they guard themselves against failure by saying that the gods are to blame.[206] All through his attack, in fact, he treats the actions of the purifiers as if they were to

---

although he is once again healed by the god after incubation; in case 7 a man is punished with marks on his face for not giving the god the money he had received from a patient for being healed).

[203] *Morb. Sacr.* ch. 1 paras. 17 and 19 (G) (L VI 356.6f, 8f). At ch. 1 paras. 33ff (G) (L VI 360.13ff) we have an outline sketch of what may have been a quite elaborate symbolic schema associating certain behaviour on the part of the patient with particular deities, e.g. 'if he utters a higher-pitched and louder cry, they say he is like a horse and blame Poseidon'.

[204] See especially ch. 1 paras. 11, 20 and 27 (G) (L VI 354.15, 356.9ff, 358.13ff).

[205] E.g. ch. 1 paras. 4, 23f, 26 (G) (L VI 352.7ff, 358.1ff, 11ff).

[206] Ch. 1 para. 20 (G) (L VI 356.9ff), cf. above p. 18.

be assessed not – or certainly not merely – in terms of their felicity or appropriateness, but in terms of the practical results that were obtained.

Although it has, in the past, often been argued that magical beliefs and practices are particularly common in relation to situations beyond the technological control of the group or society concerned,[207] here too our Greek evidence provides grounds for caution. First, it is not the case that the help of the gods was invoked only, or even mainly, for particularly difficult or intractable cases. On the contrary, to judge from the cures claimed,[208] it seems that the god was consulted on what the Greeks themselves considered straightforward cases (such as injuries from wounds) as well as on more difficult 'acute' diseases (such as consumption).[209] Conversely, and more

---

[207] This was Malinowski's view and it is one that figures prominently in Evans-Pritchard's study of the Zande (Evans-Pritchard 1937). One may compare, more recently, Horton on the Kalabari ('Sometimes, however, the sickness does not respond to treatment, and it becomes evident that the herbal specific used does not provide the whole answer. The native doctor may rediagnose and try another specific. But if this produces no result the suspicion will arise that "there is something else in this sickness"...It is at this stage that a diviner is likely to be called in...Using ideas about various spiritual agencies, he will relate the sickness to a wider range of circumstances – often to disturbances in the sick man's general social life', Horton 1967, p. 60) and Tambiah ('Although we should not judge their *raison d'être* in terms of applied science, we should however recognize that many (but not all) magical rites are elaborated and utilized precisely in those circumstances where non-Western man has not achieved that special kind of "advanced" scientific knowledge which can control and act upon reality to an extent that reaches beyond the realm of his own practical knowledge', Tambiah 1973, p. 226, with a reference to Evans-Pritchard's conclusion that Zande rites were most 'mystical' 'where the diseases they dealt with were the most acute and chronic'), but cf. also the critical remarks of Thomas 1971, pp. 774ff, 785ff.

[208] See above, p. 46. Similarly Aelius Aristides invokes divine assistance for every kind of medical problem.

[209] The question of whether a condition is beyond cure – even beyond treatment – is, however, one that occupied several of the Hippocratic writers. *De Arte* even makes it one of the defining characteristics of the art of medicine 'to refuse to undertake to cure cases in which the disease has already won the mastery, knowing that everything is not possible in medicine' (ch. 3, *CMG* I, I 10.21ff). Cf. *Fract.* ch. 36, L III 540.9ff on the dangers attending the reduction of the thigh and upper arm ('one should especially avoid such cases if one has a respectable excuse, for the favourable chances are few and the risks many'). Finally *Prog.*, too, is aware of the problem: 'by realising and announcing beforehand which patients were going to die, he would absolve himself from any blame' (ch. 1, L II 112.10f). Yet at no stage do any of these writers suggest that in difficult, or hopeless, cases their patients should have recourse to temple medicine. The one passage that has been taken to be an exception to this rule is in *Morb. Sacr.* itself, ch. 1, paras. 41ff (G) (L VI 362.10ff) where the writer says that what the charlatans should have done is not to treat the epileptics as if they had committed sacrilege, but to 'take the sick into the temples, there by sacrifice and prayer to make supplication to the gods', not to bury their καθαρμοί or throw them into the sea, but to take them into the temples as offerings. Herzog 1931, p. 149, concluded from this that the author himself actually *approved* of temple medicine: yet he is, rather, merely arguing that his opponents are *inconsistent*. What they should have done, *if* the god had been responsible for the disease, is to take the patients to the temples. But that

importantly, the testimony of *On the Sacred Disease* would tend to run counter to any thesis to the effect that the undermining of magical beliefs follows an increase in the control that could be exercised over the areas of experience to which the beliefs in question related. It is striking that our chief critical text deals with a topic – epilepsy – where the author himself, so far from having any effective means of treating the disease, was – *we* should have said – just as helpless as the charlatans he attacks. True, the writer *states* that epilepsy, like every other disease, is curable.[210] Yet we have only to consider how he intended to treat it – that is, principally, by the control of the temperature and humidity of the body by variations in the diet – to appreciate that, as with the 'purifiers' he was attacking, such comfort as his patients derived from his ministrations must have been very largely of a psychological nature, and thanks to their confidence in his ability or authority, rather than the result of his having, in this case, any real means of cure at his disposal.

No straightforward account, in which 'science' and 'philosophy' together and in unison stand opposed to 'magic' and the 'irrational', can be sustained in the face of the evident complexities both *within* and *between* the theory and practice of medicine on the one hand and those of the investigation concerning nature on the other. Our next task is to go back to the two key concepts of nature and of cause to examine what the different strands of speculative, rationalistic inquiry owed to pre- or at least non-speculative thought, as a first step towards determining how far the former should be seen as marking a radical break with the latter.

THE NOTIONS OF 'NATURE' AND 'CAUSE'

The idea of nature as implying a universal nexus of cause and effect comes to be made *explicit* in the course of the development of Presocratic philosophy, though we have emphasised the dangers of representing the Presocratic philosophers as having a uniform set of beliefs and attitudes on the subject. Yet an assumption of the regularity of natural phenomena is *implicit* in much of human behaviour. Whatever other factors the farmer may believe he has to

argument is based on a premiss – that the god is responsible for the disease – that the Hippocratic writer himself rejects. As we have seen (p. 26), the only sense in which he is prepared to say the disease is divine is that in which all diseases are divine – because the whole of nature is.

210 *Morb. Sacr.* ch. 18 paras. 1ff, especially 6 (G) (L VI 394.9ff, 396.5ff), see above pp. 21f.

take into account in order to insure a good crop of wheat, he knows that he will have no crop at all unless he sows seed. The hunter takes it that his arrows will normally fly straight: they will not be deflected from their course; it is not as if his chances of making a hit are as good if he points his bow in *any* direction and takes no aim at all, as if he takes careful aim at his target. We all take it for granted that stones fall, fire and smoke rise, however imprecise our ideas about 'heavy' and 'light' may be. To understand, let alone to learn from, experience at all presupposes *some* idea of the regularity of phenomena, although that idea may well be neither explicit nor universalised.[211] It may be believed, for instance, that that regularity is subject not just to exceptions,[212] but to interference from divine powers. Indeed the notion of what takes place normally or regularly may be, and often is, the basis of inferences that such an interference has taken place. A clear instance of such an inference in Homer is Teucer's reaction when his bow-string snaps when he aims at Hector at *Iliad* xv 458ff.[213] That a bow-string that he fitted new that morning (νεόστροφον, πρώϊον, 469f) should have snapped, is taken as a sign that there must be some δαίμων thwarting him, since new bow-strings are not expected to break – and similar inferences that the hand of heaven is at work can, naturally, be paralleled extensively throughout Greek literature.[214]

[211] Although, in Homer, what *we* should call natural phenomena are often associated with the gods, divine beings are not always invoked in their description, especially in the similes, e.g. *Il.* v 864f, xiv 16ff, xv 618ff, xvii 263ff.

[212] One should distinguish cases where what is regular corresponds to what is always the case (for example that the sun rises in the East) from others where it admits of exceptions (for instance the growth of a crop of wheat). 'Nature' for the Presocratic philosophers and Hippocratic writers encompasses both types of phenomena, but they do not distinguish explicitly between them as Aristotle was to do with the principle that nature is what happens 'always *or for the most part*' (see above, p. 36 n. 129).

[213] Other notable occasions when the exceptional character of an event is used as the basis of an inference that the gods are at work are *Il.* viii 139ff, xiii 68ff, xv 290ff, xvi 119ff and xxiv 563ff (where Achilles infers that a god brought Priam through the Achaean camp, since without divine help he would not have dared to come) and even more commonly the general run of the battle is cited as evidence of whom the gods are favouring. Cf. also *Od.* xvi 194ff, xx 98ff (Zeus, asked for a sign, thunders from a cloudless sky).

[214] To cite an example from the classical period, at i 174 Herodotus notes that in the digging of the canal across the isthmus at Cnidus the workforce suffered an exceptional number of injuries, particularly in the eyes, from which they concluded that they should consult Delphi to find out what was hindering them. τέρατα, portents or monsters, were, of course, generally interpreted as signs from heaven or expressions of divine anger, though what was believed to be an exceptional phenomenon varied with the state of knowledge of the individuals concerned at the time. Archilochus expresses consternation at an eclipse of the sun (Fr. 74, D): but the famous case of the fatal hesitation of the Athenian army under Nicias when an eclipse of the moon occurred in their retreat from Syracuse in 413 B.C. (Th. vii 50) shows that such eclipses were still generally feared in the late fifth century.

This serves to illustrate both a connection, and a difference, between natural philosophy and pre- or non-philosophical thought. The connection is that the notion of φύσις may be said to build directly on ordinary experience of the regularities of nature:[215] in particular inferences to divine interventions based on the breaching of those regularities presuppose a firm idea of those regularities themselves. But the difference lies in the fact that the idea that every physical phenomenon has a natural cause is neither stated – nor, it would appear, assumed – as a universal rule before philosophy. As we saw, some idea of nature does not, by itself, exclude all beliefs in personal divine interventions,[216] but once the notion of nature as a universal principle is grasped, then those interferences must be seen either as 'miracles' – the suspension of nature – or as cases of 'double determination' – where the god works through physical causes. The explicit expression of a universalised concept of nature involves a corresponding development or clarification in the notion of marvels or miracles: the category of the 'supernatural' develops, in fact, *pari passu* with that of the 'natural'.[217] Even in the philosophers, indeed, as we noted when discussing Empedocles, quite intensive investigations of nature may be combined with a belief in the possibility of wonder-working – although the exact status of the marvellous effects that Empedocles claimed could be produced is not clear.[218]

Yet if there is a distinct ambivalence in the position of some philosophers and of some medical writers, in others the emphasis is more clearly[219] on the all-embracing character of the principle that every physical event has a determinate natural cause. While the idea of what is natural in the sense of what is usual permits exceptions, the notion of what is contrary to nature, παρὰ φύσιν, comes to be used in that sense (the unusual, the irregular) not in a sense that implies that such events either have no physical cause or have causes that lie outside the domain of nature. It is the conception of a domain of nature encompassing *all* physical phenomena that is – eventually – developed by *some* philosophers and that in *some* medical writers becomes the cornerstone of the rejection of the belief in the possibility of divine intervention in physical conditions. 'Marvels'

---

215 Cf. Vlastos 1975, ch. 1.
216 See above, pp. 29ff, especially 30, on Herodotus.
217 Not only is the category of the 'supernatural' the correlative of that of the 'natural', but what are treated as 'marvellous' phenomena come to be more clearly defined once the senses of 'nature' are distinguished.
218 See above, pp. 33ff.
219 In some cases, however, reservations are in order: cf. above, p. 32 n. 109 on Leucippus.

(θαύματα) and 'monsters' (τέρατα) then pick out phenomena that are unusual but in principle intelligible, even if not yet understood:[220] and on such a view 'double determination' is otiose.

The second, related, key notion that we identified as underlying the attack on the purifiers in *On the Sacred Disease* is that of cause. Here too there are certain apparent connections, as well as differences, between the philosophical and medical writers and earlier thought. It is obvious that in the context of human behaviour, especially, the questions of who initiated or performed an action, of what human or indeed non-human agent was at work and thus in some however imprecise sense responsible or to blame for it, are of universal human interest and concern, although the assumptions made about the notion of 'responsibility' may differ profoundly from one society to another,[221] and in particular the idea that an event is due to some god or to fate may well be combined with – rather than thought of as alternative to – the notion that a human or humans are to blame. In the context of the development of Greek views on causation, it has long been recognised that much of the terminology, and some of the key ideas, originate in the human sphere. Of the words that came to be applied to causation in general, αἰτία and the cognate adjective αἴτιος are originally used primarily in the sphere of personal agency, where αἰτία may mean 'blame' or 'guilt'.[222]

Mythological 'aetiologies' are explanations only in a quite restricted sense. To attribute earthquakes to Poseidon is, from the point of view of an understanding of the nature of earthquakes, not to reduce the unknown to the known, but to exchange one unknown for another. While Poseidon's motives can be imagined in human terms (providing an answer of a kind to the question 'why?'), *how* an earthquake occurs is not thereby explained nor indeed at issue. If there is no question of assigning a historical origin to an interest in causal explanations of *some* kind, the deliberate investigation of how particular kinds of natural phenomena occur only begins with the

---

[220] By the time we come to Aristotle, at least, where τέρατα are seen as failures of the final cause (*Ph.* 199b4), they are said to be contrary to nature not in its entirety, but as what occurs in the generality of cases. 'As for the nature which is always and by necessity, nothing occurs contrary to that: unnatural occurrences are found only among those things which happen as they do for the most part, but which may happen otherwise...Even that which is contrary to nature is, in a way, in accordance with nature' (*GA* 770b9ff).

[221] The slow development of a coherent notion of responsibility in Greek thought has been traced by Adkins 1960.

[222] E.g. Pi. *O.* 1.35, cf. αἴτιος in the sense of 'culpable' in *Il.* 1 153. Some of the residual social and political associations of Greek terms for causes are discussed briefly in Lloyd 1966, pp. 230f.

philosophers: it was they who first attempted to explain what thunder, lightning, eclipses and the like are in terms of more familiar phenomena and processes.

Nevertheless to document the development of ideas about causation *as such*, we have, once again, to supplement our meagre evidence for the Presocratic philosophers from our other sources. The questions of establishing responsibility for an action, and of motivation and intention, are of recurrent concern both in the orators and, in certain contexts, in the dramatists, and passages in Herodotus and Thucydides show a developed interest in the problems of isolating the causes not just of historical events,[223] but also of certain physical phenomena.[224] On the latter question, however, it is again the medical writers who provide our richest mine of information.

The topics of what brought about a particular illness or was responsible for the amelioration in a patient's condition – and more generally of the causes, and cures, of particular types of disease – are repeatedly discussed in the Hippocratic Corpus. *On Regimen in Acute Diseases* is one work that draws attention to the fact that the same condition may have different causes.[225] *On Regimen* III remarks that 'the sufferer always lays blame – αἰτιῆται – on the thing he may happen to do at the time of the illness, even though this is not responsible – οὐκ αἴτιον ἐόν'.[226] *On Ancient Medicine* also notes that 'if the patient has done something unusual near the day of the disease such as taking a bath, or going for a walk, or eating something different (when such things are all rather beneficial than otherwise), I know that most doctors, like laymen, assign the cause (αἰτίη) [of

---

[223] Thucydides' views and comments on historical causation can be studied in many other passages besides those that deploy the terms αἰτία and πρόφασις (as in the famous and much discussed text I 23, on which see most recently De Ste Croix 1972, pp. 52–8, and Rawlings 1975). One passage of special interest in relation to the question of the survival of traditional beliefs is II 17, where he remarks that the oracle given to the Athenians that 'it were better for the Pelasgian ground to be unoccupied' came true in the opposite sense to what was expected. It was not because of the unlawful occupation of the sanctuaries that the city suffered calamities, but rather it was because of the war (and its calamities) that the sanctuaries were occupied.

[224] Thus in Herodotus' discussion of the Nile's flooding (II 20ff) he argues against the theory that the Etesian winds are responsible (αἴτιοι) on the grounds (*a*) that the Nile floods even when the Etesian winds do not blow, and (*b*) other rivers are not affected in a similar way by the Etesians.

[225] *Acut.* ch. 11, L II 314.12ff (cf. Theophrastus, *HP* IX 19.4). Cf., e.g., *Fract.* ch. 25, L III 496.11ff, on the harmful effects of bad bandaging – which the physicians in question do not recognise as the cause, αἰτίη, 500.10, of the exacerbations – and the more general discussion in *de Arte* chh. 4ff, *CMG* I, 11.5ff, concerning what is brought about by the art, and what is merely fortuitous, in disease and the recovery of health. *Flat.* ch. 1, *CMG* I, I 91.16ff is one text that points out the importance of knowing what is responsible (τὸ αἴτιον) for diseases for determining effective remedies.

[226] *Vict.* III ch. 70, L VI 606.20ff.

the disease] to one of these things, and in their ignorance of the responsible factor (αἴτιον), they stop what may have been most advantageous'.[227] The same treatise attacks the hypothesis that 'the hot', for example, is an important cause of diseases by suggesting that it is not 'the hot' itself, but the other powers it is compounded with, that brings about illnesses,[228] and the writer states the criteria that he believes a cause must fulfil: 'We must, therefore, consider the causes (αἴτια) of each condition to be those things which are such that, when they are present, the condition necessarily occurs, but when they change to another combination, it ceases.'[229] The idea of a necessary condition is first expressed in the form of the 'that without which' (ἐκεῖνο ἄνευ οὗ) in Plato's *Phaedo*.[230] But without any special terminology,[231] the author of *On Ancient Medicine* certainly has a working notion of the distinction between causal and merely con-comitant factors and conceives the former in terms of a set of factors that (as we should say) are together both necessary and sufficient conditions of the disease.[232] We should, however, add first that, like most Hippocratic writers, he is, in practice, both vague and dogmatic in his pronouncements on the causes of diseases, and, secondly and more importantly, that neither he nor any other Hippocratic writer engages in systematic testing in this context, varying the conditions of the patient or his treatment in an attempt to isolate the causal factors at work.

[227] *VM* ch. 21, *CMG* I, 1 52.17ff. *Epid.* II sec. 4 ch. 5, L v 126.10ff is one text that implies a distinction between treating the symptom and treating the underlying cause.

[228] E.g. *VM* ch. 15, *CMG* I, 1 46.18ff. In chh. 16ff he argues that hot and cold have little 'power' in the body partly on the grounds that heat is readily countered by cold and vice versa (48.10f, 49.16ff, 50.9ff), and in ch. 17 he concludes that heat is merely a concomitant (συμπάρεστι) in fevers (48.21ff, 49.2).

[229] *VM* ch. 19, *CMG* I, 1 50.7ff: cf. the insistence, in ch. 20, on knowing not merely what a pain is but also why it comes about (διὰ τί, 51.24, cf. αἴτιος, 52.3).

[230] *Phd.* 99ab, where Socrates denies that the 'that without which' can truly be said to be an αἴτιον, for the αἴτιον of an event must state why it occurs in terms of the good aimed at.

[231] Rawlings, following Weidauer 1954, has argued that the Hippocratic writers develop πρόφασις as a special term (a lexeme from φαίνω, not from φημί) for the pre-condition of a disease: 'a *prophasis* is by its very nature...visible,...it is...from outside,...it precedes a disease and can be useful in predicting the course of the disease' (Rawlings 1975, p. 43). In this sense it is close to σημεῖον (and in certain contexts to αἰτίη) but to be firmly contrasted with αἴτιον (the term for a necessary or primary cause). Reservations must, however, be expressed both about how far the two lexemes remained distinct, and about the extent to which the Hippocratic use was standardised and specialised. Generally used for an external sign or accessory cause, as opposed to necessary cause, it is sometimes a synonym for αἴτιον in the latter sense (as Rawlings recognises to be the case at *Mul.* 1 ch. 62, L VIII 126.14ff, though he explains this as a later development).

[232] It is perhaps not too far-fetched to see the principle stated in *VM* ch. 19 as a remote ancestor to Mill's Canons of Agreement and Difference or at least of Bacon's *Tabula Essentiae et Praesentiae* and his *Tabula Declinationis sive Absentiae in proximo*.

Evidence of reflection on the nature of causation can be cited from a number of medical texts. The importance of this for the criticism of traditional beliefs is clear when we turn back to some of the arguments that the author of *On the Sacred Disease* brings against his opponents. At one point he maintains that if the purifiers prohibit certain foodstuffs and the wearing of certain clothes on the grounds that these are relevant to the sacred disease, then this conflicts with their claim that the gods are at work: 'If contact with or eating of this animal generates and exacerbates the disease while abstinence from it cures the disease, then no god can be blamed (αἴτιος) and the purifications are useless: it is the foods that cure and hurt, and the idea of divine intervention comes to naught.'[233] From the Hippocratic writer's point of view it is what is regularly associated with the disease that must be held responsible for it. Now the purifiers themselves might well remain unmoved by this argument, and maintain that divine causation operates *in addition to* the physical factors they pick out as significant. Moreover there is an even greater difficulty in positively excluding supernatural causes when the main 'evidence' adduced that they are at work is the very events they are supposed to bring about – when the causes are not known independently of the effects.[234] Nevertheless the more that *regular observable* connections of physical causes and effects can be established in diseases, the easier it will be for any doctor who chooses to do so to argue that the invocation of other factors is unnecessary and unjustified, and that this is so whether the gods or divine beings are imagined as acting according to moral principles or quite capriciously, and whether the divine is cited as the sole, or an additional, explanation of diseases. The Hippocratic writer has an *ad hominem* argument against the purifiers, that if eating certain foods brings about the disease and abstention its cure, then to appeal to the gods is superfluous and mistaken: and in general he evidently hopes or assumes that his audience at least – if not his opponents themselves – will agree that whatever explanation is offered, it must consist in physical factors to the exclusion of any reference to divine or supernatural agencies.

We may now try to take stock of some of the conclusions from this

---

[233] *Morb. Sacr.* ch. 1 para. 23 (G) (L VI 358.1ff).

[234] We may contrast the appeal to divine causes to explain a class of phenomena (such as all cases of epilepsy) with invoking such causes to account for exceptional individual events (where what happens is unusual or abnormal, and where that fact may even be cited as evidence that the gods are at work, as in the case of Teucer's bowstring, cf. p. 50 above).

first inquiry. We have found that a number of popular beliefs and practices come to be challenged not only in the context of religion (from at least the sixth century) but also (from the late fifth) in the domain of medicine. 'Magic' and 'magician' are among the terms employed to disparage some such practices and their practitioners. The connotations and denotations of these terms are not fixed (any more than those of 'charlatan', ἀλαζών, were): rather they are used of what particular writers happen to disapprove of, the association with a little known foreign tribe no doubt contributing to their derogatory undertones.[235] Nevertheless the grounds for the rejection of one set of such beliefs are made plain enough in the main text that engages in a sustained polemic against them. The writer of *On the Sacred Disease* has a conception of nature, and a view of what constitutes a causal explanation, that rule out supernatural intervention in diseases.

The background of debate to which the discussion in this treatise belongs is an intricate one of complex relations both within medicine and within philosophy – and between the two. Although it is in the context of the philosophers' inquiries that the key move – the explicit expression of the idea of nature as a universal principle – is made, it is out of the question to represent all the Presocratic philosophers as sharing precisely the same views on this topic – and out of the question, too, to see them all as having adopted a uniformly sceptical and critical attitude towards traditional beliefs. Equally on many theoretical and practical issues the dividing lines that separate healers of different kinds are anything but clear-cut.

The weaknesses and vulnerability of the position of the Hippocratic rationalists are striking. This is firstly a matter of the insecure demarcations between different kinds of medical practitioners that we have just mentioned. Healers of very varied persuasions shared many therapeutic and diagnostic practices and often used the same terms to describe their aims. Secondly, there is the inexact and fanciful nature of the actual anatomical and physiological 'knowledge' that the Hippocratic writers generally claimed: we have illustrated this from *On the Sacred Disease*, and the Hippocratic collection is full of similar examples. Thirdly, although many commentators have connected the rejection of magic with an increase in the effective technological control that could be exercised over the phenomena in question, we have seen reason to doubt this. In the case of epilepsy, at least, the claims that *On the Sacred Disease* makes concerning the possibilities of cure are wishful thinking.

---

[235] Cf., e.g., Mauss (1950) 1972, p. 31, on the connection between sorcery and foreigners.

Nevertheless it is on (among other things) the question of practical effectiveness that the Hippocratic writer finds a weak spot in his opponents' position. He clearly represents them – and equally those who set up the Epidaurus inscriptions represent their god – as making claims concerning the cures effected, and this gives his attack a purchase it would not otherwise have had. Once battle was joined on that ground, the Hippocratic writers and the purifiers were, and could be seen to be, in direct competition with one another.

But if the issue in medicine was partly a matter of results, the reasons offered for success or failure varied with the individuals or groups concerned. If neither the Hippocratic writers nor – we may imagine – the temple healers were unduly deterred by failures, this was because each side had some confidence in the *kind* of explanation they proposed. Against the purifiers, the Hippocratic rationalists insisted on aetiologies, and on treatments, that referred exclusively to physical factors (though there was, as we have seen, plenty of disagreement about what came under that head). How far they persuaded their own contemporaries was another matter. Temple medicine, after all, not only continued to flourish, but actually expanded, after the fifth century B.C. The Hippocratic writers certainly had no knock-down refutation of double determination, particularly as a stubborn opponent might always multiply *ad hoc* explanations. Moreover the element of over-optimism – or pure bluff – in the Hippocratics' own position is clear: many of their treatments were ineffectual and many of the correlations and causal connections they announced as fact (such as restriction of epilepsy to those of phlegmatic constitution) were imaginary. Yet what they could and did do was – negatively – to undermine their opponents' doctrines by arguing that appeals to the gods are arbitrary and superfluous, and that secondary elaborations were indeed just that, excuses or screens for failure, and – positively – to offer an alternative explanatory framework. If some awareness of the determinate characters of things and of the regularities of natural causes and effects is part of all human experience, the plausibility of the Hippocratic rationalists' view rested partly on the fact that it was an extension or extrapolation of that awareness, now made explicit, universalised, and treated as the sole valid explanatory principle. Finally while many of their proposed correlations might be challenged and overthrown, they could hope that their overall position would be strengthened as more observable regularities were established and successful explanations achieved.

The problem of the social conditions that may have furthered or allowed the developments we have described will be discussed in our final chapter. The topic of the growth of observation and research – of the extension of the empirical base of Greek science – will occupy us in chapter three. We have found that the strength of some of the writers who were at the centre of the debate we have considered in this chapter lies in the modes of argument, both constructive and destructive, that they deployed, and this aspect of the development of Greek science will form the subject of our next study.

2

DIALECTIC AND DEMONSTRATION

There can be few societies that do not, in some degree, prize skill in speaking, and the variety of contexts in which it may be displayed is very great. Apart from in the arts of the poet or story-teller[1] and of the seer or prophet, eloquence may be exhibited in a number of other more or less formalised situations, including eulogies of the powerful[2] and contests of abuse such as the song duels reported from the Eskimos.[3] Good speaking and good judgement - and the two are often not sharply distinguished - need to be shown wherever groups of individuals meet to discuss matters of consequence concerning the running of the society, its day-to-day life and internal affairs and its relations with its neighbours.

In the context of law and justice, especially, the members of some non-literate societies are considerable connoisseurs of the speaking skills of litigants and judges, of, for example, their ability to present a case, to cross-examine witnesses and to give judgement. Thus in his study of Barotse law Gluckman reports a rich vocabulary of terms used in Lozi to 'describe different modes of expounding arguments, judicial and other'. They include separate single words for being 'able to classify affairs', for being 'clever and of prompt decision', for 'a judge who relates matters lengthily and correctly', for 'a judge who has good reasoning power and is able to ask searching questions', and again, among terms of disapproval, for 'to speak on matters without coming to the point', for 'to wander away from the subject when speaking', for 'a judge who speaks without touching on the important

---

[1] The poets may, but need not be, specialists: see, for example, Finnegan 1977, pp. 170ff.

[2] See, for example, Finnegan 1977, pp. 188ff on Zulu praise poems.

[3] See Hoebel 1964, p. 93: 'Song duels are used to work off grudges and disputes of all orders, save murder. An East Greenlander, however, may seek his satisfaction for the murder of a relative through a song contest if he is physically too weak to gain his end, or if he is so skilled in singing as to feel certain of victory. Inasmuch as East Greenlanders get so engrossed in the mere artistry of the singing as to forget the cause of the grudge, this is understandable. Singing skill among these Eskimos equals or outranks gross physical prowess.'

points at issue', for 'a person who gets entangled in words' and so on.[4] The Lozi are alert to the question of consistency in evaluating the testimony of witnesses, they distinguish between different types of evidence (direct and hearsay), and they have an operational notion of proof beyond reasonable doubt, although the ideas of consistency and proof are not explicit, let alone made the subject of self-conscious analysis.[5]

While reason and argument are, in one form or another, universal, we are primarily concerned with their role in one particular domain, that of natural philosophy and science, and here it is naturally more difficult to cite parallels to our Greek evidence from other ancient, or from non-literate, societies. Yet contests between wise men are attested in many societies,[6] and some of these contests relate to the discussion of what may loosely be called cosmological topics. Thus Ruben, in particular, has drawn attention, in his studies of Indian religion,[7] to the debates between wise men that occur in some of the older Upaniṣads, notably the Chāndogya Upaniṣad and the Bṛhadāraṇyaka Upaniṣad.[8] In the former there is a discussion between three wise men in which they give their views on – among other matters – the origins of things, including the origin of this world,[9] and in the latter there is a competition for the title of wisest in which the claims of a sage named Yājñavalkya are examined at great length by a series of questioners who pose problems on such topics as the relations between the worlds of the earth, the sun, the moon, the stars and deities of various kinds.[10] In these meetings the contest continues until either the questioner has no more questions to put,

---

[4] See Gluckman 1967, pp. 276f.
[5] See Gluckman 1967, pp. 107ff, 137.
[6] The literature relating to contests of riddles is extensive: see, for example, Huizinga (1944) 1970, pp. 127ff (and for a suggested connection with philosophy, pp. 137ff and 170ff) and Dundes 1975, e.g. pp. 95ff. We may compare the evidence for such contests between mantic seers in ancient Greece, for example that between Mopsus and Calchas, Strabo XIV 1.27.642 (Hesiod Fr. 278 Merkelbach and West), and cf. the material collected in Ohlert 1912 and Schultz 1914, cols. 88ff.
[7] See Ruben 1929 and cf. 1954, ch. 8.
[8] While it is generally agreed that these two are among the oldest extant Upaniṣads, the question of their absolute dates remains highly disputed. See Keith 1925, II p. 502: 'Certainly it is wholly impossible to make out any case for dating the oldest even of the extant Upaniṣads beyond the sixth century B.C., and the acceptance of an earlier date must rest merely on individual fancy', and cf., e.g., Ruben 1954, pp. 83f and Hume 1931, pp. vii f and p. 6 ('the best that can be done is to base conjectures upon the general aspect of the contents compared with what may be supposed to precede and to succeed. The usual date that is thus assigned to the Upanishads is around 600 B.C., just prior to the rise of Buddhism').
[9] Chāndogya Upaniṣad I 8–9 (Hume 1931, pp. 185f).
[10] Bṛhadāraṇyaka Upaniṣad III 1–9 (Hume 1931, pp. 107–26, in part also in Zaehner 1966, pp. 49–60).

or the answerer cannot reply: the winner is the person who has the last word. But although in one exchange the questioner refuses to be satisfied with a reply that simply states a claim to know and he insists that he is answered directly,[11] the ground on which some answers are accepted, but others rejected, are not in general made clear.[12] These are sages who claim to have access to sacred, esoteric wisdom, and although *they* are cross-examined by one another, the content of that wisdom itself is not called in question nor its validity tested.

This evidence shows that confrontation and debate on issues of 'cosmological' significance can be documented in the ancient world outside Greece, even if, in the example we have taken, the scale of such discussions is limited, and, more importantly, the criteria for the acceptance or rejection of statements are not public: although such criteria may be clear to the participants themselves, they are not so to an uninitiated audience. By contrast, our Greek material is far richer and it includes cosmological and natural scientific debates of a quite different order, if only because the evaluation of the strengths and weaknesses of opposing positions is now, in principle, quite open: indeed in some cases the contest was adjudicated by a lay public. In the period from the sixth to the fourth centuries B.C. there are striking, and in some respects unprecedented, developments both in the practice and in the theory of argumentation. Our task in this chapter is to describe these in so far as they relate to the growth of philosophy and science. What principal methods of argument are employed in the early stages of the development of philosophy and science? How far does that development appear to depend on the deployment of new techniques of argument, or on the self-conscious analysis of such techniques? Within what is a very extensive field of investigation,[13] we shall concentrate on two topics of particular importance, namely first the development and use of dialectical methods,[14] and secondly the formulation and application of a rigorous notion of demonstration.

---

[11] Bṛhadāraṇyaka Upaniṣad III 7.1 ('Anyone might say "I know, I know". Do you tell what you know', Hume 1931, p. 114, cf. Zaehner 1966, p. 53).

[12] In particular, what appear as incompatible answers to the same question are allowed, as when seven different answers to the question of how many gods there are are given, each of which is then expounded in turn (Bṛhadāraṇyaka Upaniṣad III 9, Hume 1931, pp. 119ff, cf. Zaehner 1966, pp. 57f).

[13] Among the more useful general discussions of the growth of rhetoric and dialectic in Greece are H. Gomperz 1912, Solmsen 1929, Radermacher 1951 and Kennedy 1963.

[14] The nature of 'dialectic' was, as we shall see, itself disputed by the Greeks. I shall use the term to cover any investigation that proceeds by a critical examination of opinions in a given field of inquiry, especially, but not exclusively, those that proceed from probable premises or commonly accepted opinions (as in Aristotle's definition of 'dialectical' syllogisms, see below, p. 62: contrast Plato's use of the term, below, pp. 101f).

ARISTOTLE'S ANALYSIS OF MODES OF REASONING

The earliest extant works entirely devoted to the analysis of modes of argument and of techniques of persuasion in general are the treatises in Aristotle's *Organon* and his *Rhetoric*, at the end of the period that chiefly concerns us. Although in the *De Sophisticis Elenchis* he claims originality for his work in the systematic analysis of reasoning in general,[15] both there and frequently elsewhere he makes clear his debts to earlier writers on rhetoric, among whom he cites several who were the authors of treatises now lost on the 'art of speaking'.[16] While Aristotle systematises the ideas of his predecessors in some areas, and modifies or goes far beyond them in others, it is useful to start by rehearsing some of the fundamental points from his analysis of argumentation.

First there is his careful and quite complex classification of the different kinds of reasoning. At the beginning of the *Topics*, 100a27ff, for instance, he distinguishes three types of syllogisms, demonstrations (proceeding from premises that are true and primary), dialectical syllogisms (based on generally accepted opinions) and eristic or contentious syllogisms, which are based on what appear to be, but are not in fact, generally accepted opinions, or which merely appear to be based on what are or appear to be such opinions.[17] Strict deductive argument is the subject-matter of the *Prior* and *Posterior Analytics*, the former providing a formal exposition of the

---

[15] *SE* 183b34–184b8.

[16] On these lost *Arts*, see further below, p. 81f. In a fragment of his lost dialogue, the *Sophist* (Fr. 65), Aristotle is reported to have called Empedocles the founder of rhetoric, Zeno that of dialectic. Elsewhere the role of Socrates in the development of dialectic is recognised (*Metaph.* 1078b25ff: cf. also 987b32 where Aristotle contrasts Plato with the Pythagoreans). One major weakness in our sources for the early development of dialectic is the lack of reliable evidence for the work of the so-called Megarians. We hear from Diogenes Laertius (II 106ff) that Euclides of Megara (a younger contemporary of Socrates and not to be confused with the mathematician) founded a school that was called first Megarians, then Eristics and then Dialecticians, acquiring the last name because they put their arguments in the form of question and answer, and one of Euclides' pupils, Eubulides, was, apparently, responsible for the first formulation of a number of important paradoxes, including the Liar. As Ryle puts it (1966, p. 112), 'We know very little about the Megarians, but we know that they had very sharp noses for logical cruces.'

[17] Elsewhere, for example at *SE* 165a38ff, there is a four-fold division of arguments, didactic, dialectical, 'peirastic' (examination-arguments) and contentious. But 'peirastic' is on other occasions said to be a part of dialectic (e.g. *SE* 169b25, 171b4, cf. also 171b9, 172a21). At *Rh.* 1355b17ff the sophist is distinguished from the dialectician by his moral purpose, rather than by his faculty (cf. also *Metaph.* 1004b22ff), and at *SE* 171b25ff 'contentious' reasoners are in turn distinguished from 'sophistic' ones in that the former aim at victory itself, the latter at reputation (with a view to making money). These broad distinctions are generally maintained, even though Aristotle's terminology shows some fluctuation.

various figures and moods of the syllogism, the latter presenting an analysis of the conditions of demonstration – where Aristotle maintains that demonstration proceeds from premisses that are themselves indemonstrable and distinguishes three kinds of such primary premisses, 'axioms' 'definitions' and 'hypotheses'.[18] Dialectic, defined as the method by which we shall be able to reason from generally accepted opinions about any problem,[19] is dealt with in the eight books of the *Topics*, which lead in turn into the discussion of 'sophistical', or merely apparent, refutations in the *De Sophisticis Elenchis*.[20]

Rhetoric, which is said to be the counterpart of dialectic, shares with it the feature that it is quite general, not confined to any particular subject-matter.[21] Interestingly enough, Aristotle suggests at *Rh.* 1354 a 3ff that all men share in both, up to a point, since all men to some degree try to examine and uphold an argument, to defend themselves and to accuse. He defines rhetoric at *Rh.* 1355 b 26f as the faculty of discovering the possible means of persuasion in relation to any subject whatever, and he distinguishes three main branches, forensic (in the law-courts), deliberative (in the public assemblies)[22] and epideictic (ceremonial oratory, where the aim is to praise or blame: here the audience does not adjudicate a legal dispute nor arrive at a political decision, but merely assesses the ability of the speaker).[23] Although rhetorical proofs or arguments naturally bulk large in his discussion, these are far from being the only important consideration in persuasion. Indeed he complains that most earlier accounts of the subject had confined themselves to such matters as the question of arousing the prejudices, compassion or anger of the judges or audience,[24] a topic he takes up for himself in

---

[18] See further below, pp. 111 and 115.

[19] See *Top.* 100 a 18ff. Dialectic is contrasted with 'philosophy' in that the former conducts examinations, the latter produces knowledge (*Metaph.* 1004 b 17–26), in that dialectic reasons to opposite conclusions (*Rh.* 1355 a 33ff), is related to opinion, rather than truth (*Top.* 105 b 30f) and is concerned with another party, while the philosopher conducts his inquiry on his own (*Top.* 155 b 7ff).

[20] *SE* is often treated as a continuation of *Top.*, as is suggested by the fact that the end of *SE*, 183 a 37ff, contains a passage that serves as an epilogue for both works.

[21] *Rh.* 1354 a 1ff. Both are faculties, δυνάμεις, not branches of knowledge, ἐπιστῆμαι (see *Rh.* 1356 a 32ff, 1359 b 12ff).

[22] *Rh.* 1359 b 18ff summarises what men deliberate about under five heads, ways and means, war and peace, the defence of the country, imports and exports and legislation. The importance, to the legislator, of knowledge of such matters as the past history of his country and the political constitutions of other states is underlined at *Rh.* 1360 a 30ff, 1365 b 22ff, even though such topics themselves belong strictly to politics, not rhetoric (cf. *Rh.* 1356 a 25ff).

[23] *Rh.* 1358 b 2ff.

[24] *Rh.* 1354 a 11ff, especially 16ff.

book II, where he gives advice about, for example, addressing audiences of different age-groups,[25] about how to make your own character look right and how to put your listeners in the appropriate frame of mind.[26]

He brings out the competitive element in rhetoric – an element which it shares with contentious reasoning, in contrast with dialectic which is a joint endeavour, not one that aims at victory.[27] But both the dialectician and the rhetorician must be able to recognise and rebut apparent, as well as real, arguments, and both the *Topics* and the *Rhetoric* contain advice about how to exploit certain tricks for your own purposes as well as about how to deal with those that may be used by your opponents.[28] The *Rhetoric* covers such subjects as the use of examples, 'enthymemes' and maxims,[29] how to bring objections to your opponents' arguments,[30] how to conduct the examination of witnesses,[31] and what to include in a peroration.[32] The *Topics* not only gives extensive rules about such matters as establishing, or objecting to, a definition, but also – in book VIII especially – discusses such aspects of the practice of dialectic as the different roles of the questioner and answerer.[33] Having briefly identified four main kinds of fallacious argument in *Topics* VIII,[34] in the *De Sophisticis Elenchis* Aristotle distinguishes no fewer than six methods of apparent refutation by fallacies that depend on language and a further seven

[25] *Rh.* II 12–14, 1388b31ff.

[26] *Rh.* II 1, 1377b24ff, 1378a15ff: the following chapters on the emotions are included in large part for this purpose.

[27] He contrasts dialectical with competitive disputes at *Top.* VIII 11 and stresses the joint aims of the former at 161a38f, cf. 159a32ff, *SE* 165b9ff, 12ff.

[28] See, e.g., *Top.* II 3 on the exploitation of ambiguities and II 5 on how to lead your opponents into assertions that are easily refuted (said to be a sophistic, but sometimes necessary, method, 111b32ff), and cf. *Top.* VIII 1, e.g. 155b23ff, 156a7ff on concealing your purposes in putting your questions (again a practice said to be contentious but necessary). *Rh.* II 24 is devoted to apparent enthymemes or rhetorical syllogisms. See Ryle 1966, p. 131, on Aristotle's 'tips in sheer eristic gamesmanship', and cf. Owen 1968, p. 107.

[29] *Rh.* II 20–26 (the 'enthymeme' or rhetorical syllogism is said to be based on probable premises or signs, *Rh.* 1355a6f, 1357a32f, *APr.* 70a10f).

[30] *Rh.* II 25, 1402a34ff, cf. *Top.* VIII 10, 161a1ff, *APr.* II 26, 69a37ff.

[31] *Rh.* III 18, 1418b39ff.

[32] *Rh.* III 19, 1419b10ff.

[33] Especially *Top.* VIII 4, 159a15ff, but cf. also, e.g., 156b18ff, 159a38ff, 161a16ff, and note the references to the audience present at debates, e.g. at *SE* 169b31 and 174a36. Cf. Moraux 1968, and on the background see also Ryle 1965 and 1966, pp. 110ff and 196ff ('the organization of the eristic moot').

[34] *Top.* 162b3ff. The four are (1) when an argument only appears to be brought to a conclusion; (2) when it comes to a conclusion but not to the conclusion proposed; (3) when it comes to the proposed conclusion but not by the appropriate mode of inquiry; and (4) when the conclusion is reached from false premises. He goes on in *Top.* VIII 13, 162b34ff, to identify, for example, five types of begging the question: cf. also *Top.* 161b19ff.

that do not.[35] While the *Prior* and *Posterior Analytics* set out the first systematic theory of deductive argument, the *Rhetoric*, *Topics* and *De Sophisticis Elenchis* together provide a comprehensive, highly professional and practical discussion of argument and persuasion where – besides showing his logical acuteness and his gifts for analysis – Aristotle is clearly drawing on a wide experience of the uses of different modes of reasoning.[36]

By the middle of the fourth century B.C. dialectic and rhetoric had both developed into sophisticated disciplines, and we may be sure that the Greeks' interest in, and appreciation of, skill in those fields were both keen and widespread. Although Aristotle was the first to undertake the systematic formal analysis of valid and invalid argument, reflection on aspects of argumentation and on the techniques of persuasion had begun long before him. Moreover he and those who wrote before and after him on the art of rhetoric had a rich store of material to draw on in early Greek literature. Homer, in particular, was represented, in this as in so many other matters, as the teacher of the Greeks.[37] Thus Aulus Gellius suggested that he provided models for each of the three main styles of speaking, Ulysses of the grand, Menelaus of the restrained and Nestor of the intermediate,[38] and Aristotle even said that Homer first taught men how 'to tell lies in the right way', for instance how to use paralogisms, illustrating this by finding an (implicit) example of the fallacy of the consequent in the conversation between Odysseus and Penelope in *Odyssey* book XIX.[39] Homer provides plenty of examples of arguments and inferences of various types, as, for instance, when individual heroes justify an inference that supernatural agencies are at

[35] The six that depend on language are equivocation, amphiboly, combination, division, accent and form of expression: the seven that do not depend on language are accident, the use of words with or without qualification, ignoratio elenchi, petitio principii, consequent, false cause and many questions, *SE* 165b23ff, 166b20ff. In several cases the term by which the fallacy is still known ultimately derives from Aristotle.

[36] The importance of training and practice is mentioned, e.g. at *Rh.* 1410b8, *SE* 175a23ff, and is a frequent theme in the *Topics*, e.g. 108a12ff, 161a24ff and especially VIII 14, 163a29ff.

[37] Thus *R.* 606ef is one of many passages in which Plato is critical of Homer's influence as educator of the Greeks. In *Ion* 540b ff the idea that one can learn from Homer how different types of person should speak is attacked.

[38] Aulus Gellius VI 14.7, cf. Cicero, *Brut.* 10.40, Quintilian, *Inst.* x 1.46ff, XII 10.64 and other passages cited by Radermacher 1951, pp. 6ff, 9f.

[39] *Po.* 1460a18ff. At *Od.* XIX 164ff Odysseus tells Penelope he is a Cretan from Cnossos who once entertained Odysseus on his voyage to Troy, and as evidence of this he describes Odysseus' dress and companions. Penelope commits the fallacy of inferring the truth of the antecedent from that of the consequent: if his story were true, he would know these details; he does know them; so his story is true.

work,[40] or in other contexts where an implicit appeal is made to some notion of what is – in general, or in particular circumstances – probable or to be expected,[41] or again where analogies, drawn from experience or from mythology, are cited either in reasoning to a conclusion or with the aim of persuasion.[42]

Yet if a variety of modes of reasoning can be documented in our earliest Greek literary sources – and there can clearly be no question of attempting to assign an origin to techniques of persuasion as such – the problems we must pose relate to the types of argument employed in early Greek philosophy and science, and the interaction of the theory of argument and its practice. Our principal questions, as we formulated them, concern how far the development of philosophy and science depends on the deployment of new techniques of argument or on their self-conscious analysis. We may divide our discussion into three main parts, first the use of argument in the earliest period of Greek speculative thought, culminating in the work of Parmenides and his followers, second the growth of rhetoric and dialectic and their influence on early Greek natural science, and third the development of axiomatics and of the theory and practice of demonstration.

EARLY PHILOSOPHICAL ARGUMENTATION

For natural philosophers before Xenophanes the limitations of our information impose severe restrictions on our discussion. Sometimes, however, our secondary sources record reasoning that has some chance of going back to an original argument, whether abstract or empirical,

[40] Two such cases which have already been cited (p. 50 n. 213) are *Il.* xxiv 563ff and *Od.* xvi 194ff. In the former, Achilles says he knows that some god guided Priam to the Achaians' ships: 'for no mortal, not even one in the prime of life, would dare to come to the camp; nor would he escape the notice of the guards, nor easily thrust back the bolt of our doors'. In the latter, Telemachus, faced with a transformed Odysseus, infers that a δαίμων is deceiving him: 'for no mortal man could devise these things with his own mind, at least...for just now you were an old man'. These cases may be compared with an indirect proof or *reductio* (see below, p. 76 and n. 87) in that the speaker justifies a conclusion by rejecting the implied alternative as being in conflict with some feature of the given situation. Schematically: *A*. For otherwise (i.e. if not *A*), not *B*. But *B*. Thus the former argument can be recast in the form of a Modus Tollens, although to do so is to reveal what is left implicit or what has to be supplied in the original statement. '(If a god did not help) no mortal would dare...(But you dared). So I know a god helped you.'
[41] See, for instance, Nausicaa's representation of what some of the Phaeacians might say if they saw her arrive in the city with Odysseus (*Od.* vi 273ff). Here there is no direct reference to the probable *as such*: we may contrast the frequent explicit use of the topic of probability in, for example, the orators (on which see, for instance, Dover 1968, p. 57 and cf. below, p. 80 nn. 104 and 105 and p. 81).
[42] The use of arguments from analogy, including *a fortiori* arguments, in Homer was considered briefly in my 1966, pp. 384ff.

and while those that chiefly rely on empirical considerations will occupy us in the next chapter,[43] two instances of abstract argumentation, both concerning Anaximander, are especially noteworthy.

The first concerns his principle, the boundless, itself. At *Ph.* 204 b 22 ff Aristotle refers to certain unnamed thinkers who suppose there to be an infinite body over and above the elements, so that the other elements are not destroyed by whichever of them is infinite. They are all characterised by oppositions, air being cold, for example, water wet and fire hot: so that 'if one of these were infinite, the rest would by now have been destroyed'.[44] There must, then, have been something different from air, water and the like, from which they – and one may add everything else – came to be. Aristotle does not mention Anaximander by name,[45] but Simplicius does in his commentary on the passage[46] and he elsewhere speaks of the Boundless in Anaximander as being something separate and distinct from such things as water, which came to be from it.[47] Both Aristotle and Simplicius have undoubtedly reformulated the argument in their own, Peripatetic, terms.[48] Yet it may well be that – whether or not he was directly stimulated by reflecting on Thales' choice of water as principle – Anaximander was influenced in his own doctrine of the Boundless by considering a difficulty that arose from any such view, namely how, if water was the origin of things, fire could ever have come to be. If he recognised that a similar objection could be mounted against any cosmogony that took one of the obvious natural substances or world-masses as its starting-point, then he had what we may describe as indirect proof of the conclusion that the origin of things must be something other than those natural substances.

The second argument is both more striking and better attested, since Aristotle himself mentions Anaximander by name. This is in his discussion of his predecessors' views on the shape of the earth and on why it is at rest in the *De Caelo* II 13. After describing the doctrines held by a number of other theorists, including, for instance, Thales'

---

[43] The use of empirical evidence by the Presocratic natural philosophers is discussed in ch. 3, pp. 139ff. Many of their theories and explanations are derived from, or supported by, fairly simple analogies: this was considered in some detail in my 1966, chh. 4 and 5.

[44] *Ph.* 204 b 28f.

[45] At *Metaph.* 1069 b 22, however, he speaks of Anaximander as one of a number of thinkers who held that everything came to be from a state when 'everything was together': cf. also *Ph.* 187 a 20f.

[46] *In Ph.* 479.33 and 480.1.

[47] *In Ph.* 24.16ff.

[48] Not only in their use of the notion of 'elements', but also, probably, in their analysis of those elements in turn in terms of the primary opposites, hot, cold, wet and dry: see Hölscher (1953) 1970 and Lloyd (1964) 1970.

idea that the earth is at rest because it floats on water,[49] Aristotle
goes on to say that some, including Anaximander, held that it was
the earth's 'indifference' that was responsible for it remaining where
it is.[50] Their argument was that that which is situated at the centre
and related similarly to the extremes has no tendency to move in one
direction rather than in any other. Moreover it is impossible for it
to move in opposite directions at the same time. Therefore it necessarily remains at rest. If this was indeed Anaximander's reasoning,[51] then we may see this as the first extant instance, in natural
science, of the use of what we may call the principle of sufficient
reason.

   Any reconstruction of the dialectical arguments used by the
Milesians must remain conjectural. But with later philosophers,
Xenophanes, Heraclitus and especially Parmenides, we are on firmer
ground. First the arguments that Xenophanes brought against
anthropomorphism provide good examples of informal attempts to
reduce that view to absurdity. To attack the idea of men picturing
the gods in their own image he draws an analogy with animals
representing the gods as like themselves: 'If oxen and horses and
lions had hands and could draw with their hands and produce works
of art like men, horses would draw the forms of the gods like horses,
and oxen like oxen, and they would make their bodies such as each
of them had themselves.'[52] If the idea of animal representations of
gods in the form of animals was absurd,[53] then – Xenophanes
implies – the same should apply, by parity of reasoning, to human
conceptions of gods in the form of humans. Then although the
extant fragments of Heraclitus are more often suggestive and
elliptical or paradoxical, rather than explicit and deductive, there
are some notable examples of hypothetical arguments in the thought
experiments of Fr. 7 ('if everything became smoke, noses would

---

[49] *Cael.* 294a28ff. He also mentions Xenophanes' view (that the roots of the earth stretch
     down indefinitely, *Cael.* 294a21ff – an idea criticised by Empedocles, Fr. 39, *Cael.*
     294a24ff) and those of Anaximenes, Anaxagoras and Democritus, *Cael.* 294b13ff.
[50] *Cael.* 295b10ff. The historicity of this report has recently been questioned, though on
     weak grounds, by J. M. Robinson 1971.
[51] Again the possibility that Anaximander was prompted to his view by reflecting on the
     difficulties present in that of Thales has long been recognised. Aristotle records as one
     of his own main objections to Thales that the same argument applies to the water
     which is supposed to hold the earth up, as to the earth itself (*Cael.* 294a32ff) – though
     we have no way of determining at what point that objection was formulated.
[52] Fr. 15, cf. also Frr. 14 and 16: and cf. above, pp. 11f.
[53] No doubt in part because the animals are thought of as responsible. It is not clear
     whether Xenophanes knew (as Herodotus certainly did) that the Egyptians often
     represented the gods in the form of animals, but gods appear frequently in the guise
     of birds in Homer (e.g. *Od.* III 371f).

discriminate them') and Fr. 23 ('they would not know the name of Justice, if these things did not exist').[54]

Such isolated examples of arguments of types much used in later Greek philosophy and science are interesting, but they pale in significance beside the developments that take place with Parmenides, who was – as all recognise – the first to produce a sustained deductive argument. The interpretation of many aspects of the Way of Truth is highly disputed, and it is not my intention to engage in a detailed analysis of the issues that relate to the substance of the arguments it contains.[55] Rather it is the form and structure of those arguments that concern us, and here there can be no doubt about two essential points, (1) that Fr. 2 sets out the starting-point for the whole, and (2) that the conclusions of Fr. 8 are established by arguments linked in a strict deductive chain.

The starting-point is the apparently incontrovertible proposition that 'it is'. This is stated, and established, in the stronger form – 'it is and it is impossible for it not to be' – in Fr. 2, which points out that the alternative – 'it is not and it needs must not be' – must be rejected on the grounds that 'you could not know what is not, at least,[56] nor could you declare it',[57] for – as Fr. 3 goes on to say – 'it is the same thing that can be thought and can be'. The sense of the verb to be, the subject we must understand, and the nature of the alternatives presented, are all controversial issues, but that does not affect my main point here, which is that in fragment 2 Parmenides was evidently attempting to establish a starting-point that all would have to accept. Thus on one plausible reading of Fr. 2.6, one point that he appeals to is that we could not expect to learn anything from any way of inquiry that began with the statement that the object it was inquiring into does not exist.[58]

The starting-point once secure, Parmenides gives an articulated argument in the long fragment 8 to show that it is ungenerated and indestructible, not subject to movement or change, and both

---

[54] See also Fr. 99 ('if there were no sun, it would be night on account of the other stars') and the testimony included as Fr. 4. Cf. also the thought-experiment in Xenophanes Fr. 38 ('if god had not created yellow honey, they would say that figs were much sweeter').

[55] Among the fundamental recent studies the following may be mentioned in particular, Owen (1960) 1975, Mansfeld 1964, Tarán 1965, Hölscher 1968, pp. 90ff, Mourelatos 1970. The nature of the alternatives presented in Fr. 2 is discussed in my 1966, pp. 103ff.

[56] τό γε μὴ ἐόν (Fr. 2.7): an alternative translation would be 'it, at least if it were not'.

[57] οὔτε φράσαις (Fr. 2.8): not evidently in the sense that you could not pronounce the words (the goddess just has), but that you could not assert it as true.

[58] Taking παναπευθέα as 'that from which nothing can be learned' (cf. Mourelatos' 'that from which no tidings ever come') rather than 'unintelligible' or 'inconceivable', and taking ἔστι as (at least primarily) existential.

spatially and temporally invariant. Again many points of detail, some of them important ones, are obscure or disputed or both: but the *main* structure of the argument is fairly well agreed, and that it has a rigorous deductive form is not in question. He first sets out what he is going to establish at Fr. 8.2ff,[59] and then proceeds to demonstrate each point in turn in an argument that falls into four main sections.[60] In the first section, the denial of 'it is not' (established already in Fr. 2) is used to show that 'it is ungenerated and indestructible'. The arguments in each of the later sections take over conclusions from the earlier ones, which are cited as premises picked out in each case by a clause introduced with ἐπεί.[61] Although there is plenty of room for disagreement about the precise force of particular stretches of argument, the overall tactics Parmenides uses in his demonstration are not in serious doubt. The fragment forms a carefully articulated whole in which the later sections build on the conclusions of the earlier in an orderly sequence of argumentaion.

The method, then, no less than the content, of the Way of Truth is revolutionary. We have not only a set of conclusions that completely overturn common assumptions:[62] but those conclusions are established by an extraordinarily tightly knit chain of arguments.[63] Moreover among the individual arguments he uses, two deserve particular notice. First we now have a far more certain example than in Anaximander of an appeal to what we may call the principle of sufficient reason. This is in the proof that 'it is ungenerated', at Fr. 8.9f, where the goddess demands: 'What need would have

---

[59] There is disagreement both about the readings of verse 4 (οὖλον μουνογενές seems preferable to ἐστι γὰρ οὐλομελές and either οὐδ᾽ ἀτέλεστον or ἠδὲ τέλειον to ἠδ᾽ ἀτέλεστον) and about whether verses 5 and 6 continue setting out what is to be proved, or begin the proof. But these issues do not substantially affect the central point that Fr. 8 begins with a statement of what is to be demonstrated.

[60] Fr. 8.5–21, 22–25, 26–33 and 42–49 (34–41 make points about the relation between thinking and being that are similar to Fr. 2.7f, Fr. 3 and Fr. 6, but fall outside the main structure).

[61] Fr. 8.22, 27 and 42. Thus the proposition that 'it is not divisible' is proved at Fr. 8.22ff using the premiss that 'it is all unqualifiedly' (taking ὁμοῖον as adverbial, with Owen): cf. 'it is completely' established in the first section at Fr. 8.11. 'It is unmoved... without beginning, without end' is shown at Fr. 8.26ff using the premiss that 'coming to be and passing away' had been ruled out: cf. Fr. 8.6ff and 19ff in the first section. Finally 'it is perfected in all directions' is shown at Fr. 8.42ff using the premiss that 'there is an ultimate boundary': cf. Fr. 8.26 and 30f in the third section.

[62] The incoherence of the opinions of ordinary men is mentioned in Fr. 6.4ff (they consider that 'being and not being are the same and again not the same') and Fr. 8.38ff (which rejects their use of the terms coming-to-be and passing away, being and not being, changing place and alteration of bright colour).

[63] The goddess who instructs Parmenides herself says: 'It is all one to me where I begin: for I shall come back there again' (Fr. 5).

raised it to grow later or earlier, starting from nothing?'[64] If no cause can be adduced, or no explanation given, to account for the universe coming to be at one time rather than at any other, then we may consider this an argument for rejecting the notion that it came to be at all.[65] Secondly, the same disproof of coming-to-be contains a clear instance of a *reductio* argument, in this case an implicit Modus Tollens where a thesis is rejected on the grounds that one of its consequents is false ('If *A*, then *B*; but not *B*; so not *A*'). 'If it came to be', the goddess says at Fr. 8.20, 'it is not.' But – as we had been shown – 'it is not' must be denied; and so, as she concludes, 'coming-to-be' must also be rejected.[66]

The immense impact that Parmenides had on subsequent Greek philosophy derives as much from his methods as from his conclusions. He is the first thinker to set up a fundamental opposition between the senses and abstract argument or reason,[67] and to express an unequivocal judgement on the relative trustworthiness of each. In Fr. 7 the goddess tells Parmenides not 'to let habit, born of experience, force you to let wander your heedless eye or echoing ear or tongue along this road': rather, he is to 'judge by λόγος – reasoned argument – the much-contested refutation, spoken by me'. What we may broadly call empirical methods and evidence are not merely not used: they are ruled out. Their relevance is restricted to the Way of δόξα – Seeming or Opinion – the 'deceitful' cosmology 'in which there is no true belief', that the goddess offers Parmenides after the end of the Way of Truth 'so that no judgement of mortals may outstrip you'.[68] The Way of Truth itself proceeds strictly deductively from a single, incontrovertible starting-point, and it served as a model of a rigorous demonstration for later philosophers. Although it was not until the fourth century that arguments came to be analysed as such, several late fifth-century writers – whether in direct imitation of Parmenides or not – exemplify patterns of reasoning that are very similar to his. The arguments in his own two

---

[64] I.e., probably, at any one time rather than at any other.

[65] The relevance of this argument is not confined, of course, to such earlier Greek cosmologists as Parmenides himself may have had in mind.

[66] Verse 20 continues: οὐδ' εἴ ποτε μέλλει ἔσεσθαι. This might be taken either as implying a similar argument to deny that it will be ('nor is it, if it is coming to be in the future'), or as part of the consequent ('if it came to be, it is not – not even if it is going to be').

[67] The question of the reliability of the senses was, however, also broached by Heraclitus Fr. 107 (but cf. Fr. 55), and cf. Xenophanes Fr. 34 on the limitations of human knowledge in general.

[68] Fr. 8.61. The status of the Way of Seeming is much disputed. But it is sufficiently clear from the goddess' statement that 'there is no true belief' in it (Fr. 1.30, cf. Fr. 8.50ff) that it is not intended to *modify* the conclusions of the Way of Truth.

chief followers, Zeno and Melissus, especially, often resemble, though in certain respects they may be said to advance on, those in the Way of Truth, and their use of *reductio* arguments and of the dilemma is particularly noteworthy.

That the whole of Zeno's principal treatise was intended to support Parmenides' position by reducing the views of his opponents to absurdity is suggested by what Plato tells us in the *Parmenides*. There (128 cd) Zeno is made to say that his work

> is in truth a kind of support for Parmenides' argument against those who try to make fun of it by saying that if there is one, then many absurd and contradictory consequences result for his argument. This book, therefore, opposes those who assert that there are many, and it repays them in the same coin – and with more – aiming to show that their hypothesis, that there are many, is subject to far more absurd consequences than the hypothesis that there is one, if one prosecutes the inquiry sufficiently.

According to this report, the strategic aim of Zeno's arguments was to provide indirect proof of 'the one' (Parmenides' monistic position) by refuting the opposite[69] hypothesis that 'there are many', and although it has often been thought that he had particular pluralists in mind, the range of application of his destructive arguments appears to be quite general and they seem to be designed to refute *any* version of a pluralist doctrine.[70]

The material from Simplicius collected as fragments 1 and 2 in Diels–Kranz, and the four arguments against motion for which Aristotle is our main (though admittedly rather unsatisfactory) source, provide us with our best opportunities to see how this strategy was implemented in practice.[71] The articulation of Frr. 1 and 2 can be reconstructed from Simplicius' introductory remarks.[72] The original argument appears to have had four main limbs. (1) The first aimed to establish that if there are many, they have no size.

---

[69] Following Ranulf 1924, I argued in my 1966, p. 107, that the hypotheses of 'the one' and 'the many' are treated by Zeno as if they were mutually exclusive and exhaustive alternatives. A clear distinction between contraries and contradictories does not antedate Plato: and the first full explicit analysis of different modes of opposition is in Aristotle.

[70] Against the views of Tannery 1887, pp. 124ff, and 1930, pp. 255ff, Cornford 1922 and 1923, Lee 1936 and Raven 1948, one may refer to Fränkel (originally 1942), Owen (1957–8) and Vlastos (1953) in Allen–Furley 1975 especially. The idea that Zeno's arguments are directed especially, if not exclusively, against the doctrine that Cornford labelled 'number-point-atomism' (the identification of integers, geometrical points and physical atoms) suffers from the major disadvantage that there is no good evidence to confirm that such a view was held before Zeno.

[71] Fr. 3 and Fr. 4 (both antinomies) take a similar form to Frr. 1 and 2, but we have less evidence on which to reconstruct the original arguments.

[72] Note especially προδείξας at *In Ph.* 139.18 and at 141.1. The reconstruction adopted here follows that of Fränkel (1942/1975) in the main.

Simplicius remarks, baldly, that this was shown from the admission that 'each of the many is the same as itself and one'.[73] The underlying argument has plausibly been suggested to have been that if they have size, they have parts, and so are not (ultimate) units, but collections of units. But the many in question cannot be (mere) collections of units: they must be the ultimate units themselves. So they can have no size.[74] (2) The second limb of the argument then shows, on the contrary, that they must have size. For if they had no size, they would not be; but they have been assumed to be; so they must have size.[75] (3) Zeno then explores the consequences of saying that they have size in an argument the conclusion of which is that 'they are so large as to be unlimited'.[76] (4) The overall conclusion is then stated that, 'if there are many, they are so small as to have no size, and so large as to be unlimited'.[77]

Many problems of interpretation remain disputed, and this is true in particular of the question of whether, when Zeno concluded in (3) that the many are 'so large as to be unlimited', this is a fallacy or merely a deliberate trick.[78] Yet certain features of the form and overall structure of the arguments are clear. First both the first and the second limbs (as reconstructed) are indirect proofs or *reductiones* that consist in implicit Modus Tollens arguments. In (1) if they have

[73] *In Ph.* 139.19.

[74] Cf. Fränkel (1942) 1975, pp. 110–11.

[75] Simplicius, *In Ph.* 141.1f states that 'If what is had no size, it would not be.' The argument is given at *In Ph.* 139.11ff (Fr. 2). If a thing has no size, then 'if it were added to something, it would not increase it', and if it were subtracted from something, it would not decrease it. But that which, when added or subtracted, causes no increase or diminution to what it is added to or subtracted from, is nothing.

[76] The argument is given by Simplicius at *In Ph.* 141.2ff (Fr. 1). 'If it exists, then each necessarily has some size and bulk and one part of it stands out (or 'is distinct', ἀπέχειν) from another. And the same argument applies to the part that is in the lead (or 'is before it', or 'projects', προὔχοντος). It too will have size and part of it will be in the lead.' Moreover this process can be continued indefinitely. So they are 'so large as to be unlimited'. On the much debated sense of προέχειν here, compare Fränkel (1942) 1975, p. 118, Owen (1957–8) 1975, p. 146, Vlastos (1959) 1975, p. 178 and most recently Booth 1978.

[77] This conclusion is stated at Simplicius, *In Ph.* 141.6ff.

[78] The key to this is the sense of the term translated 'unlimited' (ἄπειρα, *In Ph.* 141.8). If (*a*), as most of the ancient commentators assumed, it means 'infinitely large', the argument is fallacious. Zeno has shown that there is an infinite number of parts, but not that what is is infinitely large – for the sum of an infinite series may be finite. If, alternatively (*b*), as Fränkel has suggested, the sense is merely 'limitless', the argument goes through, but only troubles the pluralists if they do not spot the trick: he has shown that the object has an indefinite number of parts, but uses a term that might suggest they have infinite magnitude. See especially Vlastos (1959) 1975 and Fränkel (1942) 1975 and cf. Furley 1967, pp. 68f, who points out (p. 77 n. 9) that on the latter interpretation (*b*), the argument still retains some force as an antinomy (since there is still a contradiction between 'of no magnitude' and 'limitless' in the sense specified), although its force as a dilemma is removed (since the pluralists could accept one of the alternatives presented).

size, they are not units: but they are units; so they have no size. In (2) if they have no size, they do not exist; but they exist; so they have size. Secondly, and more strikingly, the four-fold argument as a whole is a carefully constructed antinomy. It reduces the assumption that there are many to absurdity by drawing incompatible consequences from it, namely that they are both so small as to have no size, and so large as to be unlimited. Schematically, the form of the argument is: if A, then both X and Y; but (it is assumed) not both X and Y; and so (by implication) not A.

The four arguments against motion reported by Aristotle[79] form a further part of Zeno's indirect proof of the one. That much is reasonably clear, although many of the specific moves in these arguments and their interrelationship are in doubt. An exposition of the set would take us well beyond our present concerns,[80] but we may comment briefly on the main aspect of their structure that is of particular interest in considering Zeno's techniques of argument, namely the use of the dilemma. On one interpretation that stems from Tannery the four arguments display a complex symmetry, the first pair, the Dichotomy[81] and the Achilles, attacking motion on the assumption that space and time are infinitely divisible, and the second, the Flying Arrow and the Moving Rows, doing so on the assumption that space and time consist of indivisible minima (atomic quanta). This view represents Zeno constructing a highly sophisticated multiple dilemma, but it runs into several major difficulties, the most damaging of which is that if the Moving Rows attacks a clearly defined notion of atomic quanta, then all the ancient commentators missed the point.[82]

[79] Chiefly *Ph*. 239 b 9ff, cf. also 233 a 21ff, 239 b 5ff, 263 a 4ff.
[80] The modern literature on these arguments is immense, but apart from the items already mentioned above (p. 72 n. 70), see especially Booth 1957*a* and *b*, Vlastos (1966*a*) 1975, (1966*b*) 1975 and Furley 1967.
[81] Also called the Stadium (from Aristotle, *Top*. 160 b 7ff): but this may be misleading as a stadium is also referred to in the fourth argument (the Moving Rows).
[82] Aristotle reports the argument at *Ph*. 239 b 33ff (cf. also Simplicius, *In Ph*. 1016.9ff, from which the usual diagram is derived). Two rows of bodies (*B*s and *C*s) are imagined as moving past one another, and past a third stationary row (*A*s), in opposite directions:

$$A \ A \ A \ A$$
$$B \ B \ B \ B \rightarrow$$
$$\leftarrow C \ C \ C \ C$$

In this situation, each member of row *B* passes two members of row *C* for every one of row *A*, and Aristotle states the conclusion as being that half the time is equal to its double. What is not clear is whether the rows comprise bodies of any kind, or whether they are thought of as indivisible units moving in indivisible units of time. If the latter is the case, the paradox is undoubtedly more interesting. Yet none of our ancient sources takes it in that way, and Aristotle in particular refutes the argument simply by pointing out that the speed of a moving object is relative to what it is measured against.

Yet even if it seems that these arguments are not the tightly knit set that they have sometimes been made out to be, it is still probable enough that – together with Frr. 1 and 2 – they exploit *dilemmas* to show the incoherence of any notion of *division*. It may be that the second pair of arguments against motion[83] does not attack a well-defined conception of atomic quanta of space and time, so much as an *unspecified* notion of discrete units of *some* kind.[84] But the first pair clearly attack the idea of the infinite divisibility of space (at least).[85]

[83] The Flying Arrow, as reported by Aristotle at *Ph.* 239b5ff, appears to hinge on a confusion between 'instant' and 'interval'. Aristotle has clearly abbreviated the argument, but the essential steps seem to be: (1) everything is always at rest when it is 'at a place equal to itself'; (2) the moving object is always at a place equal to itself 'in the now' (ἐν τῷ νῦν); therefore (3) the object moving is always at rest (ἕστηκεν, *Ph.* 239b30: contrast *Ph.* 239b7 where the conclusion is that it is unmoved, ἀκίνητον). Whether or not Zeno himself used the expression ἐν τῷ νῦν, 'in the now' in step (2) is ambiguous, for it can be used either (*a*) of the durationless instant (the analogue, in the continuum of time, of the geometrical point) or (*b*) of an indefinitely small interval or period of time. The argument can then be read in two quite different ways, depending on which sense we take. (A) If we take it to mean 'durationless instant', then proposition (2) is true, but the conclusion that follows is that the arrow is neither in motion nor at rest. It is, we should say, meaningless to talk of either motion or rest taking place *in* an instant, though we need to be able to talk of motion or rest measured *at* an instant. But if (B) the now refers to an interval, however small, then proposition (2) is false. The plausibility of proposition (2) depends on taking 'now' strictly as an instant: but the conclusion in step (3) only follows if it is taken in the sense of minimal interval. It is, however, far from clear whether the confusion here was one that Zeno's real or imaginary opponents made or one that he himself did. It is possible that he was exploiting an incoherence in the position of anyone who tried to explain motion on the assumption that time is made up of 'nows' without distinguishing the idea of a durationless instant from that of a minimal interval. But we cannot rule out the possibility that Zeno himself simply failed to distinguish the two. The distinction was only made explicit for the first time by Aristotle, and once that distinction was available, anyone who postulated atomic quanta of time could block the paradox by pointing out that such quanta are not of zero magnitude (and so denying proposition (2)), although they would still face the difficulties raised by Zeno's Frr. 1 and 2. It should, however, be stressed that we have no reliable independent evidence that such a notion of atomic quanta of time was held in the fifth century.

[84] Furley 1967, pp. 74f, suggests that the Moving Rows is directed against the distinction between moving and static: it depends not on indivisibles, but only on the idea that a length can be divided into sections such that to traverse the length is to 'be opposite to' each section in turn for a period of time. The essential (fallacious) step is that since a *C* and a *B* are opposite the same *A* for the same time, they are opposite each other for that time. Cf. also Owen (1957–8) 1975, p. 164 n. 22.

[85] On what seems the more likely interpretation of the Dichotomy, it uses an infinite regress to suggest that motion cannot begin. Before a moving object reaches any given point, it must first reach a point half way to it, and again to reach that point, it must first reach a point half way to *it*. Since this process can be continued indefinitely, a point can always be found between the point to be reached and the starting-point, and so motion cannot begin. On another interpretation (see Vlastos (1966b) 1975) the argument is designed to show not that no movement can begin, but that no movement can be completed: to reach the end point in a movement, one must first reach a point half way to it, then a point half way between that point and the end, and so on. On that version the argument would reduplicate the Achilles, which similarly attacks motion on the assumption that space is infinitely divisible. Here Achilles and his opponent (it is the commentators who identify this as a tortoise) are imagined as moving, and at

And the alternative notion of a division that terminates in ultimate units is dealt with and ruled out in Frr. 1 and 2, which show (as we have seen) that no coherent account of such units can be given: they cannot have size, for they are ultimate units; but no more can they have no size, for they exist.

In Melissus, Parmenides' other main fifth-century follower, we have a further rich haul of deductive arguments. The surviving fragments[86] provide one instance after another where theses (including some that conflict with Parmenides' own doctrines) are established by the method of indirect proof or *reductio*. Strictly, a *reductio* proceeds by first assuming the contradictory of the thesis to be proved, and then refuting this by deducing *either* its contradictory, *or* some pair of contradictory propositions (as in an antinomy), *or* some other statement known to be false (as in the general schema of Modus Tollens).[87] Bearing in mind that Melissus does not distinguish between different modes of opposition, and that, like both Parmenides and Zeno before him, he often treats other pairs of opposites – not just strict contradictories – as mutually exclusive and exhaustive alternatives,[88] we can cite examples of different kinds of *reductio* from his fragments.

Melissus general strategy in Fr. 8 follows the first of these three types. He there aims to establish that there is one by considering not (it is true) the contradictory, but the contrary proposition, that there are many. After a series of arguments he arrives at what he no doubt took to be the self-contradictory conclusion that 'if there were many, they would have to be such as the one is'.[89] He proceeds in a similar

---

different speeds. Even though Achilles is much faster than the tortoise, he will never catch him. He must first reach the point from which the tortoise starts. When he has reached that point, the tortoise will have advanced a certain distance, and Achilles must then reach *that* point. Again the process can be continued indefinitely, and so, as Aristotle puts it (*Ph.* 239 b 17f), 'the slower necessarily always has a lead'. Although Aristotle sometimes reports the Dichotomy as if it too attacked the notion of traversing a distance (e.g. *Ph.* 233 a 21ff, 263 a 4ff, *Top.* 160 b 7ff), in introducing the argument at *Ph.* 239 b 9ff he speaks of it as concluding that there is no movement.

86 Our secondary sources, especially Simplicius' paraphrase and the pseudo-Aristotelian *De Melisso*, contain many further examples of similar arguments, though we cannot tell how far these have been reformulated.

87 Schematically, to prove a proposition A, first assume not A. Then show that, if not A, then either (i) A (contradicting not A); or (ii) both B and not B; or (iii) C (some other false statement). (i), (ii) and (iii) provide different means of refuting the assumption not A, and so of establishing A.

88 This aspect of Eleatic argumentation was documented in my 1966, pp. 103ff.

89 Fr. 8 (2)–(6). Although Melissus evidently thought of this conclusion as absurd, it has often been pointed out that (whether or not they were consciously reacting to Melissus' arguments) the Atomists adopted a position close to that rejected here, in that they postulated a plurality of entities, each of which has the characteristics that the Eleatics attributed to the one alone.

fashion in Fr. 1, where he establishes that it is 'ungenerated' and that 'it always was what it was and always will be'. The first part of that conclusion, that it is ungenerated, is shown by considering the thesis that it is generated, and then arguing that this thesis is self-contradictory. 'If it came to be, it is necessary that, before it came to be, it was nothing. But if it was nothing, nothing could in any way come to be from nothing.'[90]

Perhaps the clearest example of a *reductio via* the contradictory of the proposition assumed comes within Fr. 8. This contains an attack on the senses where he first assumes (Fr. 8 (2)) that 'we see and hear correctly'. He argues against this by first stating that what the senses report is that things change (Fr. 8 (3)). 'But if it changed, then what is was destroyed and what is not came to be.' Here, as often elsewhere, the link between the antecedent and the consequent, and the (implicit) rejection of the consequent, involve Parmenidean assumptions that are open to challenge. But the conclusion he arrives at is that 'we did not see correctly' (Fr. 8 (5)). Again the schema of argument is: if A, not A; and Melissus signals the self-contradiction with the remark: 'so these things do not agree with one another' (Fr. 8 (4)).

Elsewhere we find many examples of arguments of the general Modus Tollens type, not that these are always fully displayed in what became the canonical form: 'if A, then B; but not B; so not A'. In particular the second premiss has sometimes to be supplied, usually from previous stretches of argument. Thus fragment 5 shows that 'it is one' from its being boundless or limitless. 'If it were not one, it will form a boundary in relation to something else.' But (as had been shown in Frr. 2, 3 and 4) it is limitless: so we must conclude that it is one.[91] Fr. 7 produces a long argument of a similar kind to show that 'it could not perish, nor become greater, nor be arranged differently, nor does it feel pain, nor distress. For if any of these things happened to it, it would no longer be one.' But once its unity is taken to have been established, we must rule out these modes of change. He argues against its being altered in Fr. 7 (2): 'for if it is altered, what is is necessarily not the same, but that which was before perishes, and that which is not comes to be'. Here the rejection of the consequent involves Parmenidean assumptions and is implicit, but later stretches of argument are more fully articulated. At Fr. 7 (4)

[90] This conclusion has to be rejected both as contrary to fact and as contrary to hypothesis (if it is generated, nothing is generated).

[91] Cf. also Fr. 6. In quoting Fr. 5, however, Simplicius also records Eudemus' complaint against the vagueness of the first premiss (*In Ph.* 110.6ff). Melissus has not specified how it would not be limitless, if it were many.

he shows that 'it does not feel pain': one of his arguments is that 'it would not be the same, if it felt pain', and he had earlier proved that it must be 'the same'. When it comes to showing that 'it does not feel distress', Melissus merely notes in Fr. 7 (6) that 'the same argument' applies. We should certainly not assume that Melissus himself had analysed, or was in a position to offer a formal analysis of, these arguments. But what these and other[92] fragments show is the increasing frequency and variety of the use of the *reductio* in philosophical contexts.

Parmenides, Zeno and Melissus are the first considerable exponents of rigorous deductive argumentation. As we noted before, it is not difficult to cite examples of various kinds of reasoning, including arguments that can be recast as indirect proofs or *reductiones*, both before and outside philosophy. Yet with the Eleatics λόγος – reasoned argument – comes to be recognised explicitly as *the* method of philosophical inquiry. Parmenides and Melissus both advance arguments to suggest the invalidity of sense-perception and all three Eleatics practise what Parmenides preaches when he advocates a method that relies on λόγος alone. All three produce not just isolated examples of indirect proofs, but stretches of sustained deductive reasoning, and the influence of the models they provided on subsequent philosophy and science would be hard to exaggerate.

There is, as yet, however, no formal analysis of different modes of argument as such, no attempt expressly to define the distinction between necessary and probable reasoning,[93] or that between proof and persuasion. There can be no question of ascribing to the Eleatics themselves any sharp or explicit distinction between strict demonstration and dialectic. If there is a case for saying that Parmenides was the first to set out to provide deductive demonstrations of a set

---

[92] Cf. also Frr. 9 and 10 which show that what is 'has no body' (for 'if it had thickness, it would have parts, and would no longer be one') and that 'it is indivisible' (for 'if it were divided, it would move, and if it moved, it would not be'). How precisely Melissus reconciled the three statements, that it is infinite in magnitude (Fr. 3), 'full' (Fr. 7) and without body (Fr. 9) is a major crux in the interpretation of his philosophy (see most recently Reale 1970, ch. 7), but it does not affect our analysis of the forms of argument he employed.

[93] Knowing for certain and only surmising were already distinguished in Homer (e.g. *Od.* 1 215f), and a contrast between clear knowledge and opinion or conjecture is used by Xenophanes (Frr. 34–35), by Alcmaeon (Fr. 1) and by Parmenides (Fr. 1.28ff, Fr. 8.50ff) for whom the Way of Opinion is a matter of what is likely (ἐοικότα, Fr. 8.60), while the Way of Truth (in which what is is held fast by necessity, ἀνάγκη, Fr. 8.30, but cf. also Fr. 10) is a way of true conviction (πειθώ, Fr. 2.4, cf. πιστόν Fr. 8.50 and πίστις Fr. 1.30). Yet not even in Parmenides is the conception of demonstration explicit, or the contrast between necessary and probable reasoning defined.

of philosophical theses, we may recall that Aristotle is reported to have described Zeno as the founder of dialectic,[94] and two of the features that characterise the dialectical method, as conceived by Aristotle, are indeed exemplified in Zeno's arguments (as also to some extent in those of Melissus), namely (i) reasoning to opposite conclusions,[95] and (ii) arguing from commonly accepted premisses.[96] The Greeks eventually drew firm distinctions between different modes of reasoning, but the developments that led to those distinctions are complex ones. In the latter part of the fifth century we have not only a rapid growth in the actual practice of argumentation both inside and more especially outside philosophy, but also the first conscious reflections on aspects of persuasion and reasoning in the context of discussions of the art of speaking. The domain of probable arguments (generally including both rhetoric and dialectic[97]) comes increasingly to be distinguished from that of demonstration, and it will be convenient to consider the two separately and in turn.

THE DEVELOPMENT OF RHETORIC

The development in both the practice and the theory of rhetoric from around the middle of the fifth century B.C. is dramatic. Although skill in speaking was already a prized quality in Homer,[98] the contexts in which such a skill could be – and even had to be – exercised were increased out of all recognition in the framework of the institutions of the city-state – particularly, but by no means exclusively, in the democracies. We shall be returning to this question later,[99] but the essential point, for our present purposes, is a simple one, that as the degree of involvement of ordinary citizens in the political life of the city rose, so too did the value and importance of the ability to persuade your fellow men by argument, whether in political debate in assemblies or on embassies, or in the various kinds of law-courts that proliferated in Athens, and to a lesser extent also elsewhere, in the fifth century.

From Antiphon around the middle of the fifth century, through Lysias, Andocides, Isocrates and Isaeus (late fifth, early fourth century) down to Aeschines and Demosthenes, we have a wealth of original material by which to judge the actual practice of argu-

---

94 Fr. 65, see above p. 62 n. 16.
95 *Rh.* 1355a33ff: cf., e.g., the antinomy of Zeno Frr. 1 and 2.
96 E.g. *Top.* 100a18ff: cf. the arguments in both Zeno and Melissus starting from the assumption that 'there are many'.
97 On the contrast between rhetoric and dialectic in, for example, Plato, see further below, pp. 100f.
98 E.g. *Il.* IX 438–43, XVIII 249ff.   99 See further below, pp. 242ff, 245ff.

mentation of the Attic orators in each of the three main modes of
speech-making – forensic, deliberative and epideictic – identified
by Aristotle.[100] My aim here is not to attempt a complete typology of
the techniques of persuasion they deployed,[101] but rather merely to
draw out the evidence that they and our other sources provide
concerning the increase of awareness of, and interest in, reasoning
and persuasion as such.

First there is the growth of professionalism in the art of speaking.
From around the middle of the fifth century there were paid speech-
writers, λογογράφοι, who composed speeches for others to deliver in
the law-courts[102] and assembly. The majority of Lysias' extant works
are just such speeches, and although he won a reputation for the
simplicity of his style, his use of what Aristotle called 'artificial
proofs'[103] is sophisticated enough.[104] Moreover some noted orators
and speech-writers wrote artificial speeches on imaginary law-suits
with the intention that these should serve as models to be imitated on
real-life occasions. We have some early examples of this in the *Tetralogies* ascribed to Antiphon, sets of four speeches, two for the prosecution and two for the defence, on fictitious cases.[105] As, perhaps,
a natural extension of this, the art of speaking came to be taught
professionally, that is by men who set themselves up as teachers of
rhetoric and who took fees from their pupils in that capacity. The
first man to charge for his instruction, at least according to Plato, was
Protagoras,[106] and most of the major sophists of the late fifth and

[100] Cf. above, p. 63.
[101] For one attempted typology, see Perelman and Olbrechts-Tyteca 1969.
[102] By custom, if not law, litigants conducted their own cases in Athens in the fifth and
fourth centuries, though they might be helped by a συνήγορος – advocate – who also
spoke on their behalf: see A. R. W. Harrison 1968–71, I p. 156.
[103] *Rh.* 1355 b 35ff: as opposed to ἄτεχνοι πίστεις, such as witnesses, evidence produced
under torture, contracts and the like.
[104] Apart from his skilful manipulation of emotive appeals – the denigration of opponents'
characters, commendation of his own side and flattery of his audience – he often attacks
the opposition on the grounds of inconsistency (e.g. III 24) or irrelevance (XII 38 draws
attention to the commonness of this expedient to mask a failure to answer a charge).
Explicit appeals to what is probable (εἰκός) are frequent, as are arguments *a fortiori*
(e.g. VII 24), disjunctive or dilemma arguments (e.g. VII 14) and, especially, indirect
proofs, as in the disproof of a point of fact or of the imputation of a motive or intention
by first considering what the likely consequences would be, if true, and then contrasting the known circumstances of the case (see, e.g., I 40–2, III 22–3, XXII 11ff).
[105] As with Lysias, a rich variety of argumentative devices can be illustrated from the
works of Antiphon. They include explicit appeals to probabilities (particularly common
in the first Tetralogy, see especially II 1.9f, 2.8, 4.8 and 10), the use of the topic of
consistency (e.g. I 10, V 36f, 49f, 54), arguments *a fortiori* (e.g. II 1.9), disjunctive or
dilemma arguments (e.g. III 3.8) and refutations that proceed by considering what the
likely consequences of a thesis would be and (explicitly or implicitly) contrasting the
known circumstances of a case (e.g. II 1.4, V 52).
[106] Plato, *Prt.* 349a, *Hp. Ma.* 282d.

fourth centuries included rhetoric among the subjects they claimed to teach.

Yet another sign of this professionalism was the production of treatises or handbooks on the subject. The reconstruction of their contents is, to be sure, problematic.[107] Of our earlier sources, Plato, especially, writes from a point of view that is not just unsympathetic, but positively hostile, to the art of rhetoric as it was generally practised, while the extent of the first-hand information available to such later commentators as Cicero is uncertain.[108] What is incontestable, however, is that such treatises began to be written in some numbers from the middle of the fifth century. The first we hear of to have done so are two Sicilians, Corax and Tisias, but they were followed by many others, including Gorgias, Thrasymachus, Theodorus, Polus, Licymnius, Evenus and Alcidamas, to mention only some of the writers on rhetoric for whom we have good early evidence in Plato or Aristotle. Several of these early *Arts* may have taken as their chief theme the elements of which the speech should be composed: at *Rh.* 1354b16ff one of Aristotle's criticisms of his predecessors implies that they attempted to lay down what each of the main parts of the speech should contain.[109] But that there were discussions of arguments, including especially those that appeal to probabilities, is also clear. Thus Plato indicates that Tisias, Theodorus and Evenus dealt with aspects of proof and refutation,[110] and according to Aristotle Corax's *Art* largely consisted of a treatment of how to manipulate arguments based on either real, or what were passed off as real, probabilities.[111]

Many aspects of the early history of rhetoric as a conscious art must remain obscure. But for one key figure at the beginning of this development we have good original evidence that throws light not only on the actual techniques of argument he used, but also on his views on persuasion and on the power of the word. This is Gorgias, who provides a bridge between those whom we conventionally classify as sophists and the philosophers.[112] First, in the work called *On What is Not or on Nature*,[113] Gorgias exploits arguments that are

---

107 The evidence is collected by Radermacher 1951.

108 It is generally agreed that most late commentators depended on such sources as Aristotle's work (now lost) the *Collection of Arts*, rather than on the original treatises themselves.   109 Cf. also Plato, *Phdr.* 266d ff.

110 See *Phdr.* 266e–267a, 273a.   111 *Rh.* 1402a7–23.

112 Isocrates x 2–3 already compares Gorgias (and Protagoras) with Zeno and Melissus.

113 Our two main sources are Sextus Empiricus (*M.* VII 65ff, DK 82B3) and the pseudo-Aristotelian *De Melisso, Xenophane, Gorgia* (*De MXG*). On the much debated question of the purpose of the treatise, see, for example, Gigon 1936, Kerferd 1955, Bröcker 1958, Sicking 1964, Guthrie 1969, pp. 192ff.

similar in kind to those we have discussed from Parmenides, Zeno and Melissus. The difference is that now Eleatic arguments[114] are deployed against the Eleatics' own theses, as well as against those of their opponents, to establish the entirely negative conclusion that 'nothing exists; but that if it exists, it is unknowable; and if it exists and is knowable, it still cannot be indicated to others'.[115] The first part of that conclusion, that nothing exists, is established by a multi-stage dilemma argument, comparable with Zeno's. 'If something exists, *either* what is exists, *or* what is not, *or* both what is and what is not.'[116] Each of these possibilities is then considered and rejected in turn, in arguments that in some cases employ a further disjunction. Thus to refute the subordinate thesis that 'what is' exists, he considers the possibilities that it is *either* eternal, *or* generated, *or* both eternal and generated,[117] and these are rejected in turn, in one instance in an argument that sets up yet another dilemma: if it is generated, then it is generated *either* from what is, *or* from what is not.[118] Although there are minor divergences between our two sources for these arguments,[119] they agree in the main in the general pattern of reasoning they ascribe to Gorgias: to refute a thesis, he first subdivides it into a number of subordinate theses one of which must be true if the thesis itself is to be true, and then demolishes each of these subordinate theses in turn.

But while in *On What is Not* Gorgias applied Eleatic-style deductive reasoning to ontological questions similar to those that the Eleatic philosophers themselves investigated, we have other evidence for Gorgias' use of techniques of persuasion in a more general, rhetorical context in the three speeches that are ascribed to him, the fragmentary *Funeral Oration*, the *Encomium of Helen*, and the *Defence of Palamedes*.[120] These are, indeed, among our earliest extensive examples of Greek oratory, and all three exhibit the elaborate style (including the studied use of antithesis, anaphora and chiasmus) for which Gorgias was famous.[121] But apart from the complex formal design

---

[114] *De MXG* 979a22ff remarks that Gorgias tries to prove some points 'like Melissus', and others 'like Zeno'. Cf. also 979b22, 25, 37.

[115] See Sextus, *M.* VII 65 and *De MXG* 979a12f.

[116] Sextus, *M.* VII 66, cf. *De MXG* 979a35ff, b9f.

[117] Sextus, *M.* VII 68. The third possibility, the conjunction both eternal and generated, is not mentioned at *De MXG* 979b20ff.

[118] Sextus, *M.* VII 71, *De MXG* 979b27ff.

[119] Notably in that Sextus sometimes mentions, while *De MXG* omits, a third possibility, consisting of the conjunction of the first two (as in VII 68, above n. 117).

[120] Although the authenticity of the *Helen* and the *Palamedes* has, in the past, often been questioned, they are now generally accepted as genuine (see, for example, Segal 1962, p. 100 and n. 10).

[121] See, e.g., Diodorus Siculus XII 53.2ff.

and literary polish of these works, the *Helen* and the *Palamedes* are both exercises in persuasion – their aim being to exonerate Helen and Palamedes respectively from any guilt for the acts they are alleged to have committed.

Two features of these speeches are particularly noteworthy. First in both Gorgias employs disjunctive arguments that are strikingly reminiscent of – though to be sure more informal than – the dilemmas of *On What is Not*.[122] In the *Helen*, for instance, he distinguishes four possible explanations of why Helen acted as she did, namely (i) she was compelled by a god or fate, (ii) she was compelled by force, (iii) she was persuaded by arguments, and (iv) she was captivated by love, and he then claims that whichever explanation is adopted, Helen should not be held guilty.[123] In the *Palamedes*, too, Palamedes uses dilemmas, putting before his judges a series of possibilities relating to his case and then showing his innocence whichever of these is envisaged.[124] In neither speech are the alternatives such as to be formally mutually exclusive and exhaustive.[125] Yet, as in the would-be strict demonstrative reasoning of *On What is Not*, so also in the more general rhetorical context of the *Helen* and the *Palamedes*, Gorgias uses arguments that derive their force from the assumption that for a given thesis to be true, one or other of a set of subordinate theses must be true.

Secondly, the speeches provide important evidence concerning Gorgias' own reflections on the topic of persuasion. In the *Helen*, paras. 8ff, especially, when he argues that if Helen did what she did because she was 'persuaded' by λόγοι,[126] then she is not guilty, he speaks about the power of λόγος in general. To illustrate how it can affect the emotions, he first adduces poetry (para. 9), referring to the terror, pity and yearning that the audience feels. 'Inspired incan-

---

[122] On the similarity between the argumentation in these speeches and in *On What is Not*, see especially H. Gomperz 1912, pp. 22ff, Gigon 1936, pp. 190f, Segal 1962, pp. 99 and 115, Solmsen 1975, p. 12 n. 3.

[123] These four possibilities are set out in paras. 6 and 20.

[124] Thus at paras. 11 if he asks whether he was supposed to have committed the crime alone or with accomplices, and suggests that both are implausible. To undertake the crime was not the work of one man. Turning to the other possibility, he introduces a second dilemma: who were his accomplices, free men or slaves? If free men, then his judges should know about it themselves, and indeed be implicated with him. On the other hand it is surely incredible that he used slaves as accomplices, for they would denounce him either willingly, in the hope of their freedom, or under compulsion. Having shown he could not betray the Greeks, he goes on to argue that even if he had been able to, he had no motive to do so (paras. 13ff).

[125] The use of pairs of opposites in the arguments in the *Palamedes* is discussed in my 1966, pp. 120f.

[126] λόγος is, throughout, ambiguous, covering 'word' 'speech' and 'argument'.

tations', too, have a 'power' that 'soothes and persuades and transports the soul by its wizardry'[127] (para. 10). To be persuaded by λόγοι, then, he claims,[128] is equivalent to being abducted by force, and to demonstrate the effects of persuasion he says (para. 13): 'That persuasion, when added to λόγος, impresses the soul as it wishes can be learnt first from the λόγοι of the meteorologists,[129] who, by removing one opinion[130] and implanting another, make what is incredible and invisible appear before the eyes of opinion; secondly, from the constraining trials that take place through λόγοι, in which a single λόγος delights and persuades a whole crowd, when it is composed with skill, not spoken with truth;[131] and thirdly, from the contests of the λόγοι of the philosophers, in which is revealed a quickness of thought that demonstrates the mutability of belief in opinion.' Finally (para. 14) he compares the power of λόγος over the soul with the effect of drugs on the body.

This is the first extended passage that attempts something approaching a general statement concerning persuasion and the role of argument within it. As in Xenophanes, Parmenides and others,[132] truth is contrasted with seeming or opinion, and here[133] the latter is the sphere of persuasion.[134] The power of λόγος to deceive and to mislead is clearly acknowledged: it is, after all, Gorgias' aim at this point in the *Helen* to argue that she should be exonerated if she did what she did because she was won over by λόγοι. At the same time the examples given to illustrate the powers of persuasion are remarkable. Whilst the second of the three examples refers to the use of argument in the assemblies and the law-courts, the first and third relate to the work of natural scientists (the 'meteorologists') and to that of the philosophers. We could hardly have clearer indication of

---

[127] γοητείᾳ. On the use of this and other terms from magical practices (e.g. ἐπῳδαί, μαγεία, φαρμακεύω and ἐκγοητεύω) to describe rhetoric in the *Helen*, see especially de Romilly 1975.

[128] The beginning of para. 12 is corrupt, but the general sense of the argument is clear enough.

[129] μετεωρολόγοι, that is those who studied things in the heavens and things under the earth.

[130] δόξα, which covers both 'opinion' and 'seeming' 'semblance'.

[131] No doubt this refers to deliberative and forensic speeches.

[132] Xenophanes Frr. 34–35, Parmenides Fr. 1.30ff, Fr. 8.50ff, see above p. 78 n. 93.

[133] We may contrast Parmenides, in whom πειθώ is more often linked with ἀλήθεια (e.g. Fr. 2.4) even though mortals are deceived in what they confidently take – πεποιθότες – to be true, Fr. 8.39.

[134] The truth/opinion distinction recurs in the *Palamedes*, especially para. 24, where Palamedes accuses his prosecutor of trusting opinion, not truth. Opinion is open to all men and is unreliable: it is truth alone that should be trusted. The recognition of the negative or destructive possibilities of persuasion in the *Helen* contrasts with the position ascribed to Gorgias in some passages in Plato, where he is made to claim the superiority of rhetoric to all other arts precisely because it is unconcerned with the truth and confines itself to the task of persuasion (e.g. *Grg.* 452e, 454e ff, 459a–c, cf. *Phlb.* 58a ff).

the recognition, on the part of one notable orator, of the role of persuasive arguments in other contexts besides those of deliberative and forensic speeches.

Our survey has been brief, but there is, in any event, no need to multiply evidence to support the uncontroversial general conclusion that there was a rapid expansion in both the practice and the theory of public-speaking in various contexts in the fifth and fourth centuries. However the question that must now be raised is the significance of these developments for our concerns, that is first for the criticism of traditional beliefs as a whole, and then more specifically for the growth of natural science.

On the first topic we may restrict ourselves to some general remarks. It is evidently no mere coincidence that the period which sees the rise of professionalism in the art of speaking is also a period of radical criticism of certain aspects, at least, of Greek traditional beliefs. This is already suggested by the fact that the same individuals sometimes figure prominently in both developments. Several of the sophists who were among the first to offer to teach rhetoric also gained a reputation as critics of received opinions.[135] It would, it is true, be absurd to suggest that *all* the sophists were such, or that their criticisms extended to *all* such opinions. Nor should we forget that in some cases their doubts or objections had been anticipated by earlier writers.[136] Nevertheless the example of Protagoras is instructive. Not only was he one of the foremost teachers of the art of public speaking, but he is said to have been the first to have suggested that there are two opposed arguments on every issue.[137] The ability to sustain *either* side of a question was part of the training of the orator, as we can observe from the sets of speeches for the prosecution and for the defence in Antiphon's *Tetralogies*.[138] But a precisely similar

---

[135] Cf. above, pp. 14f, on the views expressed by Prodicus and Critias on the origins of beliefs in the gods.

[136] Thus Protagoras Fr. 4, which disclaims knowledge about the gods ('concerning the gods I have no means of knowing whether they are or are not or what they are like in form: for there are many hindrances to knowing, the obscurity of the subject and the shortness of human life'), may be compared with Xenophanes Fr. 34 ('there never was a man, nor will there ever be, who knows the certain truth about the gods and all the other things about which I speak'), and cf. also Heraclitus Fr. 86.

[137] Diogenes Laertius IX 51: cf. the title of a work in two books, the *Antilogiai*, and the tradition that he was one who claimed to be able to make the worse cause appear the better (e.g. Aristotle, *Rh.* 1402 a 23ff).

[138] See above, p. 80. If, as Morrison 1961 cf. 1963 has argued, Antiphon the Orator is one and the same man as Antiphon the Sophist, this would provide another example of the connection between rhetoric and sophistic. The contents of the Sophist's work on *Truth*, as revealed by *POxy*. XI 1364, are, however, sufficiently dissimilar from the

procedure was also deployed in the sphere of moral and political beliefs.[139]

Here, apart from Protagoras himself and many texts in Plato and Aristotle,[140] we may cite the *Dissoi Logoi*, or *Double Arguments*, a document from some time at the end of the fifth or the beginning of the fourth century that sets out a series of opposed theses on such questions as good and evil, the beautiful and the shameful, the just and the unjust, truth and falsehood and so on.[141] Again although Socrates is, in many respects, a quite exceptional figure,[142] certain aspects of his practice of *elenchus* may be viewed as, in part, a more systematic and sustained version of techniques that already had a wide general application in the context of deliberative and forensic oratory. The cross-examination of witnesses and the evaluation of evidence in general, the insistence that an account should be given – λόγον διδόναι – and the assessment of arguments put forward on either side of a disputed issue were all familiar enough to his interlocutors and audience from their experience of the law-courts and assemblies.[143] Socrates' distinctive contribution was to turn the searchlight of his scrutiny on current moral and political assumptions to expose – as he saw it – their shallowness and incoherence.

RHETORIC AND THE DEVELOPMENT OF NATURAL SCIENCE

But it is one thing to relate the development of rhetoric to that of the criticism of popular beliefs in general: it is quite another to attempt to argue that the former is directly relevant to the growth of natural science. While many scholars would agree that certain rhetorical tricks and stylistic traits[144] can be found in some writers

ideas we find expressed by, or attributed to, the Orator, to make the identification a difficult one.

[139] See the discussion of arguments 'in utramque partem' in Solmsen 1975, ch. 1, and cf. Moraux 1968, pp. 300ff, on their use in the Academy and by Aristotle's school, and Kudlien 1974 on their application in medicine.

[140] Thus Aristotle notes (*SE* 173a7ff) that reference to the twin opposed moral standards of 'nature' and 'convention' was a favourite commonplace for generating paradoxes (which 'all the ancients thought held good'), and at *Top.* 163a36ff he mentions the usefulness of examining arguments both for and against *any* thesis as a general training in dialectic.

[141] DK II 405ff: for a brief discussion, see Guthrie 1969, pp. 316ff.

[142] As a major point of difference from the sophists, Socrates received no fees for his instruction. Plato further insists that Socrates' *elenchus*, unlike the contentious reasoning of the sophists, was directed at discovering the truth, not at scoring victories nor at mere entertainment: see further below, pp. 100f.

[143] See below, pp. 249f, 252f.

[144] The question of how precisely these are to be defined and identified is, to be sure, a difficult one. But among the traits that may be taken to characterise rhetorical pieces

who are relevant to the development of early Greek natural science, most would want to maintain that there is no more to the connection than that. In particular it would commonly be assumed that there is a fundamental contrast between the aims of natural science on the one hand and those of rhetoric on the other, in that the former seeks the truth in its domain, the investigation of natural phenomena, while the latter is concerned only with persuasion – in its, quite different, field. Yet, as we shall see, too absolute a contrast between 'truth' and 'persuasion' oversimplifies and distorts a complex situation.

Our chief difficulty here, as so often elsewhere, stems from the nature of our evidence, the lack of extended original texts in two of the three main areas that we might hope to investigate. First, too little remains of the later Presocratic natural philosophers for our inquiry there to get far.[145] Secondly, although we know that several of those whom we customarily think of primarily as 'sophists' rather than as 'natural philosophers' also engaged in investigations concerning nature[146] and in the field of mathematics,[147] the dearth of original texts again means that our inquiry draws a blank. It is, then, to the third main area of pre-Aristotelian speculative thought that we must turn, namely to the substantial remains of fifth- and fourth-century medicine.[148]

are a concern for the main divisions of the work (including introduction and recapitulation or peroration), the heavy use of antithesis and the apostrophising of an opponent or objector.

[145] The fragments do, however, contain some notable examples of arguments of a kind similar to those we have found used in Parmenides and the later Eleatic philosophers. Thus we may cite examples of indirect proofs from Empedocles (Fr. 17.31ff), Anaxagoras (Fr. 12) and Diogenes of Apollonia (Fr. 2) – though this is not to say that these arguments were always modelled directly on those of the Eleatics themselves.

[146] Due caution must, it is true, be exercised in evaluating the testimonies of some later writers that certain sophists were natural philosophers or wrote treatises *On Nature*: the fact that Gorgias' work *On What is Not* had as its subsidiary title *On Nature* shows that a treatise with that name might well not contain any discussion relevant to the investigation of particular natural phenomena. We do, however, have good evidence for an interest in physical questions for no less than four major sophists, namely Gorgias himself, Hippias, Prodicus and Antiphon. (1) Gorgias' interest in certain specific problems connected with vision is attested by Plato (*Men.* 76c ff) and by Theophrastus (*Ign.* 13.73). (2) The range of subjects that Hippias was prepared to teach was comprehensive: they certainly included both music and astronomy according to Plato (*Hp. Ma.* 285b ff, *Prt.* 315c, 318d ff). (3) Galen ascribes not only a treatise on the nature of man, but also a specific theory of phlegm to Prodicus (*Nat. Fac.* II 11, *Scr. Min.* III 195.17ff Helmreich, K II 130.4ff, cf. K xv 325.11ff). (4) The second book of Antiphon's *Truth* evidently dealt with a variety of cosmological topics: the fragments we have cover a wide range of meteorological, geological and biological questions (Frr. 22–39).

[147] Both Hippias and Antiphon did original work in mathematics and Protagoras was evidently concerned to challenge the status of mathematics in Fr. 7 (see below, p. 116 n. 299).

[148] The roles of rhetoric and dialectic in the Hippocratic writers have been a rather neglected topic. See, however, Bourgey 1953, pp. 109ff ('les médecins sophistes et théoriciens'), Laín Entralgo 1970, ch. 4, and Kudlien's pioneering paper, 1974.

Now it is generally recognised that some of the treatises included in the Hippocratic Collection are in the nature of sophistic ἐπιδείξεις, exhibition pieces composed by men who were almost certainly not themselves medical practitioners. The two outstanding examples of this are *On the Art*, which offers a defence of the art of medicine against its detractors, and which has sometimes been thought to have been composed by Protagoras or Hippias,[149] and *On Breaths*, which sets out to prove that all diseases have a single cause, namely air.[150] Not only is the medical knowledge shown in these works superficial, but both show considerable stylistic elaboration, for example in announcing the subject of the present discourse in a proem or general introductory section,[151] in the studied use of antithesis,[152] and in dealing with the objections of imaginary opponents.[153] Their aim is merely to put a plausible thesis before some evidently quite inexpert audience, and in doing so they have no compunction in ignoring or drastically oversimplifying complex medical problems.[154]

But these two examples might, perhaps, be taken to be very much the exceptions that prove the rule: it might be argued that it is precisely because of the contrast between *On the Art* and *On Breaths* on the one hand, and the bulk of the Hippocratic treatises on the other, that we should conclude that the latter show no important signs of having been influenced by the development of rhetoric and of the techniques of persuasion and argument that we outlined above. That certain works are substantially free from such influences should be acknowledged immediately: this is true of two groups of treatises, in particular, namely the books of *Epidemics* and the aphoristic works.[155]

---

[149] In his edition, Gomperz argued that Protagoras is the author (T. Gomperz 1910, pp. 22ff), but W. H. S. Jones, 1923–31, II p. 187, put it that 'almost as good a case could be made out for considering the author to be Hippias'.

[150] To these two Jones, for example, would add *Decent.* and *Praec.*, as well as *Nat. Hom.*, on which see below, pp. 92ff. Cf. Bourgey 1953, pp. 114ff.

[151] The author of *de Arte* speaks of the treatise as an ἀπόδειξις (*CMG* I, I 10.18f, 11.2f, cf. ἐπίδειξις used of his opponents in ch. 1, 9.4 and of those who know the art ch. 13, 19.6f). After an introductory section that includes a definition of medicine (ch. 3, 10.19ff), the writer begins his demonstration in ch. 4, 11.5ff, by saying that 'the beginning of my speech is a point that will be agreed by all' (cf. Diogenes of Apollonia, Fr. 1). *Flat.* ch. 1 contains introductory remarks before the writer turns to 'the discourse that is to come' (*CMG* I, I 92.12). This writer too says that he will show (ἐπιδείξω, ch. 5, 94.6) that all diseases come from air, and he ends his treatise by saying (ch. 15, 101.17ff) that this has now been shown, ἐπιδέδεικται, and that his 'hypothesis' is true.

[152] Particularly prominent in *de Arte*, but present also in *Flat.*

[153] E.g. *de Arte* ch. 5, *CMG* I, I 11.20, ἐρεῖ δὴ ὁ τἀναντία λέγων and *Flat.* ch. 10, *CMG* I, I 97.10, ἴσως ἄν τις εἴποι, cf. 94. 15.

[154] The breathtaking hypothesis of *Flat.* – that all diseases come from air – has already been mentioned. *De Arte* ch. 7, *CMG* I, I 13.10ff claims that the reasons for failures to achieve a cure lie rather with the patient not obeying the instructions of the doctor, than with the doctor himself.   [155] E.g. *Aph.*, *Coac.*, *Prorrh.* I.

Yet there is, on the other hand, a large body of treatises which include some that no one would doubt to be the work of experienced medical practitioners, and indeed that have been accounted among the foremost examples of medical science in the fifth and fourth centuries, where such influences are unmistakable. The treatises include *On Ancient Medicine, On Regimen in Acute Diseases, On the Nature of Man, On Diseases* I and some of the principal surgical and embryological works,[156] as well as a group of minor, later writings.[157] Although *On the Art* and *On Breaths* are extreme cases, it can be argued that the influence of the development of rhetoric and dialectic on some of the finest instances of Hippocratic medical literature is widespread.

We may approach this topic first from the point of view of certain aspects of the doctors' medical *practice*. Two features of this clearly involved the Hippocratic physician in a task of *persuasion*. First he had to prevail upon his patients to entrust themselves to his care – whether or not he considered himself in direct competition with other doctors or healers. Secondly there was a further opportunity – and need – for persuasion in the context of a joint consultation between doctors about a case.

The problem of winning a clientele is often mentioned by Hippocratic writers. Several of the so-called deontological treatises issue warnings against attempts to attract clients by ostentatious behaviour of various kinds. *Precepts*, for instance, tells the doctor to avoid extravagant dress,[158] and *Decorum* implies a similar lesson.[159] Then *Prorrhetic* II censures extravagant claims for diagnosis and cure.[160] The surgical treatises, too, criticise the use of elaborate mechanical devices to impress patients, as also the practice of fancy bandaging. Thus *Joints* comments that attempts to reduce hump-back by succussions on a ladder are useless:

Succussions on a ladder never straightened any case, so far as I know, and the practitioners who use this method are chiefly those who want to make the vulgar herd gape, for to such it seems marvellous to see a man suspended or shaken or treated in such ways; and they always applaud these performances, never troubling themselves about the result of the operation, whether bad or good.[161]

---

[156] *Art., Fract.,* and *Morb.* IV will all figure below.

[157] E.g. *Prorrh.* II and the deontological works, especially *Praec.* and *Decent.*

[158] *Praec.* ch. 10, *CMG* I, I 33.32ff (reading θρύψις for τρῖψις) which refers particularly to the use of luxurious headgear and strange perfumes.

[159] *Decent.* ch. 2, *CMG* I, I 25.17ff, and ch. 3, 25.20ff.

[160] *Prorrh.* II ch. 1, L IX 6.1ff, see above, p. 45 n. 195.

[161] *Art.* ch. 42, L IV 182.14ff, translation Withington in W. H. S. Jones 1923–31, III; cf. ch. 14, 120.15ff, ch. 44, 188.13ff, and ch. 62, 268.3ff. Yet although the author criticises mechanical devices used for effect, he sometimes has recourse to them himself: see especially ch. 48, 212.17ff and ch. 70, 288.11ff (a description of a mechanical method

The same author also castigates doctors who achieve nothing but a display – the verb used is ἐπέδειξεν – in producing complex, but actually quite useless, bandages.[162] Several writers deplore the fact that the public is taken in by new-fangled cures,[163] or that they do not look beyond the technical jargon to the realities of a therapy.[164] What *Joints*,[165] *Precepts*[166] and *Decorum*[167] all inveigh against is the habit of turning a consultation into a public lecture – an occasion to make a display of skill – and it is clear that this was sometimes done.

Yet although several writers thus point to the excesses that doctors sometimes went to,[168] the task of persuasion remained. *Prognosis* is explicit that one of the essential aims of prognosis is to persuade the patient to entrust himself to the doctor's care.

If he is able to tell his patients in advance when he visits them not only about their past and present symptoms, but also to tell them what is going to happen, as well as to fill in the details they have omitted, he will increase his reputation as a medical practitioner and people will have no qualms in putting themselves under his care.[169]

Among the qualities that *Decorum* suggests the ideal doctor should possess are that he should be 'severe in encounters' (that is, no doubt, in controversy), 'ready to reply, harsh towards opposition...silent in the face of disturbances and resolute in the face of silences', and that he should be able to set forth clearly and gracefully what he has

---

of reducing a dislocation of the hip that is said to be both natural and ἀγωνιστικόν). Cf. also *Fract.* ch. 16, L III 476.8ff on a technique that is described as 'plausible for the layman, and without blame for the doctor', though it is 'less in conformity with the art', and *Medic.* ch. 2, *CMG* I, I 21.11ff.

[162] *Art.* ch. 35, L IV 158.4ff, cf. *Medic.* ch. 4, *CMG* I, I 21.32ff, *Fract.* ch. 2, L III 418.8ff.

[163] See, e.g., *Acut.* ch. 2, L II 234.2–238.1, *Fract.* ch. 1, L III 414.7ff, and cf. *Praec.* ch. 5, *CMG* I, I 31.26ff.

[164] Cf. the observations of *Acut.* ch. 2, L II 238.1ff, and the defensive remarks in *Prog.* ch. 25, L II 190.6ff, on the absence of the names of certain diseases in his account.

[165] E.g. *Art.* ch. 1, L IV 78.9ff (referring to what was evidently an open debate, with both physicians and laymen present, on what was thought to be a case of forward dislocation of the shoulder).

[166] *Praec.* ch. 12, *CMG* I, I 34.5ff, with the particular, ironic, injunction to avoid quotations from the poets.

[167] See *Decent.* ch. 2, *CMG* I, I 25.15ff on the crowds that quacks gather round them. Cf. from a later period Polybius XII 25d–e which laments the way in which rhetorical skills may count for more, with the public, than practical medical experience.

[168] *Epid.* VI sec. 5 ch. 7, L V 318.1–4, even refers to concealing a wad of wool in the palm of the hand and pretending to remove it from the patient's ear, in cases of ear-ache, so that he believes that it has been discharged. The author concludes the chapter with the single word ἀπάτη, 'deceit' 'trickery'. But the passage is evidence (if evidence is needed) that such tricks were used: nor is it clear whether the author himself condemns, or whether he would condone, this deception.

[169] *Prog.* ch. 1, L II 110.2ff (translations after Chadwick and Mann 1978), cf. also 112.6ff ('in this way one would justly be wondered at...' θαυμάζοιτο...δικαίως).

demonstrated[170] – a picture that the writer states to be appropriate not just to medicine, but to wisdom in general.[171] The Greek doctor was given instruction not only about the questions he should put to his patient,[172] but also on how to withstand the cross-examination to which he himself would be subjected.[173] In a situation where several doctors were present, it was up to each to sustain his point of view very much as if they were contending advocates.[174]

The treatise *On Diseases* I, especially, which contains a detailed pathological theory, begins with a passage that explains that the whole purpose of the work is to provide guidelines both on how to put, and answer, questions and on how to meet objections – where it seems clear that the objections do not simply come from prospective patients, but arise in the context of a semi-formalised debate. 'He who wishes to ask questions correctly, and to answer the questioner, and to debate (ἀντιλέγειν) correctly, on the subject of healing, must bear in mind the following things.'[175] These turn out to include not

---

[170] *Decent.* ch. 3, *CMG* I, 1 25.25 to 26.6. It is noteworthy that the author states that 'many cases require not reasoning – συλλογισμός – but help – βοηθείη' (ch. 11, 28.20f). This chapter recommends that the doctor should know what has to be done before going in to the patient: 'You should forecast what will happen from your experience; for that will add to your reputation and it is easy to learn' (ch. 11, 28.18–22).

[171] *Decent.* ch. 1, *CMG* I, 1 25.2ff.

[172] This is a recurrent theme in many different treatises, e.g. *Acut. (Sp.)* ch. 9, L II 436.8ff, *Aff.* ch. 37, L VI 246.16ff, *Morb.* II ch. 47, L VII 66.4ff, *Nat. Mul.* ch. 10, L VII 326.3ff, *Mul.* I ch. 21, L VIII 60.15ff, *Steril.* ch. 213, L VIII 410.14f and ch. 230, 440.13f, *Prorrh.* II chh. 27, 34, 41, 42, L IX 60.1ff, 66.8ff, 70.22ff, 74.4ff. Whether the patient is telling the truth or not is queried, for example, at *Epid.* IV ch. 6, L V 146.11f (cf. 160.6 and 162.5), cf. *Decent.* ch. 14, *CMG* I, 1 29.3f. *Mul.* I ch. 62, L VIII 126.12ff, draws attention to the problem that doctors are sometimes deceived because women are inhibited from speaking about conditions about which they are ashamed (and the cases just mentioned from *Epid.* IV were all women). Conversely *Praec.* ch. 2, *CMG* I, 1 31.6ff, insists that the doctor should not himself, from mistaken professional pride, hold back from asking laymen questions relevant to the case. In later Greek medicine, the topic of questioning the patient was further developed. One notable example of a treatise devoted to the subject is Rufus' *Quaestiones Medicinales* (*CMG* Suppl. IV Gärtner).

[173] See, for example, *Decent.* ch. 12, *CMG* I, 1 28.25, which speaks of the need for the doctor to show an ability to respond to objections, and *VM* ch. 15, *CMG* I, 1 46.26ff, which considers the questions that a hypothetical patient might put to a doctor.

[174] The frequent references to incorrect medical practices that are such a feature of such treatises as *Fract.*, *Art.*, and *Acut.*, should, no doubt, be seen in this light, as – in part at least – polemical or agonistic in purpose (cf. further Ducatillon 1977, pp. 229ff). As *Art.* ch. 1, L IV 78.1ff, 9ff, makes clear, the lay public, as well as his colleagues, to contend with, and *Aff.* ch. 1, L VI 208.16ff, promises advice to the layman on what he can contribute in his discussions with doctors. The fact that doctors were, for one reason or another, sometimes inhibited from calling in other doctors may be inferred from *Praec.* chh. 7 and 8, *CMG* I, 1 32.22ff, 33.5ff. In later medicine, as Kudlien 1974, pp. 187ff, has pointed out, the topic of arguing on both sides of the question is common: see, for example, Soranus, *Gyn.* 1 7.30ff and 1 11.42, *CMG* IV 20.2ff, 29.17ff, and the Δικτυακά of Dionysius of Aegae (Dulière 1965), and cf. Galen's criticisms, K VIII 56.4ff.   [175] *Morb.* I ch. 1, L VI 140.1ff.

only the origins and causes of diseases, which are of doubtful outcome, their transformations and so on, but also

what is said or done conjecturally by the doctor in relation to the patient, and by
the patient in relation to the doctor; also what is done or said exactly in the art,
and what is correct in it and what is not correct; what its beginning, its end and its
middle are and anything else that has correctly been shown to exist or not to exist
in it, both small things and great, and many and few; and that everything in the
art is one and that one is everything;[176] and the things that are possible for one to
think and say and, if need be, do, and those that are not possible for one either to
think or to say or to do.[177]

Guidance is also promised on which other arts it is like and which it
is not like,[178] as well as on which parts of the body are hot or cold or
dry or wet or strong or weak or dense or rare. The writer concludes:

one must bear these points in mind and retain them in discourses. Whatever
mistake in these matters anyone makes *either in speaking or in asking questions or in
answering* – if he describes what are many as few, or the great as small, or the
impossible as possible, or whatever other mistake he makes in his speech – one
may, bearing these things in mind, *attack him in reply* (ἐν τῇ ἀντιλογίῃ) in this
way.[179]

Moreover as this last example already indicates, it was not just in
connection with the discussion of certain medical practices that the
Hippocratic doctor sometimes found himself in a competitive
situation that tested his skills in debate. This was also the case on
some occasions, at least, where what was at issue was not a matter of
diagnosis or therapy, but general theoretical questions in such fields
as pathology, physiology or embryology. The best known evidence
for this comes from *On the Nature of Man*. This opens with a dis-
claimer: 'He who is used to hearing speakers talk about the nature
of man beyond its relevance to medicine will not find the present
account suitable to listen to.' The author immediately explains why:

I am not going to assert that man is all air, or fire, or water, or earth, or anything
else that is not a manifest constituent of man. But I leave such matters to those who
wish to speak about them. However those who make such assertions do not seem to
me to have correct discernment. For they all have the same opinion, but they do
not all say the same thing: yet they give the same reasoning for their opinion. For
they say that what is is a single thing, and this is the one and the all, but they do
not agree on their names for it. For one of them says that this thing, the one and
the all, is air, but another says fire, another water, another earth, and each adds to
his own speech evidences and proofs which amount to nothing. For when they all
have the same opinion, but do not all say the same thing, it is clear that they do

---

[176] ὅτι ἅπαν ἐστὶν ἐν αὐτῇ ἕν, καὶ ὅτι ἓν πάντα. The Heraclitean and Eleatic associations of
the author's use of such opposites as one and many, is and is not, great and small, are
obvious.　　　　　　　　　　　　　[177] L VI 140.10–19.
[178] L VI 142.1ff: cf. the references to other arts in *de Arte*, ch. 1, *CMG* I, 1 9.14ff, and in
Flat. ch. 1, *CMG* I, 1 91. 2ff.　　　　[179] L VI 142.7–12.

not know what they are talking about. One can discover this most easily by being present at their debates – αὐτέοισιν ἀντιλέγουσιν. When the same men debate with each other in front of the same audience, the same speaker never wins three times in succession, but now one does, now another, now whoever happens to have the glibbest tongue in front of the crowd. Yet it is right to expect that the person who says he has correct discernment about these matters should always make his own argument prevail, if he does know the truth and sets it forth correctly. But such men seem to me to undo themselves in the terminology of their arguments through their ignorance, and to establish the theory of Melissus [that is, that the one is unchanging].[180]

This text clearly indicates that even such, as we might suppose, specialised or technical topics as the ultimate constituents of man were the subject of public debates between contending speakers in front of a lay audience in the late fifth or early fourth century B.C. The particular speakers the writer has in mind in chapter 1 are not doctors: that becomes clear from the contrast he draws at the beginning of chapter 2.[181] But as that chapter goes on to show, there were not only natural philosophers, but also medical men, who adopted similar monistic theories, even though they selected not air, fire, water or earth, but such things as blood, bile and phlegm as the basis of their doctrines: 'they too add the same reasoning, that a man is one – whatever each of them chooses to name it – and that this changes its form and power, being compelled by the hot and the cold, and it becomes sweet and bitter, white and black, and every other kind'.[182] Now the criticisms that the author of *On the Nature of Man* levels against the lecturers he attacks might lead us to expect a radically different, certainly a less abstract and superficial, approach in his own discussion of the question of the constitution of man. Up to a point this may be so, in that he evidently makes some effort to bring empirical data to bear. Yet the contrast between this Hippocratic writer and the theorists he criticises is not as great as might at first be thought likely.

His principal complaint is that they are monists.[183] Although he promises 'evidences' and 'proofs' of his own pluralistic doctrine,[184]

---

[180] *Nat. Hom.* ch. 1, L VI 32.1–34.7.
[181] *Nat. Hom.* ch. 2, L VI 34.8ff.
[182] Ch. 2, L VI 34.10ff.
[183] At the end of ch. 1 (L VI 34.6f) he complained that the monists 'establish the theory of Melissus': in ch. 2 (34.17ff) we have the converse of an argument in Melissus Fr. 7 (cf. also Diogenes of Apollonia Fr. 2). Whereas Melissus had argued that if it felt pain, it would not be the same (but it is the same, so it does not feel pain, see above, p. 78), *Nat. Hom.* ch. 2 argues that if man were one (that is, consisted of a single element), he would not feel pain, as part of an indirect proof of the conclusion that man is a plurality: since he evidently does feel pain, he cannot be one.
[184] His own view is that man consists of four humours, phlegm, blood, yellow and black bile, each of which he associates with two of the four primary opposites, cold and wet,

these bear a strong resemblance to the reasoning he attributes to his opponents. He believes they are influenced by seeing what a man is purged of when he dies: sometimes he evacuates bile or, it might be, phlegm, and they then conclude that the human body consists of this one thing.[185] Yet his chief demonstration of his own four-humour theory takes a precisely similar form – only he identifies not just one, but four different substances (each of which is purged at different times or by a different drug) as the fundamental constituents of the body.[186] Although on certain purely medical matters we may imagine that the writer could lay claim to a certain expertise – or at least to some experience – that distinguished him from the out and out layman, on the topic of the constitution of the human body his own approach is not markedly different from that of the lecturers whose theories he dismisses as sterile. Although he castigates their contentiousness and lack of proof, his lecture[187] too is a similar exercise in persuasion, relying on some plausible (but quite arbitrary) arguments supported rather thinly by actual empirical evidence.

On the Nature of Man is not the only text that throws light on the open debates that were held particularly on topics where the interests of the medical men overlapped with those of the natural philosophers. Among the points that *On Diseases* I promises to cover in its advice on how to play the roles of questioner and respondent are which parts of the body are hot or cold or dry or wet, which are strong or weak, dense or rare.[188] Other treatises, too, reflect a context of debate when, for example, they envisage, and deal with, objectors. Thus the writer of *On Regimen in Acute Diseases* breaks off at one point to consider what might be said in favour of the opposite argument.[189] In *On Fleshes*, when the writer argues that the unborn embryo suckles, he considers what should be said in reply to someone who asks how anyone can know about the behaviour of the embryo in the womb.[190] The author of *On Regimen* I bewails the fact that many

hot and wet, hot and dry, and cold and dry respectively, and each of which he says comes to predominate in turn in the body according to the four seasons, winter, spring, summer, autumn, chh. 4, 5 and 7, L vi 38.19ff, 40.15ff, 46.9ff.

[185] Ch. 6, L vi 44.3ff.

[186] See further below, pp. 149f on ch. 5, L vi 42.8ff, ch. 6, 44.11ff, ch. 7, 46.17ff, 48.10ff, 50.9ff: that the substances evacuated are the elementary constituents of the body is simply assumed.

[187] Note ἀκούειν in ch. 1, L vi 32.3.

[188] *Morb.* I ch. 1, L vi 142.2ff, see above, p. 92.

[189] *Acut.* ch. 11, L ii 302.6. The beginning of the treatise engages in polemic with the authors, and revisers, of the *Cnidian Sentences*, where, for once, it is clear that the attack is directed at a written text, rather than a spoken discourse.

[190] *Carn.* ch. 6, L viii 592.16ff. Cf. also ch. 19, 614.10ff, where he says that if anyone wishes to scrutinise (ἐλέγξαι) what he has said about the seven-month embryo, it is easy to do so – though this turns out to be just a matter of consulting the midwives.

people, when they have heard one exposition on a topic, thereafter refuse to listen to anyone else speak about it,[191] and the writer of *On Diseases* IV says that he has expatiated on the subject of whether drink goes to the lungs (a much debated issue) because of the difficulty of persuading the listener to change his opinion with your own arguments.[192]

The general point can be made particularly clearly from a consideration of *On Ancient Medicine*, a treatise that concerns itself especially with the problem of the correct approach to the study of medicine. The author's polemic against various kinds of opponents, both 'sophists' and 'doctors', envisages lectures as much as written works. As he puts it in both ch. 1 and ch. 20, his subject is what is or has been *said or written* about medicine and about its relations with natural philosophy:[193] at the end of ch. 1 he refers explicitly to the *audience* who – he says – cannot tell whether what is said on the basis of a hypothesis is true or not.[194] Again in ch. 2 he insists that those who speak about the art of medicine should do so in a way that is clear to the lay public,[195] and he concludes that chapter with the remark that 'if anyone departs from what is popular knowledge and does not make himself intelligible to his audience, he will miss the truth. Therefore for this reason we have no need of a hypothesis.'[196] Evidently the vehicle of communication he has in mind is the lecture as much as, or even more than, the written text. Moreover much of the later part of the treatise is taken up with an imaginary debate with those who employed 'hypotheses'. At one point he claims that he has presented his opponent with a considerable ἀπορίη (quandary) with the question that he puts to him.[197] At another he meets an objection that his opponent might raise against him.[198] It is clear that these are not just empty stylistic traits: rather they reflect the author's experience of, and they show that he remains close to, a situation of live dialectical debate.

---

[191] *Vict.* 1 ch. 1, L VI 466.16ff.
[192] *Morb.* IV ch. 56, L VII 608.14–21, reading ἀκούοντα, with R. Joly, at 608.20, rather than Littré's ἀκόντα (which stresses the reluctance of the author's opponents to give up their opinion).
[193] *VM* ch. 1, *CMG* I, 1 36.2, ch. 20, 51.12f.
[194] Ch. 1, *CMG* I, 1 36.18–21 (note τοῖς ἀκούουσι at 19). Cf. further below, p. 135, on the subject of 'hypotheses' in this work.
[195] Ch. 2, *CMG* I, 1 37.9ff: he adduces as a reason for this that medicine is about what any human being suffers.
[196] Ch. 2, *CMG* I, 1 37.17–19.
[197] Ch. 13, *CMG* I, 1 44.27f: οἶμαι γὰρ ἔγωγε πολλὴν ἀπορίην ἐρωτηθέντι παρασχεῖν. Cf. also ch. 15, 46.26f, where he considers the question that an imaginary patient might put to the doctor who bases his theory on hot, cold, wet and dry.
[198] Ch. 17, *CMG* I, 1 48.21.

In the one area of Greek natural science for which we have extensive original texts from the late fifth and early fourth centuries, there is a good deal of evidence to suggest the role of open, and often competitive, public debates. Whether or not we think of a direct influence from the sophists or others who were instrumental in bringing about the developments in rhetoric we outlined earlier, it is clear enough that what we may broadly call sophistical or rhetorical elements – not just stylistic traits, but also techniques of argument and question and answer – can be found in a substantial body of medical writings. Not only such topics as the art of medicine itself, but also the origin of diseases, the constitution of the human body and a wide range of other problems in what we should call physiology, biology and indeed physics, were, it seems, sometimes debated openly. In such circumstances, it is not hard to see that the distinction between the exposition that a professional medical practitioner might give, and the ἐπίδειξις of a professional sophist, might be a fine one. The 'professional medical practitioner' and the 'professional sophist' were not, in any case, so readily identifiable as those terms might suggest. Although there is a general distinction between those who earned their living largely by medical practice on the one hand, and those who taught such subjects as the art of speaking on the other, neither category was at all sharply delineated. Their membership might overlap, and in two respects they found themselves in analogous situations. A teaching function was common to both – though some medical practitioners no doubt confined their instruction to those who were going to practise the art themselves.[199] Secondly, both had to attract a clientele. As we have seen, some of the medical authors write of the need to exercise restraint in this regard. Yet the very fact that they did so indicates where the temptations lay: any visit to a patient might be turned into an occasion for a display of skill or learning; a joint consultation might degenerate into a dispute between contending experts.

Finally evidence from outside the Hippocratic Corpus confirms that speculation on medical topics was far from confined to medical men and illustrates the overlap in interest not just between the 'medical practitioners' and the 'sophists', but also between the former and the natural philosophers. Aristotle remarked that it is the business of the student of nature (φυσικός) to 'inquire into the first principles of

---

199 The Hippocratic *Oath* specifies: 'I will hand on precepts, lectures and all other learning to my sons, to those of my master and to those pupils duly apprenticed and sworn, and to none other' (*CMG* I, I 4.9ff). This should not, however, be taken to be uniform practice among all the doctors represented in the Hippocratic Corpus.

health and disease... Most of those who study nature end by dealing
with medicine, while those of the doctors who practise their art in a
more philosophical manner take their medical principles from
nature',[200] and elsewhere he distinguished three kinds of persons who
had a claim to be able to speak on medical matters, that is not just
the ordinary practitioner and the 'master-craftsman', but also the
'man educated in the art' – the man who has studied medicine but
does not necessarily practise it.[201] The place that medicine might
occupy in the interests of others besides doctors can best be demonstrated by the fact that, following several other philosophers who are
known to have tackled the topic of the origin of diseases,[202] Plato
himself devoted a detailed six-page account to the subject in the
Timaeus,[203] while Aristotle too covered, or certainly aimed to cover,
the question in a treatise *On Disease and Health*.[204]

The Hippocratic doctor thus found himself in a complex and
competitive situation that often called for the exercise of skills in
persuasion and debate. The importance of this for our understanding
of the early development of Greek natural science is two-fold. First it
helps to explain the rather arid quality of much of the extant literature. The superficiality of some discussions of intricate and difficult
questions, and a certain dogmatism – the tendency to argue single-
mindedly for one particular thesis and against all others – may sometimes reflect the nature of the audience and the competitive character
of the *agon*. The remarkable proliferation of theories dealing with the
same central issues may well be considered one of the great strengths
not only of Hippocratic medicine, but also of Presocratic natural
philosophy. Yet while the critical examination of other doctrines is
sometimes well developed, this fertility in speculation is often not
matched by a corresponding *self*-criticism. In a situation of competi-

---

[200] *Sens.* 436a17–b1. The point is repeated in substantially the same terms at *Resp.*
480b22ff.

[201] *Pol.* 1282a1ff. Cf. Thucydides (II 48) who, after noting that anyone, whether doctor
or layman, might speculate about the causes of the plague at Athens, sets out to give
a detailed description of its course himself.

[202] The general theories of disease put forward by Hippon and by Philolaus are reported
in Anon. Lond. xi 22ff (where the name is an almost certain restoration), and xviii 8ff.
Particular pathological doctrines are attributed to Anaxagoras (Aristotle, *PA* 677a5ff)
and to Democritus (Soranus, *Gyn.* iii 17, *CMG* iv 105.2ff), who may, indeed, have
written works on 'prognosis', dietetics and 'medical judgement': the titles of such
works are certainly recorded by Diogenes Laertius, ix 48, although their authenticity
has been called in question.

[203] *Ti.* 81e–87b. In what is preserved of Meno's history of medicine in the papyrus
Anonymus Londinensis, more space is devoted to Plato's theory of diseases (xiv 11–
xviii 8) than to that of any other writer, including Hippocrates.

[204] No such work is extant, but Aristotle refers prospectively to such a treatise at *Long.*
464b32f, cf. *PA* 653a8f and Bonitz 1870, 104a47ff.

tive debate, however, this is readily understandable. The speaker's role was to advocate his own cause, to present his own thesis in as favourable a light as possible. It was not his responsibility to scrutinise, let alone to draw attention to, the weaknesses of his own case with the same keenness with which he probed those of his opponents. Given an interested but inexpert audience, technical detail, and even the careful marshalling of data, might well be quite inappropriate, and would, in any event, be likely to be less telling than the well-chosen plausible – or would-be demonstrative – argument.

Secondly, we must recognise the role of such debates in providing a framework of discussion on a variety of natural scientific problems. Aristotle's Lyceum was the first ancient institution that began to act as something like a centre of research in the natural sciences,[205] although Plato's Academy – on which the Lyceum was in part at least modelled – certainly anticipated it in fostering interest in certain areas of what we may call advanced studies, including the exact sciences. Yet in the late fifth or early fourth century B.C. those who engaged in scientific inquiry, when not quite isolated individuals, belonged, at most, to such loosely structured associations as the medical schools, such as those at Cos and Cnidus,[206] or the Pythagorean fraternities of Magna Graecia. Moreover, although by the end of the fifth century the manufacture and production of books had begun to develop, and literacy was by then established at least in a small section of the population,[207] the chief, even if no longer the only, medium for the propagation of scientific, as for other kinds of, knowledge was still, and was for long to remain, the spoken, not the written, word. In these circumstances – where no significant institutional support for natural science existed, and in an intellectual milieu which was still essentially oral, small-scale and face-to-face, the sophistic-type ἐπίδειξις and the kind of public debate alluded to in *On the Nature of Man* undoubtedly provided important vehicles for the exchange and dissemination of scientific ideas.

THE CRITICISM OF RHETORIC

It has long been recognised that the advance of rhetoric and of the sophistic movement contributed to a greater awareness of the use of arguments and helped to lead, eventually, to their formal analysis.

[205] This will be discussed in ch. 3, below, pp. 201ff.
[206] Both the extent of the doctrinal uniformity of such 'schools', and the nature of their organisation, are matters of conjecture (see most recently Lonie 1978). While there is no reason to doubt that teaching was one of their functions, there is no evidence that they attempted to initiate programmes of research on a corporate basis.
[207] See further below, pp. 239f.

Yet, as again is well known, this advance met with a hostile reception in some quarters. There is, to be sure, an element of exaggeration in some aspects of the reaction the sophists provoked. Thus Aristophanes' humour and abuse are directed somewhat indiscriminately at a wide variety of targets, ranging from politicians and poets, through astronomers, sophists and seers, to sexual deviants,[208] and he had no compunction in assimilating Socrates to the new learning. Yet some of the basis of the resentment felt towards the sophists is understandable enough. In that, in general, they moved from city to city in search of pupils, they were on the periphery of city-state society and were easily, and to some extent correctly, identified as a focus for the criticism of traditional morality. Moreover in the terms that he used to describe the power of rhetoric Gorgias himself invites a comparison with magic and witchcraft,[209] and so it is hardly surprising that the reception the sophists had was sometimes marked with an ambivalence similar to that which can be detected in popular attitudes towards the purifiers whom we discussed in chapter 1.

The key charge was that of making the worse (or weaker) argument appear the better (stronger). Our sources report that Protagoras claimed to be able to teach this;[210] Aristophanes makes great play with the theme in the *Clouds*, where he stages a mock *agon* between the just and the unjust λόγος;[211] and according to Plato it was one of the accusations brought against Socrates.[212] Xenophon's Ischomachus is careful to distinguish practising the art of speaking from learning how to make the worse argument appear the better,[213] and Isocrates too protests indignantly at the calumny levelled at him.[214] The charge evidently became a standard one, and though often, no doubt, distorted, it has at least this much foundation, that orators were expected to be able to support either side of a case: and as we have seen this applied not only in straightforwardly forensic contexts (where Antiphon's *Tetralogies* provided a model) but also in more

<hr>

208 *Nu.* 331 ff is one typical passage where we find sophists grouped together with Θουριομάντεις Ιατροτέχνας σφραγιδονυχαργοκομήτας, κυκλίων τε χορῶν ᾀσματοκάμπτας ἄνδρας μετεωροφένακας.

209 See above, p. 84 n. 127. The comparison recurs, but is turned against the sophists in Plato, e.g. *Sph.* 234e f, *Plt.* 303c, *Euthd.* 289e f, though Plato also describes Socrates' effect on his interlocutors as a bewitching, *R.* 358b.

210 D.L. ix 51, cf. Aristotle, *Rh.* 1402a23ff.

211 *Nu.* 882ff.

212 *Ap.* 18bc, 19bc (followed by a reference to Aristophanes) and 23d (where Socrates implies this was a stock, glib, charge made against anyone who philosophised).

213 *Oec.* 11. 23–25. At *Cyn.* 13. 4 (cf. 8) Xenophon represents the sophists as practising the art of deception.

214 xv 15f, cf. 259ff. xiii is a general warning against the pretensions and deceits of the sophists, cf. also x 1ff.

general, moral ones (as in the *Dissoi Logoi*).²¹⁵ Suspicion of rhetoric
was clearly widespread enough to make it advisable for the professional speech-writer to conceal his traces, and early on it became
a commonplace to insist on your own lack of skill in speaking and to
represent your opponents as dangerously misleading, and un-
scrupulous, manipulators of arguments.²¹⁶

As in Aristophanes, moral disapproval is the primary element in
Plato's more intense and complex reaction to the sophists. In such
dialogues as *Protagoras*, *Gorgias*, *Euthydemus*, *Phaedrus* and *Sophist*
especially he develops a series of contrasts between rhetoric, sophistic
and eristic on the one hand,²¹⁷ and dialectic on the other. The
sophists take pay for their instruction: but true philosophy is not to
be bought or sold; rhetoric is not an art, but flattery aimed at
gratification;²¹⁸ it seeks victory, not truth, and it deals with probabilities, intent on persuasion, not on knowledge.²¹⁹ Given that his
own dialectical method – throughout his life – owed much to
Socratic *elenchus*, it was clearly essential for Plato not only to under-
line the differences between dialectic and rhetoric or sophistic,²²⁰
but also to distinguish what he represented as the true method of
elenchus from the false. Thus in the *Gorgias* he explicitly contrasts the
correct method with the type of rhetorical *elenchus* practised in the
law-courts.²²¹ The latter depends largely on the number of witnesses
you can muster against your opponents: the former is not a matter of
counting heads, but of gaining the agreement of one man – the
person whose ideas are under examination;²²² it is not directed at the
man himself, however, but at the subject under discussion.²²³ If there
is a competitive element in the true method, it is a rivalry to get to the
truth.²²⁴

---

²¹⁵ See above, pp. 85f.
²¹⁶ See, for example, Antiphon v 1–7, Lysias xii 86, xvii 1, Isocrates xv 42, cf. Dover
    1974, pp. 25ff.
²¹⁷ Though Plato distinguishes between rhetoric and sophistic, for example, he also, on
    occasion, assimilates them or emphasises their similarities as at *Grg.* 465c, 520a.
²¹⁸ As most famously at *Grg.* 464b ff, 502de: contrast Isocrates xv 197ff.
²¹⁹ See especially *Grg.* 452e ff, 458e ff, *Phdr.* 259e ff, 272d ff, *Phlb.* 58a ff, cf. *Phd.* 91a.
²²⁰ The contrast between a sophistic ἐπίδειξις and dialectic is pointed up in both *Prt.*,
    e.g. 328d ff, and *Grg.*, e.g. 447b ff. The great sophists are represented as claiming to
    be able not only to produce set speeches, but also to engage in question and answer
    (*Grg.* 449bc, cf. 458de, *Prt.* 329b, cf. 334e–335d), though their inadequacies in the
    latter are exposed by their confrontations with Socrates.
²²¹ *Grg.* 471e ff. One may compare Aristotle, *Pol.* 1268b41ff, who refers to what he
    already considers an archaic system of law, where what counts is the number of
    witnesses brought into court (cf. Gernet (1948–9) 1968, p. 245 and 1955, pp. 61–81).
²²² *Grg.* 471e, 472bc, 474a, 475e–476a, 522d.
²²³ *Grg.* 457e, cf. 453c, 454c: at 473b we are told that the truth is never refuted.
²²⁴ See *Grg.* 505e, cf. 457d and Thrasymachus' accusation at *R.* 336c.

Moreover this general characterisation is supplemented – though admittedly unsystematically – with many specific remarks concerning dialectical procedure. Thus in different contexts it is pointed out that it is useless to talk about a subject which you have not defined;[225]each person must take his turn at asking, and at answering, the questions,[226] not evading the questions by play-acting,[227] nor by descending to verbal abuse;[228] one must guard against ambiguous[229] or compound[230] questions, and not specify in advance the kind of answer required.[231] Above all there is an insistence on the need to follow the argument wherever it leads and to accept its conclusions, however unpalatable,[232] and on the primary importance of consistency:[233] indeed the principal type of refutation is that in which it is claimed that a self-contradiction has been shown.[234] Although Plato was certainly not the first person to be interested in establishing what we may call the correct rules of procedure for conducting a dialectical inquiry, we have fuller observations on the subject from him than from any earlier writer.[235]

In these respects – in incorporating *elenchus* and in making recommendations concerning the technique of question and answer – Plato belongs to a long line of critical inquiry that stretches far back into earlier philosophy and science. 'Dialectic' is now redefined to exclude rhetoric and sophistic.[236] His chief criticisms of the latter rest on fundamental distinctions between the sphere of the 'probable' and that of the 'true', and between 'persuasion' and 'proof'.[237]

---

[225] See *Grg.* 448de, *Phdr.* 263d, 270c ff, 277b6.

[226] See *Prt.* 338de, 348a, *Grg.* 458ab, 462a, 474b, 506c, *R.* 350c–e.

[227] *Grg.* 500bc. The seriousness of the enterprise (although it is treated by the young as a form of sport) is stressed at *R.* 539b–d, cf. 336e and *Tht.* 167e in Protagoras' 'defence'. It is the kind of purely verbal trickery indulged in by Euthydemus and Dionysodorus that is a mere game, *Euthd.* 277d ff.

[228] See *Grg.* 457d, *R.* 343a ff. *Grg.* 473de points out that neither attempts to frighten your opponent, nor mocking him, constitute a refutation.

[229] E.g. *Euthd.* 275d, cf. 277d ff, 295bc. Ambiguity in general is a frequent source of fallacious reasoning in Plato, although the extent to which he was conscious of fallacy *as such* is disputed, see R. Robinson (1942) 1969, pp. 16ff, Sprague 1962. In *Euthd.*, Socrates' complaint against the sophists is primarily a moral one: he rejects their arguments as much for their triviality as for their fallaciousness.

[230] E.g. *Grg.* 466cd.   [231] E.g. *R.* 337ab.

[232] See, from *Grg.* alone, 454c, 479bc, 480a, 480de, 497b, 498e–499b, 503cd, 509e.

[233] A particularly prominent theme in *Grg.*, e.g. 457e, 460e–461a, 482bc, 487b, 499b, and in *Prt.*, e.g. 333a, 339b–d, 361a–c.

[234] On the different modes of *elenchus* (not always a refutation) see especially R. Robinson 1953, chh. 2–3.

[235] Cf. R. Robinson 1953, Ryle 1965 and 1966, ch. 4, especially.

[236] Cf. R. Robinson 1953, p. 85, who remarks: 'The reason why Plato constantly pillories eristic and distinguishes it from dialectic is that in truth his own dialectic very closely resembled eristic.' Cf. also Ryle 1966, pp. 126ff.

[237] In certain contexts and for some purposes, however, Plato has to content himself with

Whilst in many of his remarks on dialectical method Plato remains close to earlier traditions, in the middle and later dialogues his criteria and requirements become stricter and more complex. The dialectician must not only be able to conduct question and answer, and to give and receive an account:[238] he must also be 'synoptic', able to grasp the connections between things and to determine the true reality – that is, in the *Republic*, the transcendent Forms – underlying appearances.[239] Dialectic comes to be seen particularly in terms of the ability to discern the similarities and differences between things,[240] and it must be not only critical but rigorous. Thus in the late *Philebus* its subject-matter is still defined as what is unchanging, and it is distinguished from the power of persuasion by its clarity, exactness and truth.[241] Whilst Plato's recommendations on the correct conduct of an *elenchus* mark a new and decisive turn in the development of views on what 'dialectic' should consist in, he also had an important, indeed crucial, contribution to make in the development of a formal notion of demonstration. It is to this, and to the related, far more problematic, question of the growth of the concept of an axiomatic system, that we must now turn.

THE DEVELOPMENT OF DEMONSTRATION

Our first points must be negative ones. If, as we have seen, the first extant sustained deductive reasoning appears in Parmenides, who is also the first to have set up a fundamental opposition between the use of the senses and abstract argument, he has no term for 'proof' and not necessarily any clear criteria for one. Although the orators, for instance, employ an informal notion of proof that is perfectly adequate for practical purposes – when, for example, a claim is made that either the facts of a case, or the motives of the agents, have been sufficiently established[242] – neither they, nor any other of the fifth and early fourth century whom we have considered provide any clear indication that the formal conditions of proof had begun to be

---

a merely probable account, as notably in that of the visible world in *Ti.* (see especially 29bc), and cf., e.g., *Grg.* 523a, 524ab, *Phd.* 108de, *Phdr.* 245c and 246a ff.

[238] These features are emphasised in the account of dialectic as the supreme study in *Republic* VII, e.g. 531e, 533c, 534b, 534d.

[239] E.g. *R.* 531cd, 532d ff, 534ab, 537cd.

[240] Especially *Phdr.* 265de, *Sph.* 253b–e, *Plt.* 285ab.

[241] *Phlb.* 58a–c, 59a–c.

[242] See especially the use of ἀποδείκνυμι (e.g. Antiphon II 3.1, IV 3.7, 4.9, V 64, 81, Lysias III 40, VII 43, XIII 49, 51, XV 11), ἐπιδείκνυμι (Antiphon II 4.3, III 4.9, IV 2.7, V 19, Lysias IV 12, XII 56, XIII 62) and ἀποφαίνω (Lysias XIII 51).

analysed.[243] In particular, although the medical writers frequently claim that they have demonstrated some theory or opinion or use the term 'necessity' to describe the causal relations they assert to hold, the 'proofs' they adduce are generally quite informal and their criteria for such are evidently far from strict.[244]

The most promising area for our inquiry is the one for which our original sources are least adequate, namely mathematics. Although commentators on Euclid's *Elements* identify for us the authors of particular theorems or groups of theorems, and texts in both Plato and Aristotle afford us some insight into the work of their contemporaries and predecessors, the reconstruction of pre-Euclidean mathematics and of the stages whereby the material that now appears in the *Elements*[245] was assembled is, inevitably, a desperately conjectural affair.[246] On one issue of fundamental concern to us, namely the suggestion that the notion of demonstration in mathematics originated from, or was the result of the influence of, Eleatic philosophy, it is as well to recognise at the outset that the direct evidence by which such a thesis might be conclusively confirmed or refuted is not forthcoming.[247] Nevertheless some relevant aspects of the development of Greek mathematics are not in doubt.

First early Greek mathematics is, already in the pre-Platonic period, remarkably heterogeneous. We can distinguish at least four main, as well as a large number of subsidiary, branches of theoretical interests.[248] First, number-theory – including the division of numbers

---

[243] Thus like the orators, the historians have no strict formal criteria for proof, even though in their assessments of differing accounts of historical events they are frequently alert to the distinction between what is merely a possible story and one that has been established beyond reasonable doubt.

[244] Thus when the author of *Nat. Hom.* promises to 'demonstrate' (ἀποδείξω ch. 2, L VI 36.12, cf. δεικνύναι 36.6, ἀποδείκνυμι ch. 5, 44.2) and to 'show the necessities (ἀνάγκας ἀποφανῶ) through which each thing is increased or decreased in the body' (ch. 2, 36.15f), this 'proof' rests mainly on the evidence of the substances drawn from the body by certain drugs (see above, p. 94 and n. 186). While at ch. 3, 36.17, for instance, ἀνάγκη is used to express the conclusion of a *reductio* argument (cf. also ch. 2, 36.1), elsewhere the same term is used of the author's view of what happens to each constituent of the body after death: 'again it is necessary that each thing returns to its own nature' (ch. 3, 38.10f, cf. ch. 4, 40.6 and 12, ch. 5, 42.6, ch. 7, 50.3, ch. 8, 50.19). Similarly causal sequences are often described as necessary in such treatises as *Aph.*, *Aër.*, *VM* and *Vict.*

[245] Euclid's *Elements* were composed – it is generally thought – some time around 300 B.C.

[246] A full survey of the problem, with extensive references to previous views, has recently been undertaken by Knorr 1975.

[247] The thesis has recently been maintained by Szabó 1964–6 and 1969.

[248] There is a pointed contrast between the utilitarian, and the non-utilitarian, justifications of the study of mathematics in different fourth-century authors. For the former, see, for example, the views ascribed to Socrates in Xenophon, *Mem.* IV 7.2–5; for the latter, Plato's account of the role of mathematics in the higher education of the guardians, *R.* 524d ff. Cf. also Isocrates XI 22f, XII 26f, XV 261ff.

into odd and even, the investigation of certain elementary propositions involving odd and even numbers,[249]the classification of
'figured' numbers and the generation of such numbers by the
application of the 'gnomon'[250] – was evidently a major preoccupation of some fifth- and fourth-century mathematicians, particularly
but not exclusively Pythagoreans. Secondly, there is the so-called
Heronic tradition of metrical geometry, characterised by its concern
with the solution of problems of mensuration, such as the determination of the areas of plane figures of various kinds.[251] Thirdly, there
is the non-metrical geometry that is represented, for example, by
several of the contributions to the three favourite pre-Platonic
special problems, (i) the squaring of the circle, (ii) the trisection of
an angle, and (iii) the duplication of the cube, and, especially, by
the work of Hippocrates of Chios.[252] Finally there is a further
distinct area of interest in the applications of mathematics to music
theory, leading up to the work of Archytas of Tarentum, a contemporary of Plato.[253]

Reflecting, in part, the heterogeneity of these areas of study
themselves, the methods that were used, and the conceptions of
'proof' that were operational, in different parts of early Greek
mathematics were far from standardised and varied appreciably in
rigour. Thus when Proclus, for instance,[254] reports that Thales is said
to have 'proved' (ἀποδεῖξαι) the proposition that the diameter
bisects the circle, we may have doubts not only about the attribution

---

[249] Several such propositions are collected in Euclid IX 21–34, e.g. that the sum of any
multitude of even numbers is even (IX 21), and that the product of an odd number and
an odd number is odd (IX 29), though how far Euclid's presentation follows earlier
models is far from clear.

[250] That is, the study of the varieties of plane numbers (triangular, square, oblong,
pentagonal, etc.) and of solid numbers (cubic, pyramidal, etc.) obtained by the
corresponding geometrical arrangements of points: see, for example, Heath 1921,
I pp. 76–84. The application of the 'gnomon' to generate series of such numbers is
attested in, for example, Aristotle, *Ph.* 203a10ff, cf. *Cat.* 15a29ff.

[251] Such metrical geometry can be extensively illustrated also in the extant remains of
both Egyptian and Babylonian mathematics (see, for example, Neugebauer 1957,
chh. 2 and 4, van der Waerden 1954, pp. 31ff, 75ff). While it is clear that the Greeks
drew on these other traditions some time before Euclid, let alone before Hero (first
century A.D.), the stages and timing of the transmission of those aspects of Egyptian
and Babylonian mathematics are largely a matter of conjecture.

[252] See further below, pp. 108ff.

[253] Archytas' study of mean proportionals is one area of overlap between his interests in
music theory and his more general mathematical work: see, for example, Heath 1921,
I pp. 213ff, pp. 246ff, Szabó, 1969 Part II.

[254] Proclus, *In Euc.* 157.10f. Elsewhere (e.g. *In Euc.* 352.14ff) Proclus claims to be
drawing on the history of Eudemus (fourth century B.C.) for his reports on Thales:
yet reservations may still be expressed about the information available to Eudemus,
when Thales may not have written anything, and when Aristotle himself is consistently
guarded in his remarks about Thales' ideas.

to Thales (about whom little definite information survived to the time of Aristotle, let alone later), but also about the nature of the proof by which such a proposition might have been established in the early stages of Greek mathematics. One simple practical technique that might well have been used in that context is that of superposition, which persists in the seventh common opinion of Euclid, the so-called 'axiom of congruence', that states that 'things which coincide with one another are equal to one another',[255] and which is applied in the proof of *Elements* I 4, asserting the congruence of two triangles that have two equal sides and the angle contained by them equal.[256] Again it is likely that the notion of proof operational in early studies of odd and even numbers was a quite informal one. Before number theory itself was put on an axiomatic foundation, propositions such as that which states that the sum of a multitude of even numbers is itself even were, we may presume, 'shown' simply by direct reference to the unit or dot representations of such numbers.[257]

The famous passage in Plato's *Meno* in which Socrates has a slave solve the problem of constructing a square double the area of a given square affords some insight into early Greek geometrical procedures.[258] Plato's own particular purpose is to establish the theory of recollection, but he does so by taking a geometrical illustration that incorporates what were, by then, clearly recognised as elementary techniques. Socrates proceeds by question and answer and claims that he is not instructing the slave, although, as has often been observed, the questions put are leading ones. The slave first tries various incorrect solutions, and at one point it is suggested to him that he might *point* to the correct line, if he is unable to *number* it.[259] This no doubt alludes to the incommensurability of the diameter and the side, but at the same time Socrates acknowledges that the slave might resolve the problem merely by referring to the diagram. The two features of the eventual solution that are particularly noteworthy are (i) that it depends entirely on the correct construction,[260]

---

[255] A similar notion may underlie the definition of straight line as 'that which lies evenly with the points on itself' (I Def. 4), cf. Plato, *Prm.* 137e.

[256] I 4 (I 11.4ff HS), cf. I 8 (I 16.11ff HS). Heath 1926, I pp. 224ff, 248ff, arguing that the seventh (or in his numbering fourth) common opinion may be an interpolation, suggested that Euclid used the method of superposition with some reluctance and may have been aware of objections to it as not admissible as a theoretical means of proving equality, only as a practical test of it. But whatever is true of Euclid, the method was, no doubt, used without qualms in parts of earlier Greek geometry. Cf. especially von Fritz (1959) 1971, pp. 430ff.

[257] For a plausible reconstruction of the types of method that may have been used in this context, see Becker 1936e, pp. 533ff.

[258] *Men.* 82b–85d.   [259] *Men.* 84a1.

[260] That is, of the square on the diameter of the original square. The close connection

and (ii) that once this construction has been carried out the solution to the problem is treated as obvious on direct inspection.[261]

In certain contexts, and for some purposes, early Greek mathematicians employed heuristic or practical methods and were content with a loose, informal notion of 'proof'. Yet the first attempts to give strict deductive demonstrations certainly go back into the fifth century. To illustrate this we need go no further than the well-known example of the incommensurability of the side and diameter of the square.[262] Whereas elsewhere it is often appropriate and necessary to distinguish between the discovery of a mathematical theorem and the discovery of its proof,[263] a feature of the incommensurability of the side and diameter is that the justification of this result can only be based on logical deduction.[264] We are not in a position to say for certain when or by whom the proof was first given, nor which of several possible methods was employed. Yet we may infer from Plato's report concerning the work of Theodorus that the result was already known to mathematicians by his time (and he may be taken to have been active in the decade or so either side of 400 B.C.) and the same text in Plato also indicates that Theodorus himself carried out further studies of incommensurables up to that of the side of the seventeen foot square (or, as we should say, $\sqrt{17}$).[265]

Moreover although the question of just how the incommensurability of the side and the diameter was originally demonstrated is hotly disputed,[266] it is clear from a text in Aristotle that one method

between 'construction' and 'proof' in Greek geometry is further suggested by the fact that the term γράφειν continues to be used to cover both: see for instance in Plato, *Tht.* 147d, Aristotle, *APr.* 65a4ff, *Top.* 158b29ff.

[261] That the diameter bisects the square is asserted (not proved) at *Men.* 84e4ff.

[262] This was the form in which Greek mathematicians treated what we should call the problem of the irrationality of $\sqrt{2}$.

[263] Cf., e.g., the report in the introduction to Archimedes' *Method* (II 430.1ff), that the relations between the volumes of the cone and cylinder, and between those of the pyramid and prism, were first discovered by Democritus, though their proofs were first given by Eudoxus.

[264] Approximations to $\sqrt{2}$ and $\sqrt{3}$ are attested in Babylonian sources, but there is no evidence that the Babylonians knew that $1 : \sqrt{2}$ cannot be expressed as a ratio between integers, or that, if they did, they had grasped the significance of that fact: see Neugebauer 1957, p. 48.

[265] *Tht.* 147d. The fact that Theodorus began with the side of the three foot square shows that the case of the two foot square was well known. The method that Theodorus himself used has been much discussed: see, e.g., Vogt 1909–10, Heath 1921, I pp. 202ff, von Fritz 1934*b*, Anderhub 1941, van der Waerden 1954, pp. 142ff, Heller 1956–8, pp. 13ff, Wasserstein 1958, Szabó 1962, pp. 69ff, Knorr 1975, chh. 3 and 4.

[266] One interpretation has it that the discovery arose from the application of anthyphairesis ('reciprocal subtraction'), the algorithm set out for integers in Euclid VII 1–3 (VII 1 states: 'Two unequal numbers being set out, and the less being continually subtracted in turn from the greater, if the number which is left never measures

proceeded by an indirect proof. At *APr.* 41 a 23ff Aristotle takes as an example of reasoning *per impossibile* the proof that the side and diameter are incommensurable that first assumes that they are commensurable and then deduces the impossible result that 'odd numbers equal even numbers'. As every student of Greek mathematics knows, a proof of this general form appears in the 'appendix' to Euclid x (App. 27, III 231.10ff HS). In this, to paraphrase the essential steps, the diagonal (AC) is first assumed to be commensurable with the side (AB) and $a:b$ is taken to be their ratio expressed in lowest terms.[267] That is, $AC:AB = a:b$. So $AC^2:AB^2 = a^2:b^2$. But (by Pythagoras' theorem) $AC^2 = 2AB^2$. So $a^2 = 2b^2$. So a^2, and therefore also a, is even, and since $a:b$ are the lowest terms, b *is odd*. Since a is even, let $a = 2c$. So (from the previous step, $a^2 = 2b^2$) $4c^2 = 2b^2$. So $2c^2 = b^2$. So b^2, and b, *are even*. So since the original assumption of commensurability leads to the contradiction that b *is both odd and even*, that assumption is false. Now it is not certain that this was the precise proof that Aristotle had in mind – let alone that it was the original method by which incommensurability was established.[268] Yet we can be sure of this, at least, that *some* proof of

the one before it until a unit is left, the original numbers will be prime to one another') and for homogeneous magnitudes in x 2–4 (x 2 states: 'If, when the less of two unequal magnitudes is continually subtracted in turn from the greater, that which is left never measures the one before it, the magnitudes will be incommensurable'). Thus, in arithmetic, anthyphairesis can be used to show that two numbers have no common factor: but when, in geometry, the process of subtraction continues indefinitely, the magnitudes are incommensurable. It has been suggested that mathematicians (usually identified as Pythagoreans) discovered incommensurability when they realised either that the anthyphairesis of a line cut in mean and extreme ratio continues indefinitely, or that the algorithm applied to side and diameter numbers does. (The former has been suggested in connection with studies of the regular pentagon: the side and the diagonal are in mean and extreme ratio, the intersection of the diagonals forms another regular pentagon and this process can be continued indefinitely: see, for example, von Fritz (1945) 1970, pp. 401ff, and Heller 1958, pp. 9ff. For the latter suggestion, see Heller 1956–8, pp. 3ff and 1958, pp. 14ff). But against this it may be, as Knorr 1975, ch. 2 (following Heath) has argued, that in this context anthyphairesis was originally used rather as a method of approximating to the length of the diameter, than in order to prove its incommensurability.

267 This is too strong a condition: all that is necessary is that not both a and b are even. The proof in Euclid also includes a redundant step to show that AC is greater than one: more correctly – given that some Greek mathematicians are reported to have held that one is both odd and even, Aristotle, *Metaph.* 986a 19f – in the alternative (and more general) version of the proof that follows in x App. 27 (III 233.15ff HS) a step is introduced to show that AB is greater than one.

268 Apart from the anthyphairesis interpretations (see above, p. 106 n. 266) two main other possibilities are explored by Knorr (1975, chh. 2 and 6). (1) The first uses the same diagram as *Meno* 84d ff, and assumes first that AG and DB both represent integers. But $AGFE$, a square number, is even (it is double $DBHI$). And $DBHI$ in turn is double $ABCD$, and so it too is even, and this process, conceived geometrically, can clearly be carried on indefinitely. But if $AGFE$ represents a finite number, its successive division in half must terminate. So the initial assumption, that both AG

incommensurability of a strict type, taking the form of a *reductio*, was well known by the fourth century.

Moreover it was not just that strict methods of proof were used in some areas of early Greek mathematics, for attempts had also begun to be made to systematise parts of geometry by the end of the fifth century. We have, to be sure, to rely in part on what the late commentators tell us, but the main point, at least, can be confirmed by good original evidence. In the general information that Proclus gives us concerning Euclid's predecessors, he reports that the first person to have composed a book of *Elements* was Hippocrates of Chios, and further that several other later mathematicians, including Archytas and Theaetetus, 'increased the number of theorems and progressed towards a more scientific arrangement of them'.[269] Meanwhile Aristotle explains how the term 'elements' itself was used: 'We give the name "elements" to those geometrical propositions, the proofs of which are implied in the proofs of all or most of the others.'[270]

Although the detailed reconstruction of the contents of these early *Elements* cannot be attempted with any degree of confidence, we can

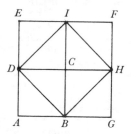

and *DB* represent integers, leads to an absurdity. (2) Secondly incommensurability might have been discovered in the study of right-angled triangles, using some simple theorems concerning Pythagorean triples, namely (*a*) that if the largest number is even, then all three numbers are even, and (*b*) that if one of the numbers is odd, then the largest number is odd, and one of the other two numbers is odd and the other even. Taking a right-angled isosceles triangle, what are the numbers represented by the sides? Let X be the number for the hypotenuse, and Y for the side. If X is even, then by (*a*) all three numbers are even. By halving each side of the triangle we obtain another right-angled isosceles triangle, and if its hypotenuse is even, the same applies. This process can, if necessary, be continued indefinitely. But if it continues indefinitely, this contradicts the assumption that X is a finite integer. Assume, then, alternatively, that it is odd (or that we arrive at an odd number for the hypotenuse after bisection). But if so, then by (*b*) one of the two other sides is odd, and the other is even, and this contradicts the assumption that the triangle is isosceles.

[269] *In Euc.* 66.7ff. Hippocrates' work is generally dated to around 430 B.C. Theaetetus, a slightly younger contemporary of Plato, died in 369.

[270] *Metaph.* 998 a 25ff. That the, or a, mathematical use lies behind the distinction between elements and complexes drawn by Plato at *Tht.* 201 e ff has been suggested, for example, by Morrow 1970. The term στοιχεῖον there, however, has its primary, literal sense, 'letter'.

gain some idea of Hippocrates' methods and of the state of geometrical knowledge at the time from Simplicius, who preserves an
extended passage from the second book of Eudemus' *History of
Geometry* recording Hippocrates' work on the quadrature of lunes.[271]
Eudemus tells us that he

made his starting-point, and set out as the first of the theorems useful for his
quadratures, that similar segments of circles have the same ratios as the squares on
their bases. And this he proved by showing that the squares on the diameters have
the same ratios as the circles.[272]

The quadratures of lunes with external circumferences that are equal
to, greater than, and less than, a semi-circle are then set out in turn,
followed finally by the quadrature of the sum of a lune and a circle.
  Several fundamental points emerge from a consideration of this,
the most complex and sustained piece of fifth-century mathematical
reasoning that has come down to us. First Hippocrates uses strict
demonstrative methods.[273] This is true not only of the quadratures
themselves,[274] but also of the preliminary constructions: it is notable
that he is not content merely to construct lunes with external
circumferences greater, or less, that a semi-circle, but in both cases
provides proof of these inequalities.[275] Secondly, he presupposes a
large number of geometrical theorems, corresponding not only to
Euclid I and II (including, for example, extensions of Pythagoras'
theorem to obtuse and acute triangles), but also to much of III and IV
(dealing with the circle and inscribed polygons) and of VI.[276] We
must repeat that we have no means of determining the precise form
or the extent of his *Elements*. But if, as is possible, it included a study
of the geometry presupposed in the quadrature of lunes, that would

---

[271] Simplicius, *In Ph.* 60.22–69.34. Diels' edition distinguished between what he took to
be direct quotations from Eudemus and Simplicius' own additions, though some aspects
of this question remain disputed. On the problems of interpretation, see especially
Bretschneider 1870, pp. 99ff, Rudio 1907, Heath 1921, I pp. 189ff, and van der
Waerden 1954, pp. 131ff.

[272] Simplicius, *In Ph.* 61.5ff.

[273] That is not to say that every stage in his original reasoning is clear. In particular how
he established the theorem that circles are as the squares on their diameters is in doubt.

[274] The quadrature of the lune with external circumference greater than a semi-circle,
omitted by Eudemus as obvious, is duly supplied by Simplicius, *In Ph.* 63.19ff.

[275] The latter example is also notable for the evidence it provides that Hippocrates was
familiar with the method of νεῦσις, inclination or verging, used at *In Ph.* 64.17ff. See
Heath 1921, I pp. 194ff.

[276] At *In Ph.* 61.12ff 'similar segments' are said to be those 'which are the same part of
the circles respectively, as for instance a semi-circle is similar to a semi-circle, and a
third part of a circle to a third part'. As Heath and many others have noted, the idea of
'part' used here corresponds to the pre-Eudoxan notion of proportionality (set out
in Euclid VII Def. 21, II 104.25ff HS = Heath Def. 20) that, strictly speaking, applies
only to commensurable magnitudes.

imply that a considerable body of geometrical theorems had already been brought into some kind of order by the end of the fifth century B.C.

The systematisation of geometry thus initiated by Hippocrates was evidently of outstanding importance: indeed the notion of strict mathematical demonstration was, as is now generally recognised, the distinctive achievement of early Greek, as opposed to Egyptian or Babylonian, mathematics.[277] But the question of the origin of or stimulus to this development becomes all the more pressing. How far should we see it as the result of an interaction of philosophy and mathematics, even, as has sometimes been suggested, the direct outcome of Eleatic influences, or how far are we dealing with an independent development? There is no question of being able to pinpoint direct debts in the original texts themselves, but it is helpful and necessary first to distinguish between two aspects of the problem, that is on the one hand the question of methods of demonstration and on the other that of the axiomatic foundations of mathematics. So far as the former goes, it seems probable enough – even when due allowance is made for the fragmentary nature of our sources – that strict deductive proofs occurred in philosophy before they did in mathematics. There is no reliable evidence for rigorous demonstrations in mathematics before the time of Parmenides, and it was the Eleatics who provided the first clear statement of the key thesis that serves as the epistemological basis for any abstract inquiry such as mathematics, namely the insistence on the use of reason (as opposed to the senses) as the criterion. There are, too, certain obvious – and natural – similarities between some of the chief modes of argument that we find in the Eleatic philosophers and those of early Greek mathematics. This applies particularly to what was to be one of the major types of Greek mathematical reasoning,[278] namely indirect proof or the *reductio*, much used – as we have seen – by the Eleatics and exemplified in the strict demonstrations of the incommensurability of the side and the diagonal. The possibility that the mathematicians were directly influenced by the Eleatic philosophers in this regard[279] is, then, there, although (i) we cannot confirm this and (ii) more importantly we must recognise that any

---

[277] See further below, pp. 229f.

[278] Thus Itard 1961, p. 72, calculates that *reductio ad absurdum* is used in 14 propositions (and sometimes several times in a single proposition) out of 39 in Euclid VII, in 6 out of 27 in VIII and in 13 out of 36 in IX.

[279] Szabó 1964–6 and 1969, pp. 328ff, for instance, argues for other influences as well, notably on the conception of unity and in the use of the idea of infinite divisibility: but I am concerned here with the question of modes of deductive argument.

such influence was a limited one, being confined principally to the notion of a strict deductive argument and not extending to such distinctive mathematical procedures as the reciprocal subtraction algorithm, anthyphairesis (see above, p. 106 n. 266) or the application of areas, let alone mechanical techniques such as the use of a straight edge in solving a problem of 'converging'.

The question of whether there were specific fifth-century philosophical influences on the development of the notion of the axiomatic foundations of mathematics is, however, another matter.[280] By the time we come to Euclid himself clear distinctions are drawn between three types of first principles – the starting-points of all mathematical reasoning – namely definitions, postulates and common opinions.[281] As is well known, these correspond in the main, though not exactly, to the three kinds of general indemonstrable starting-points of demonstrations identified by Aristotle in his formal logic, namely definitions, hypotheses and axioms.[282] Yet a good deal of obscurity surrounds the earlier history of the notion of such principles. Plato's emphasis on the need for exact definitions in mathematics, as elsewhere, is not in doubt, and, as we shall see, his discussions of hypotheses are of fundamental importance in this development. Moreover the mathematical excursus in the *Theaetetus* authorises us to attribute a concern for the careful definition of mathematical terms to Theaetetus.[283] But the evidence for any conception of the first principles of reasoning in the fifth century, whether in mathematics or in philosophy, is very limited.

First it is doubtful whether the Eleatics themselves had a clear conception of such starting-points. Although, beginning with Plato,[284] later commentators spoke of their notion that 'there is one', and of

---

[280] Cf., e.g., the discussions in Scholz (1930) 1975, pp. 50ff, Lee 1935, Einarson 1936, Szabó 1960–2, 1964–6 and 1969, Mueller 1969, and J. Barnes (1969) 1975, pp. 69ff.

[281] Roughly speaking, the common opinions are self-evident principles that apply to the whole of mathematics (thus the third states that 'If equals be subtracted from equals, the remainders are equal'), while the postulates are the fundamental geometrical assumptions underlying Euclid's geometry: the first three relate to the possibility of carrying out certain constructions (e.g. 'to draw a straight line from any point to any point') and the other two assume certain truths concerning geometrical constructions, namely that all right angles are equal and that non-parallel straight lines meet at a point.

[282] One of Aristotle's axioms corresponds to Euclid's third common opinion (see above, n. 281). Aristotle's hypotheses differed from the definitions as the assumptions of the existence or non-existence of the objects defined.

[283] *Tht.* 147 d ff: the problem tackled there is that of distinguishing and defining commensurable and incommensurable lengths, or more strictly those that are commensurable in length from those that are so only in square. Cf. Aristotle's remark, *Top.* 158 b 29ff, that in mathematics some things are not easily proved for want of a definition.

[284] E.g. *Prm.* 128 d 5f, 136 a 4f.

the alternative idea of their opponents that 'there are many', as hypotheses, there is no evidence that the Eleatics themselves used that term, and indeed even if they had, they would probably have meant no more by it than some roughly defined notion of an assumption. The fact that Aristotle singles out the search for general definitions as one of Socrates' distinctive contributions[285] suggests a relative lack of concern in this regard on the part of earlier philosophers. Meanwhile there can be no question of dating any attempt to formulate general laws of reasoning, such as the law of contradiction or that of excluded middle, before the fourth century.[286]

So far as the fifth century goes, we have little reason to suppose that philosophical influences were at work on the development of an explicit conception of the axiomatic foundations of mathematics. But how far had the formulation of that notion progressed in mathematics itself before Plato? It is notable that in his report on Hippocrates' quadrature of lunes Eudemus says that Hippocrates took as his starting-point (ἀρχή) the theorem that 'similar segments of circles have the same ratios as the squares on their bases', though he goes on to add that Hippocrates proved this by showing that the squares on the diameters have the same ratios as the circles.[287] We should not necessarily suppose that ἀρχή was Hippocrates' own term, but even if it were, it is clearly not used here of an ultimate principle of mathematical reasoning, but rather of the starting-point of a set of demonstrations: the ἀρχή is a proposition that is used in proving certain conclusions, but one that is itself established independently. Yet the systematisation of a body of geometrical theorems in books of *Elements* by Hippocrates and his successors itself implies a distinction between more and less fundamental propositions, and if Aristotle is anything to go by, the term 'elements', στοιχεῖον, itself was used of primary principles or starting-points of *some* kind – even though their nature may not yet have been clearly specified. Particular assumptions, conceptions and indeed definitions used in early Greek mathematics are reported often enough in our sources, which sometimes comment on their status as *fundamental* assumptions. Yet the very variety of competing definitions of such terms as point and number, for instance, is striking,[288] and on the all-important question

285 *Metaph.* 1078b27ff, cf. 987b1ff.
286 See, e.g., Lloyd 1966, pp. 139–41, 162–4.
287 See above, p. 109.
288 Commenting on Euclid's definition (I Def. 1) of a point as that which has no part, Proclus (*In Euc.* 95.21f) says that 'the Pythagoreans' defined the point as a monad having position, a definition that certainly antedates Aristotle since he refers to it at *de An.* 409a6, as also does the definition as the 'extremity of a line' (*Top.* 141b19ff,

of how such starting-points were combined to form an adequate and consistent set of ultimate principles to serve as the basis of mathematics, our sources are generally uninformative.[289]

The evidence in Plato, in particular, throws important light on the extent, and the limitations, of the idea of the axiomatic foundations of mathematics before him. In two of the well-known and much discussed texts in which he treats of the nature of hypotheses he refers to the use made of them by mathematicians. When the term is introduced in the *Meno* (86 e ff) it is explained by referring to the procedure of geometers. They are represented as tackling such a problem as the inscription of a given area as a triangle in a given circle 'by means of a hypothesis' which states that if certain conditions are fulfilled, then one result follows, but if not, not.[290] Here, then, the hypothesis is in no sense an ultimate assumption, but in the *Republic* (510 c ff), when Plato again uses a mathematical example to illustrate a (though not the same) use of hypothesis, he says that the mathematicians 'hypothesise the odd and the even, and the figures and the three kinds of angles, and other things like these in each inquiry'. They 'do not give an account of these either to themselves or to others, as being clear to all, but beginning from

cf. *Metaph.* 992 a 19ff an obscure passage where Plato is said to have rejected the existence of points and *either* (according to Heath 1949, p. 200) to have called points the 'beginning of a line' and to have posited indivisible lines, *or* (Ross) to have applied the name 'the beginning of a line' to indivisible lines). For a convenient survey of the rich variety of definitions of 'number' recorded in our sources from Aristotle onwards, see Klein 1968, pp. 51ff.

[289] On the other hand, the effect that the discovery of incommensurability had on subsequent Greek mathematics has often been exaggerated (see Tannery 1887, p. 98, Hasse and Scholz 1928, von Fritz (1945) 1970, but *contra*, for example, Freudenthal 1966, Burkert 1972a, pp. 455ff) and in particular the view that it effectively paralysed work in mathematics for several decades cannot be sustained. The proponents of such a view have generally dated the discovery early in the fifth (if not in the sixth) century, but that is quite unlikely (Plato can still refer to a general ignorance of incommensurability in the fourth century, *Lg.* 819d ff). Thus the idea of a connection between Zeno's arguments and incommensurability (Hasse and Scholz, 1928, pp. 10ff) has rightly been challenged (van der Waerden 1940–1, and cf. Owen (1957–8) 1975, pp. 153ff): those arguments evidently ignore incommensurability and tell, rather, for the conclusion that it was not known. Nor is it easy to explain that, in turn, by appealing to Pythagorean secretiveness (the subject of many fanciful stories in our late sources: see further p. 228 and n. 5). Incommensurability certainly posed a problem for the Pythagorean belief that 'all things are numbers': but so far as we can tell, it was far from bringing mathematical activity to a halt. From the time of Hippocrates (*c.* 430) at least, we have good evidence of continuous mathematical investigations of a high order, some, but not all, of which were directed to resolving what we may call foundational problems.

[290] On the hypothetical method in the *Meno* and the disputed questions of its relations (*a*) to διορισμός (the determination of the conditions under which a problem is capable of solution, and of those under which it is not) and (*b*) Greek geometrical analysis, see especially Cornford (1932) 1965, pp. 64ff, R. Robinson (1936) 1969, and 1953, ch. 8, Bluck 1961, pp. 75ff, and Mahoney 1968–9.

them they go through the rest consistently and end with what they set out to investigate'.

It is, once again, not certain whether either of these passages reflects an actual use of the term ὑπόθεσις in earlier mathematics.[291] But whatever the original terminology, Plato clearly indicates the use of certain basic assumptions, treated as the self-evident starting-points of mathematical deductions. His testimony is all the more convincing in that he is critical of their use, being concerned here to point out the inferiority of this procedure (διάνοια) to that of νόησις, which takes nothing for granted, but proceeds upwards to an 'unhypothesised beginning', on the basis of which the whole of the intelligible world – including the geometer's hypotheses them-selves – can be known.[292] But while this account of διάνοια points to the role of certain primary principles in mathematics, the precise character of those principles is still left, in certain respects, unclear. Plato mentions 'odd and even', the 'figures' and the 'three kinds of angles'. But it is an open question[293] whether for example, these are to be construed as – or to include – definitions (like the definitions of odd and even in Euclid vii Def. 6 and 7), or existence assumptions (corresponding to Aristotle's hypotheses)[294] or – as seems possible in the case of the figures – as assumptions concerning the possibility of carrying out certain constructions (like the first three of Euclid's postulates). Plato thus provides confirmation of what the existence of books of *Elements* might already lead us to infer, namely the recognition of the need for incontestable principles of some kind as the foundation of mathematics, but he is also evidence for a con-tinuing indeterminacy in the conception of those foundations.

Although our picture of Plato's work would, no doubt, be modified if we had more information about that of the galaxy of mathe-maticians, chief among them Eudoxus,[295] who were associated with

---

[291] The care with which ὑπόθεσις is introduced and explained with the geometrical example in the *Meno* suggests both that the term was generally unfamiliar and that it may well have had some semi-technical use in mathematics (though on the latter point doubts have been expressed by, for example, Bluck 1961, p. 92). We know from the Hippocratic treatise *VM* that ὑπόθεσις was also used in medicine in the sense of postulate, but the date of its introduction in that context is disputed, see H. Diller 1952, Lloyd 1963.

[292] In this process the hypotheses are said to be treated not as starting-points, but 'really as ὑποθέσεις' (literally what is placed under something) – steps to the unhypothesised beginning (*R.* 511 b). For a survey of competing interpretations of the procedure of the 'upward path', see, for example, R. Robinson 1953, pp. 16off.

[293] See, for example, Cornford (1932) 1965 and von Fritz (1955) 1971, pp. 361ff.

[294] In which case we must contrast Plato's statement with the view in Aristotle, *APo.* 76 b 6ff, where he insists that while arithmetic assumes the definitions of odd and even, it has to prove that they exist.

[295] Eudoxus, who was probably some 20 years younger than Plato, was certainly one of

him, usually as members of the Academy,[296] we can hardly doubt that Plato's own contribution was an important, perhaps even a crucial one, not only directly in the development of the notions of proof and of hypothesis, but also indirectly in stimulating interest in the problem of foundations. Even so, it was not until Aristotle that we have a full analysis of the conditions of demonstration: arguing against Plato, or at least against the apparent implications of the *Republic*'s notion of an 'unhypothesised beginning', Aristotle insisted that not all true propositions can be demonstrated, and that the starting-points of demonstrations are principles that are themselves indemonstrable but known to be true;[297] and, as we said, he provides us with our first extant typology of such indemonstrable starting-points. We have no reason to deny the importance of the interaction of philosophy and mathematics in the development of the notion of the axiomatic foundations of the latter, but this is a fourth-, rather than a fifth-century phenomenon, a matter of the impact of the work of Plato and Aristotle, rather than of that of the Eleatics.

INTERACTIONS OF DIALECTIC AND DEMONSTRATION

The strict notion of proof deployed in both mathematics and philosophy was formulated, in part, by means of a contrast with merely probable arguments. Yet it would be a mistake to place Greek mathematics, even in the fourth century, completely on the side of

the greatest mathematicians in the history of Greek science. He was responsible, as is generally agreed, for the theory of proportion developed in Euclid, *Elements* v (see also Proclus, *In Euc.* 67.2ff: this has the advantage over all earlier attempts at such a theory that it is applicable to incommensurable, as well as to commensurable, magnitudes), and for the method of exhaustion, based on *Elements* x 1, a type of indirect proof which was one of the most powerful and effective techniques used in Greek mathematics. One simple application was to determine an unknown area by inscribing successively greater regular polygons in it, as in the determination of the area of a circle in Euclid, *Elements* xii 2, and cf. Archimedes, *On the Measurement of the Circle*, Prop. 1, 1 232.1ff, which also circumscribes successively smaller polygons: the figure whose area is to be found is 'exhausted' only in the sense that the difference between it and the inscribed or circumscribed figure can be made as small as desired.
[296] Thus apart from Archytas, Theaetetus and Eudoxus, Proclus (*In Euc.* 67.8ff) mentions particularly Amyclas, Menaechmus, Dinostratus, Theudius, Athenaeus of Cyzicus and Philip of Mende (or Medma – sometimes identified with Philip of Opus, who is thought by some to have been the author of the *Epinomis*). The gradual extension and systematisation of the elements are clear from the remarks that Proclus makes on the work of Leodamas, Archytas and Theaetetus (*In Euc.* 66.14ff, cf. above, p. 108), on Leon (66.20ff), Eudoxus (67.2ff), Amyclas, Menaechmus and Dinostratus (67.8ff) and Theudius (67.12ff) especially. *In Euc.* 72.23ff further shows that the term element, στοιχεῖον, itself was a subject of discussion, for Menaechmus distinguished between two senses.
[297] *APo.* 1 chh. 1–3 especially.

'demonstration' as opposed to that of 'dialectic' in the Aristotelian sense.[298] Plato provides some of our best evidence for the distinction between necessary and merely persuasive arguments, but some of the texts in question indicate that the latter included not just rhetorical and sophistical, but also some mathematical, examples. Two passages are particularly revealing. In the *Phaedo* (92 d) Simmias is made to say: 'I am well aware that accounts that base their proofs on what is probable (διὰ τῶν εἰκότων) are impostors: unless one is on one's guard against them, they deceive one very badly, *in geometry and in everything else*'. The mention of geometry here is striking as it appears to be occasioned by nothing in the context (a discussion of images used to describe the nature of the soul). Secondly when, in the *Theaetetus*, Socrates, having criticised Protagoras' Man Measure doctrine, comes to offer a defence on Protagoras' behalf, he has Protagoras complain (162 e): 'There is no proof nor necessity whatsoever in what you say, but you rely on what is probable (τῷ εἰκότι): but if Theodorus or any other geometer were to rely on that in doing geometry, he would not be worth anything at all.' It is a nice touch that Protagoras[299] – the teacher, *par excellence*, of 'double arguments' – should here be made to insist on a contrast between proof and mere probability, and again the reference to geometry was – we may infer – not merely otiose.

In Aristotle, too, the *Organon* is rich in mathematical examples used to illustrate not just features of demonstrative science in the *Posterior Analytics*, but also dialectical procedures in the *Topics*, for instance points concerned with the correct or incorrect way of giving definitions.[300] Moreover although Aristotle's own distinction between 'dialectical' and 'demonstrative' syllogisms is, in general, a firm one, it is well known that on the thorny question of how the primary starting-points of demonstrative science are to be obtained, there is a tension between the points of view expressed in different parts of the *Organon*. The famous last chapter of the *Posterior Analytics* answers the question of how we apprehend the principles in terms of the exercise of νοῦς ('reason') in ἐπαγωγή ('induction'), a process that is said to

---

[298] In contrast to the sense that Plato sometimes gives it, where 'dialectic' stands for the supreme, and most rigorous, study, see above, p. 102.

[299] At *Tht.* 164e (cf. 168e) Theodorus is called a 'trustee' of Protagoras and this is usually taken to imply that he was taught by him. Whether that teaching extended to mathematics, however, might be doubted. Aristotle (*Metaph.* 998a2ff, Fr. 7) reports that Protagoras tried to refute the geometers on the grounds that a circle does not touch a ruler at a point, but we have no reason to suppose that Protagoras was interested in the practice of mathematical inquiry, as well as in the question of its ontological or epistemological basis.

[300] The most important passages are conveniently set out by Heath 1949, pp. 76ff.

begin with perception.[301] But at *Top.* I 2 the claim is made that dialectic is useful in relation to the primary bases of each science, the argument being that these cannot be discussed from the principles peculiar to each science, since they are primary to them all. Here the emphasis is not on perception or experience,[302] but more generally on the critical examination of the generally accepted opinions.[303]

Finally when we consider the specific primary principles adopted by Euclid himself, it is evident that some relate to questions that had been the subject of dialectical disputes in some cases going back into the fifth century. Thus both the inclusion of the eighth common opinion ('the whole is greater than the part')[304] and the assumption of the indivisibility of the one[305] are probably to be explained against a background that includes the arguments on whole and part and on the one and the many between the Eleatics and their opponents.[306] The most interesting example is the famous parallel postulate itself. We know from Aristotle that there were attempts to prove an assumption concerning parallels in the fourth century, though the proofs were open to the charge of circularity.[307] Euclid himself (if not one of his predecessors) must presumably have deliberately chosen to adopt the proposition that non-parallel straight lines meet at a point *as a postulate*, and although this move was bitterly criticised by the ancient commentators,[308] more recently the wisdom of treating it as a postulate in the context of the geometry he constructed on its basis has been acknowledged. There is no evidence that Euclid or any other Greek mathematician envisaged the development of non-Euclidean geometries: yet his *Elements* are not merely an axiomatic, but also an explicitly hypothetical system, in this sense at

---

[301] *APo.* 99b32ff. On the much disputed questions of the extension of 'principles' (primary propositions or primitive terms), the meaning of νοῦς and its relation to ἐπαγωγή, see most recently von Fritz (1964) 1971, pp.623ff, Kosman 1973, Lesher 1973, J. Barnes (1975), pp. 248ff, Hamlyn 1976.

[302] Cf. also *APr.* I 30, 46a17ff.

[303] *Top.* 101a36–b4, see, for example, Owen (1961) 1975, pp. 113ff, (1965) 1975, pp. 29f, J. D. G. Evans 1977, pp. 31ff, and cf. Weil (1951) 1975, pp. 88ff.

[304] The fifth common opinion in Heath's numbering, I 6.4 HS.

[305] See VII Def. 1, the definition of the unit as that by virtue of which each of the things that are said to be one.

[306] Cf. Zeno's Moving Rows argument, the conclusion of which is reported by Aristotle as that 'half the time is equal to its double' (p. 74 n. 82 above), and cf. the reference in Plato, *R.* 525de, to mathematicians who refuse to allow the one to be divided 'lest it should appear to be not one, but many parts'.

[307] *APr.* 65a4ff.

[308] Thus it was often objected that the postulate should be proved, and both Ptolemy and Proclus were among those who attempted this. It was, as is well known, an attack on this problem that eventually led, in the nineteenth century, to the development of non-Euclidean geometries.

least that it was one based on postulates and common opinions which include propositions that he must have known to have been questioned or denied by other Greek thinkers.

The nature of dialectic and its relations with demonstration were, then, as we have seen, themselves the subjects of considerable disagreement, and both Plato's position, and Aristotle's,[309] underwent certain modifications in their lifetime. Though Aristotle often emphasised the contrast between dialectic and demonstration – particularly in contexts where he was advancing an alternative to Plato's view of dialectic as the supreme study – even in Aristotle the distinction is not as absolute as might appear from some passages, since demonstration may be said to depend on dialectic for its starting-points.[310] Moreover the dialectical examination of current opinions turns out to be a good deal more important and prominent in Aristotle's actual scientific practice than the techniques for the cogent presentation of knowledge (especially the theory of the syllogism) that he set out in his analysis of demonstration.[311]

Nevertheless the formulation of rigorous criteria for demonstration was of fundamental significance for the subsequent development of Greek science. From the fourth century on, a comprehensive theory of an axiomatic, deductive system was available in Aristotle's logic, and soon after 300 B.C. Euclid's *Elements* provided an instance of such a system[312] applied in practice in mathematics. Beginning with a statement of the definitions, postulates and common opinions,[313] he proceeds to the demonstration of a massive array of mathematical propositions, with very little redundancy and high standards of consistency.[314] The aim was not merely the orderly presentation of a body of knowledge, but also the certainty to be secured by an axiomatic, deductive method.

As the prime example of such a system, the *Elements* had an immense influence not just on later mathematics, but also on physical science. The mathematisation of branches of physical inquiry begins

---

[309] On the development of Aristotle's views on the question of the supreme science, see, especially, Owen 1960, pp. 175f, 1965 and (1965) 1975.

[310] It is, too, through dialectic that trans-departmental laws of reasoning are to be established. Though the law of contradiction, for example, cannot be demonstrated directly, an opponent who denies it can be refuted: see *Metaph.* 1006a5ff.

[311] See further below, pp. 136ff and 202ff.

[312] This is not to assert that Euclid was necessarily directly influenced by Aristotle (as well as by earlier mathematicians) though it is possible that he was.

[313] Those in *Elements* I are added to when fresh starts are made in v and vii.

[314] The simplest instance of redundancy is that of the inclusion of certain definitions (such as those of 'oblong', 'rhombus' and 'rhomboid') not thereafter used in the *Elements*. But there are also more revealing discrepancies, such as that between the notions of proportion used in v and vii.

before Euclid, to be sure. Indeed explicit discussion of the relationship between mathematics and sensible phenomena goes back before
Plato. As we noted, Protagoras insisted that the circle does not touch
the ruler at a point, a topic that was probably the subject of a treatise
by Democritus,[315] who also raised – though we do not know how he
resolved – the problem of whether the contiguous surfaces of a
horizontal section of a cone are equal or unequal.[316] Plato in turn
had divorced knowledge from true opinion and claimed (in his
middle period) that sensible phenomena are the domain of the latter,
not of the former.[317] While Aristotle developed a complex, and not
entirely consistent, notion of the hierarchy of the sciences,[318] he
distinguished μαθηματική from φυσική primarily on the grounds that
while physical bodies contain volumes, surfaces, lines and points,
mathematics studies these in abstraction from physical bodies.[319] Yet
whatever views they held on this important on-going controversy,
practising scientists in the fourth, if not already in the fifth, century
had achieved a mathematisation of physical inquiry in two areas
especially, namely harmonics[320] and astronomy. From the time of
Eudoxus, if not before, the notion that some *geometrical* model is to
be used to explain the movements of the heavenly bodies was
common ground to all theoretical astronomers, even while they

---

[315] *On the Difference of Cognition or on the Contact of Circle and Sphere*: cf. Aristotle's report
(*Cael.* 307a16f, with Simplicius' comments, *In Cael.* 662.10ff) that Democritus treated
the sphere as a 'kind of angle' (or 'all angle', *Cael.* 307a2).

[316] Democritus, Fr. 155. Another fifth-century thinker who may have discussed the
relation between mathematics and sensible phenomena is Antiphon, whose attempt to
square the circle by treating it as equivalent to an inscribed regular polygon with an
indefinitely large number of sides (perhaps an anticipation of Eudoxus' method of
exhaustion) was held by Aristotle to be contrary to geometrical principles (see *Ph.*
185a16f).

[317] See, e.g., *Phd.* 65b ff, *R.* 476d ff, 523b, cf. *Ti.* 51d ff, *Phlb.* 59a, and see further
below, pp. 131ff.

[318] Thus at *Ph.* 194a7ff he calls optics, harmonics and astronomy 'the more physical of
the μαθήματα', yet he allows that certain questions in astronomy must be the domain of
the natural scientist or φυσικός (193b25ff). On the hierarchy of the sciences, see most
recently McKirahan 1978.

[319] E.g. *Ph.* 194a9ff. Owen 1970, pp. 256f, explores the tension that exists between
Aristotle's use of mathematical abstractions (for example, of the assumption that the
centre of the universe may be regarded as a point, *Cael.* II 14) and his insistence that
mathematics should be applicable to physical phenomena.

[320] Our somewhat limited information concerning Philolaus (e.g. Boethius, *Mus.* III 5,
276.15ff, cf. Fr. 6) and Archytas (Frr. 1 and 2, Ptolemy, *Harm.* I 13, 30.9ff, Boethius,
Mus. III 11, 285.9ff) can be supplemented by the reports in Plato (*R.* 530d ff, see
below, pp. 144f) and in Aristoxenus (e.g. *Harm.* II 32f, cf. 38ff). Aristoxenus, in particular, contrasts those who try to do without demonstration with those, on the other
hand, who shun sense-perception as not exact and who construct intelligible causes. On
the authority of Aristotle, *Ph.* 194a7ff, optics may be added to harmonics and astronomy
as a third early example of mathematisation: but our earliest extant treatise on this
subject is by Euclid.

disagreed both on what model to adopt, and on the conditions under which the 'phenomena' could be said to have been 'saved'.[321]

Euclid's *Elements* provided a model of rigour and systematisation that was imitated or echoed more or less closely not only in what we should call pure, but also in applied, mathematics and in physics. The initial statement of such definitions and assumptions as were necessary, followed by the orderly deductive proof of theorems, became the canonical form in many domains of inquiry. It appears, for example, in such works of pure mathematics as Archimedes' *On the Sphere and Cylinder* and Apollonius' *Conics* (mid and late third century B.C. respectively), in mechanical treatises such as Archimedes' *On the Equilibrium of Planes*, in optics in Ptolemy's work on that subject (second century A.D.), and in astronomy in Aristarchus' *On the Sizes and Distances of the Sun and Moon* (early third century B.C.).[322] Even when a Euclidean presentation was not adopted, the aim was often a similar exactness and certainty, as we see very clearly, for instance, from the contrast that Ptolemy draws at the beginning of the *Syntaxis* between mathematics (including mathematical astronomy) on the one hand, and theology and 'physics' on the other.[323] Moreover such was the prestige of mathematics and the mathematical sciences that proof *more geometrico* was sometimes represented as a desideratum in other fields of inquiry as well, such as cosmology or physiology, as we can see from the repeated remarks on the need for such proofs in Proclus (fifth century A.D.),[324] and from the claims that Galen (second century A.D.) made for the relevance of strict logical demonstrations to medicine.[325]

[321] See further below, pp. 173ff, 198f.

[322] The pattern continues in many post-Classical works, from Jordanus' *De Ratione Ponderis* (*c.* 1250), for example, through Bradwardine's *Geometria Speculativa* (fourteenth century) and Tartaglia's *Nova Scientia* (1527), to Newton's *Principia* (1687) itself.

[323] *Syntaxis* I 1, ii 6.11ff: both theology and 'physics' are conjectural, the one because of its obscurity, the other because of the instability of its subject-matter: mathematics alone yields unshakeable knowledge, proceeding as it does by means of indisputable arithmetical and geometrical demonstrations. In practice, however, in the body of the *Syntaxis*, Ptolemy expresses more doubts about aspects of his investigations than one might expect from these remarks in the Proem.

[324] A recurrent theme in *In Ti.*, especially, e.g., I 226.26ff, 228.27, 236.15ff, 258.12ff, 346.31ff.

[325] Only fragments of Galen's major work *On Demonstration*, in 15 books, remain, and many other logical treatises, such as that showing the superiority of geometrical analysis to Stoic argument, are also lost (but see, for example, *Inst. Log.*). But he frequently stresses the need for the doctor to be trained in, and to apply, strict logical methods, including demonstration (e.g. *Med. Phil.* 3, *Scr. Min.* II 6.10ff Müller, K I 59.15ff), though he also draws attention to some of the dangers, pointing out, for instance, that large mistakes may stem from small errors in the principles (*Mixt.* I 5, 17.22 Helmreich, K I 536). Moreover a famous passage at the beginning of the *Ars Parva*, K I 305.1ff, confirms that he believed that the – characteristically geometrical – methods of analysis and synthesis are applicable to medicine.

The model of an exact science was, as has always been recognised, one of the chief achievements and legacies of Greek science. Plato and Aristotle contributed much to the theory and it was in the practical applications that many of the greatest successes of Greek science were won, preeminently in astronomy (by Apollonius, Hipparchus and Ptolemy), but also in mathematical geography (Eratosthenes), in statics and hydrostatics (Archimedes), and in optics (Euclid, Archimedes, Hero, Ptolemy). The notion of the supremacy of pure reason may thereby be said to have promoted some of the triumphs of Greek science. But that same notion is also related to some of its predominant weaknesses. The aim was often indisputability, rigour, exactness: but these were sometimes bought at the price of a certain arbitrariness and dogmatism,[326] and in some contexts of a certain impoverishment of the empirical content of the inquiry. Time and again Greek scientists interpret their subjects as far as possible as branches of pure mathematics. This is true, for example, of Archimedes' treatises on statics and hydrostatics,[327] and of Aristarchus' work *On the Sizes and Distances of the Sun and Moon*: it is notorious that Aristarchus there takes as one of his hypotheses a grossly inaccurate value for the angular diameter of the moon, and the reason for this is not, in all probability, that he was incapable of the crude observation necessary for a roughly correct approximation, but simply that he was less interested in arriving at concrete results for the sizes and distances (which in any case he expresses not as values but as proportions) than in the pure geometry of his problem.[328] Finally the

---

[326] Even before Aristotle insisted on the varying degrees of exactness of different inquiries (e.g. in a famous passage in *EN* 1094 b 25ff), several fifth- and fourth-century medical writers had pointed out the impossibility, or inappropriateness, of exactness in medicine: see e.g. *Prog.* ch. 20, L II 168.16ff, *Fract.* ch. 7, L III 440.2ff, *Art.* ch. 69, L IV 286.7ff, *Morb.* I ch. 16, L VI 170.2ff, *Vict.* I ch. 2, L VI 470.13ff, *Vict.* III ch. 67, L VI 592.1ff, 594.1f, and especially *VM* chh. 9–12, e.g. *CMG* I, I 41.20ff, the author of which is particularly critical of what he represents as dogmatic tendencies in medicine (see further below, p. 135).

[327] Thus in *On the Equilibrium of Planes* most of book I and the whole of book II deal with the problems of determining the centres of gravity of various plane figures, such as the parallelogram, the triangle and (especially) parabolic segments. The law of the lever is proved in I 6 and 7, II 132.14ff, 136.18ff HS, where it is shown first for commensurable, and then for incommensurable, magnitudes that two magnitudes balance at distances that are reciprocally proportional to the magnitudes: but the question of how far Archimedes' initial postulates are themselves independent of experience or how far the law of the lever is presupposed in this proof has been much debated, see Mach 1893 and Duhem 1905–6, I pp. 9ff especially.

[328] The value taken is 2°, that is about four times the correct figure – and indeed Archimedes ascribes to Aristarchus the value of $\frac{1}{2}$° for the (approximately equivalent) apparent size of the sun (*Sand-Reckoner* II 222.6ff HS). For varying interpretations of the discrepancy, see, for example, Tannery 1912–43, I pp. 375f, Heath 1913, pp. 311ff, Wasserstein 1962, pp. 57f and Neugebauer 1975, II pp. 634ff, 642f.

Greek preoccupation – some have said obsession[329] – with demonstration is apparent also in the emphasis, in works of pure and applied mathematics alike, on the synthesis or deductive proof of the theorems to the almost total exclusion of the antecedent analysis or their method of discovery: in our chief text where, exceptionally, problems associated with discovery are discussed, namely Archimedes' *Method*, it is well known that he draws a sharp contrast between heuristic and demonstrative procedures and insists that the theorems that he has discovered by the method he sets out must thereafter be given a strict geometrical demonstration, using *reductio* and the method of exhaustion.[330]

By the time we reach Archimedes' *Method* or his strict Euclidean presentation and demonstration of a body of propositions in statics or hydrostatics, we have clearly come a long way from anything that can be paralleled in non-literate or early literate societies. But we may now pause to reflect on the character of the developments we have been tracing. Although the end-products are modes of reasoning of considerable sophistication, they were arrived at after long and complex processes of development which have both 'externalist' (that is, broadly, sociological) and 'internalist' (intellectual) features.

In part the development of argumentation within philosophy and science can be related to more general developments, the growth of a certain professionalism in the art of speaking as a whole, which itself must be understood at least in part as a response to certain changes within Greek society. We shall be considering some of the social and political aspects of the problem of the development of philosophy and science in chapter 4. But meanwhile we may note that the rise of rhetoric had both a direct and an indirect influence on that development, first directly on the actual practice of argument in natural science (for which our main evidence is provided by the medical writers) and then also indirectly, in that – in Plato's case at least – the development of a rigorous notion of demonstration owed

---

[329] See Mach 1893, pp. 18 and 82, on what he called the Greek 'mania' for demonstration, instancing Archimedes in particular.

[330] See *Method* II 428.18ff, 438.16ff HS. The method in question involves treating a plane figure whose area is to be determined as composed of a set of parallel lines indefinitely close together and then thinking of these lines as balanced by corresponding lines of the same magnitude in a figure of known area. The reason why Archimedes contrasts this method with the subsequent geometrical proof (despite the fact that the method yields results that are correct) is, no doubt, not that it uses mechanical concepts, but that it depends on infinitesimals, that is the assumption that areas and volumes may be treated as composed of their line and plane elements respectively.

something to a hostile reaction to the deployment of merely persuasive argumentation.

Even when we make allowances for the fragmentary nature of our sources, the types of argument used by the earliest cosmologists, scientists and even mathematicians appear quite elementary, and several of them (such as analogical arguments and appeals to probability) can readily be paralleled in pre- or non-philosophical contexts. We can trace the extension of the use, within philosophy, of the principle of sufficient reason, of *reductio* arguments, of Modus Tollens, the antinomy and the dilemma, but the extension is a gradual one: nor are such arguments confined to the philosophers. Eleatic argumentation marks, however, an important turning-point, both in raising directly the epistemological problem of the relation between reason and perception, and in practice in the development of strict deductive arguments. Plato and Aristotle introduce further major changes both in the actual use of arguments, and more particularly in their theoretical analysis, in, for instance, the formulation of rules of procedure in dialectical debate, in the distinction between necessary and probable arguments, and especially in the development of the notion of an axiomatic system. Fifth- and fourth-century mathematicians, too, have their contribution to make, for example to the development of specifically mathematical techniques such as anthyphairesis and the method of exhaustion.

Yet evidently – as these two last examples show – not all the types of argument actually employed in the various strands of Greek speculative thought in the fifth and fourth centuries can be directly compared with instances or anticipations in non- or pre-philosophical texts from Greece or anywhere else. The question we must turn to in conclusion is, then, this: do the radical developments that occur in either the practice or the theory of reasoning in Greek thought imply any shift or transformation in the underlying logic or rationality itself? Or rather what light do those developments throw on the problem of what it would mean to talk of any such shift or transformation? Certainly new modes of argument, some of them quite technical, can be said to have been *invented*. Important new concepts – hypothesis, postulate, proof, axiom, definition itself – come to be defined, and fundamental distinctions are drawn between, for example, valid and invalid arguments, and between necessary and probable ones. Moreover the notions of consistency, self-contradiction[331]

---

[331] Thus R. Robinson 1953, pp. 26ff, has pointed out the looseness with which the idea of 'contradicting oneself' (αὐτὸς αὑτῷ ἐναντία λέγειν) is used in the early and middle dialogues of Plato. At *Grg.* 482 bc, for instance, 'contradicting oneself' is treated not so

and even truth[332] itself undoubtedly come to be grasped more clearly – and it is in terms of truth and consistency that logic and rationality themselves have to be defined. Yet it is not the case that the logic itself is *modified* by being made explicit, *except insofar as it is made explicit.*

In the instances we have taken we may speak of an increase in clarity, in explicitness and in self-consciousness: there is a corresponding increase in confidence in the handling of certain types of argument in certain contexts (though that point should not be exaggerated: fallacies continue to be committed after their various types have been identified and defined). But the formalisation of logic consists – at least initially – in making explicit rules that are already contained in language and that are presupposed by intelligibility in communication. Even the axiomatisation of mathematics is a matter of achieving greater explicitness in setting out deductive relations between mathematical propositions. The developments we have been dealing with involve a change in the level of awareness of aspects of reasoning: and how such changes come about – how it is that individuals within a given society can come to raise fundamental questions concerning the basis of their own knowledge and the mechanisms of their own reasoning – poses a major problem we shall attempt to discuss in chapter 4. But the problem is one of trying to understand how *that* occurred – that is the conditions under which such second-order questions come to be asked – not one of trying to explain the substitution of one logic, or rationality, for another.

much from the point of view of the relationship between the propositions asserted, as from that of the psychological disorders set up in the soul. At *R.* 436b8ff Plato points out that 'it is clear that the same thing will never submit to doing or suffering opposite things, in the same respect, at least, and in the same relation and at the same time', but contradiction is not explicitly defined as a matter of the relationship between assertion and denial until Aristotle, who coined the term ἀντίφασις in this sense, e.g. *Int.* 17a33ff. But the deliberate use of apparently self-contradictory statements in writers before (or indeed after) Aristotle does not imply an alternative logic in which the law of self-contradiction is suspended: any such suspension rules out intelligible communication. Rather, like riddles and paradoxes, such statements must be understood as challenges to the reader/listener to decipher the author's hidden meaning.

[332] The semantic changes that occur in the use of the term ἀλήθεια in Greek have been studied by Boeder 1959, Heitsch 1962, 1963, Krischer 1965 and Detienne 1967, for example: thus the suggestion that initially the chief antonym was not falsehood, but λήθη, forgetfulness, receives some backing from an etymology the Greeks themselves proposed. But although in the development that ends with Aristotle's clear statement that truth and falsehood relate to propositions (e.g. *Int.* 16a9ff, cf. Plato, *Sph.* 261d ff) fundamental epistemological and logical issues are clarified (for instance the criteria that might be used to support a claim for knowledge, and the difference between truth and validity), there is no question of Aristotle thinking – or of it being the case – that the conditions under which statements *are* true or false were themselves altered during the course of this development.

The argumentative weaponry that the Greek scientist eventually had at his disposal was impressive and was put to good effect both destructively – to undermine and refute his opponents' views or common assumptions – and constructively, to advocate or attempt to prove his own ideas. We have suggested, however, that rigour was sometimes achieved at the cost of empirical content. We have considered some aspects of the development of reason and argument: we have now to turn to observation, research and experiment, to examine the factors stimulating or inhibiting the extension of the empirical base of Greek science.

3

THE DEVELOPMENT OF EMPIRICAL
RESEARCH

OBSERVATION AND RESEARCH

In our study of *On the Sacred Disease* we remarked on the accuracy of the writer's brief description of an epileptic fit, but also found that he checked few, if any, of his anatomical theories by dissection, despite the fact that he refers, at one point, to the possibility of carrying out a post-mortem examination on the brain of a goat in order to throw light on the causes of epilepsy.[1] This suggests, as a problem for investigation, the use and development of empirical research in early Greek science, a topic on which a wide range of divergent opinions has been expressed. Where Burnet, for example, wrote that 'the idea that the Greeks were not observers is ludicrously wrong',[2] Cornford saw the philosophers (at least) as dogmatists, innocent of 'the empirical theory of knowledge', although that theory was, in his view, developed by Alcmaeon and the medical writers.[3] More recently there has been a similar confrontation between Popper, who put it that most of the ideas of the early Presocratics, and the best of them, 'have nothing to do with observation', and Kirk, who saw one of the prime characteristics of Greek thought as its common sense, including its respect for observation and experience,[4] and many others have attempted general evaluations of the empirical work of Greek scientists.[5]

---

[1] See above, ch. 1, pp. 21–2 and 23–4.

[2] Burnet (1892) 1948, p. 26. Faced with the lack of actual observations recorded for the Presocratic philosophers, Burnet argued that this reflects the bias of our sources: 'We are seldom told why any early philosopher held the views he did, and the appearance of a string of "opinions" suggests dogmatism. There are, however, certain exceptions to the general character of the tradition' (Burnet goes on to refer to the 'biological and palaeontological observations' of Anaximander and Xenophanes); 'and we may reasonably suppose that, if the later Greeks had been interested in the matter, there would have been many more'.

[3] Cornford 1952, especially ch. 3. Cornford's thesis was extensively criticised by, for example, Matson 1954–5 and Vlastos (1955) 1970 and 1975.

[4] See Popper (1958–9) 1963, Kirk (1960) 1970 and 1961 and cf. Lloyd 1967.

[5] See, for example, W. H. S. Jones 1923–31, IV pp. xxii ff, H. Gomperz 1943, Edelstein (1952b) 1967, pp. 402ff, Bourgey 1953 and 1955, Sambursky 1956, pp. 1ff, 233ff, Clagett 1957, pp. 28ff, Farrington 1957, Verdenius 1962, Guthrie 1962, pp. 37f, De Ste Croix 1963, pp. 81ff, von Staden 1975.

Two preliminary conceptual points are essential for our investigation. First we must distinguish between observation and deliberate research, although, no doubt, the one shades into the other. Acuteness in observation, whether of the natural world or of human behaviour, is frequently reported from non-literate societies and is often reflected, for example, in complex classification systems.[6] But such taxonomies, incorporated into natural languages, are not – at least not generally – the products of deliberate research. The latter presupposes a particular motivation, a desire to extend knowledge. The same point can be amply illustrated from pre-scientific Greek literary texts. Many of Homer's descriptions of aspects of animal behaviour,[7] or of the effects of different types of wounds,[8] for example, are both vivid and largely true to life. Yet it would be absurd to suggest that what lies behind them is deliberate zoological and anatomical investigation. Again, Sappho's famous account of her feelings when looking at the girl she loves[9] is as detailed as some descriptions of a patient's symptoms in the Hippocratic Corpus. But such elements in that account as are not purely imaginary are the product of observation, not of research – not, that is, of observations carried out purposefully whether to obtain new data or to support or undermine hypotheses. Greek literature of all periods is full of graphic descriptions that testify to an ability to observe: our subject here is, rather, the development of deliberate observation in the rise of natural science. While our chief concern is with Greek material, we shall also refer to the evidence for sustained observations carried

---

[6] See especially Lévi-Strauss 1966.

[7] Especially in such similes as that in which Odysseus, clinging to a rock, is compared with an octopus, dragged from its lair, with pebbles sticking to its suckers (*Od.* v 432ff). The knowledge of animals shown in the Homeric poems has been discussed by Buchholz (1871–85), I part 2, Keller 1909–13, and Körner 1930, especially.

[8] One passage that Aristotle quoted in his description of the great vein (vena cava) (*HA* 513b26ff) is the account of the wounding of Thoon at *Il.* XIII 545ff, where Antilochus' spear 'severs the vein right through, which runs along the back to reach the neck'. Another notable passage is *Il.* XVI 503f which refers to the prolapse of the lungs on the withdrawal of a spear from the chest. The descriptions of wounds in Homer have been studied by, for example, Daremberg 1865, Buchholz (1871–85), II part 2, pp. 247ff, Körner 1929 and A. R. Thompson 1952.

[9] Poem 31.7ff is thus translated by Page: 'For when I look at you a moment, then I have no longer power to speak, but my tongue keeps silence; straightway a subtle flame has stolen beneath my flesh, with my eyes I see nothing, my ears are humming; a cold sweat covers me, and a trembling seizes me all over, I am paler than grass; I seem to be not far short of death.' Page 1955, p. 27 went to to note 'the accurate description of [the] physical symptoms [of her passion]': 'the symptoms, one after another, of a complex emotional state are delineated with exactitude and simplicity'. One should, however, distinguish between the status of remarks concerning humming ears, on the one hand, and that of talk of a subtle flame beneath the flesh on the other.

out in such fields as astronomy and medicine by the Babylonians and Egyptians long before Greek science began.

Our second preliminary point concerns the relationship of observation to theory. Discussions of the role of observation in Greek science have often reflected unexpressed, and in some cases incoherent, philosophical assumptions. In particular inductivist or positivist beliefs concerning a realm of 'raw' pre-theoretical data – there for the observer to observe – have been, and in some quarters continue to be, highly influential, although they should by now be recognised to be open to fundamental objections. Observation-statements use predicates that are all more or less well-entrenched in a network of theoretical assumptions.[10] There is no such thing as an observation-statement that is not to some extent theory-laden, though of course the degree of theory-ladenness varies. We may contrast, to use some examples from Hesse,[11] talk of 'particle-pair annihilation' in a cloud chamber with talk of 'two white streaks meeting and terminating at an angle', or, again, the use of the term 'epileptic fit' with the type of description of loss of voice, foaming at the mouth, clenching of the teeth and convulsive movements of the hands that we have in *On the Sacred Disease*.

We can, as these examples illustrate, retreat from greater theory-ladenness to lesser, but this retreat does not lead us to an ultimate refuge of an entirely theory-free vocabulary in which raw sense-data can be set down innocently of all assumptions. Nor is it only the explicit inductivist or positivist who may make similar presumptions concerning the pre-theoretical status of what there is to observe. Discussing certain differences between the 'mythical', and our own, perception of the world, Frankfort, for example, wrote that 'we would explain, for instance, that certain atmospheric changes broke a drought and brought about rain. The Babylonians observed the same facts but experienced them as the intervention of the gigantic bird Imdugud which came to their rescue.'[12] But the sense in which the Babylonians can be said to have 'observed the same facts' is quite question-begging.

The distinction between observation and theory is a relative, not an absolute, one, and our problem is not one of how the Greeks extended their mastery over a set of data that were always there to observe, so much as one of studying first the varying *inter*action of

---

[10] A recent clear statement of the issue is in Hesse 1974, ch. 1. Cf. also, e.g., Putnam (1962) 1975, pp. 215ff, Suppe 1974, pp. 80–6, 192–9.
[11] Hesse 1974, p. 24.
[12] Frankfort 1949, p. 15.

observations and theories or assumptions, and secondly the increasing awareness of that interaction and the gradual, though still in certain respects quite limited, realisation of the need to collect and extend the range of observational data. The inquiry falls into two main parts, first what the Greeks themselves have to say about the role of observation and research, and secondly their actual practice, where we must try to illustrate, even if highly selectively, both the strengths and weaknesses of their observational work.

THE EPISTEMOLOGICAL DEBATE

The terms in which the ancients discussed some of the issues are, in important respects, different from those of the modern debate. In the classical period there is no exact equivalent, in Greek, for our 'observation':[13] the regular late Greek word, τήρησις, is not used in that sense before the Hellenistic period.[14] But although writers of the fifth and fourth centuries B.C. do not refer to 'observation' as such, they express a variety of views about αἴσθησις, perception, about φαινόμενα, 'things that appear', about evidence (for which they develop an extensive vocabulary)[15] and about inquiry or research, ἱστορία (which, like our own 'research', may include a good deal more than what we should call *empirical* research).[16] Finally πεῖρα and ἐμπειρία, from πειρᾶσθαι 'make trial', are used generally for 'experience'.

The range of meaning of two of the main terms, αἴσθησις and φαινόμενα, is wide. Both the verb αἰσθάνεσθαι and the noun αἴσθησις cover much more than 'sense-perception', being general words for 'feel' and 'feeling', including consciousness and self-consciousness. φαινόμενα means not so much 'phenomena' in our sense, as 'the things that appear'. Sometimes it carries some of the

---

[13] There is, however, a rich vocabulary of words for 'see' 'look' 'remark' 'attend to', such as ὁρᾶν, βλέπειν, σκοπεῖν, σκέπτεσθαι, θεᾶσθαι, θεωρεῖν, ἀθρεῖν, λεύσσειν, δέρκεσθαι, νοεῖν and their compounds. The last, which covers both 'see' ('notice') and 'understand', is connected with νοῦς, 'mind' 'reason', and in certain contexts bridges the reason-sensation dichotomy: see von Fritz 1943, 1945 and 1946.

[14] The original, fifth- and fourth-century, sense of the noun is guarding or protection, as, e.g., in Aristotle, *Pol.* 1308a30, *PA* 692a7: but the verb τηρεῖν is regularly used for watching closely (e.g. Ar. *Eq.* 1145) and appears in Aristotle (*Cael.* 292a7–9, *GA* 756a33) in connection with astronomical and zoological observations.

[15] The chief terms are σημεῖον ('sign', cf. σημαίνειν 'signify'), τεκμήριον and μαρτύριον ('testimony', cf. τεκμαίρεσθαι 'infer', μαρτύρεσθαι 'call to witness', μαρτυρεῖν 'bear witness'), the same vocabulary being used in natural scientific inquiry as in such other contexts as legal proceedings and historical research, cf. below, pp. 252f.

[16] Cf. also other terms for 'search', such as ζήτησις (cf. ζήτημα what is sought, both from ζητεῖν seek), ἔρευνα (cf. ἐρευνᾶν), σκέψις (cf. σκέπτεσθαι look) and cf. εὕρημα, εὕρεσις, 'discovery' 'invention' from εὑρίσκειν (find).

undertones of mere appearance or illusion, corresponding to the use of φαίνεσθαι with the infinitive ('appears but is not'), as opposed to the use with the participle ('appears and is'), although that distinction is not, at any period, a hard and fast one. Moreover as Owen demonstrated in his analysis of the Aristotelian usage,[17] φαινόμενα may include much else besides what appears directly to the senses. It can and often does refer to what is commonly thought or believed, the ἔνδοξα or received opinions, as it does notably in a famous passage in the *Nicomachean Ethics* (1145b27f) where Aristotle criticises Socrates' paradox that no one does wrong willingly for being in plain contradiction with the φαινόμενα – that is not (despite some translators) 'what is observed', but what appears in the sense of what is commonly believed to be the case.

The pejorative undertones in some uses of these terms are reflected in the epistemological debates in which the chief issue was initially presented as one between the senses on the one hand, and reason and argument on the other. It is, however, important to recognise the variety of views maintained already in the Presocratic period. Heraclitus expresses his contempt for 'much learning', πολυμαθίη, in Fr. 40,[18] and the four individuals he names show that 'much learning' must be taken to comprise a good deal apart from a curiosity in regard to natural phenomena: the four are the mythologist and didactic poet Hesiod, the historian Hecataeus, and the philosophers Pythagoras and Xenophanes.[19] But although Heraclitus proudly proclaims that he 'sought himself' (Fr. 101), he does not condemn the senses outright, but adopts, rather, an attitude of guarded, critical acceptance of them. Fr. 55 says that 'things which can be seen, heard, learned, these are what I prefer', and Fr. 101a notes that the eyes are more accurate witnesses than the ears, although Fr. 107 warns that 'eyes and ears are bad witnesses for men if they have souls that cannot understand their language'.

Far more radical attitudes towards the senses are expressed first – as we saw in chapter 2[20] – by Parmenides and Melissus, and then by Plato. In the Way of Truth, the goddess instructs Parmenides not 'to let habit, born of experience, force you to let wander your heedless

---

[17] Owen (1961) 1975.

[18] 'Much learning does not teach sense: for otherwise it would have taught Hesiod and Pythagoras, and again Xenophanes and Hecataeus.' Cf. also Frr. 35 and 129: the latter fragment, whose authenticity is doubtful, states that Pythagoras practised ἱστορίη, and identifies his 'wisdom' (σοφίη) with 'much learning' πολυμαθίη and 'bad art' (κακοτεχνίη).

[19] The extent of the Pythagoreans' involvement in empirical research is discussed below, pp. 144ff. On Xenophanes, see below, pp. 133 and 143.

[20] See above, ch. 2, pp. 71 and 77.

eye or echoing ear or tongue along this road' (Fr. 7), and in Fr. 8 Melissus develops a *reductio* that starts from the assumption that we may accept the evidence of sight and hearing. But since it seems to us, when we use our senses, that things change,[21] and yet it is clear (on the usual Eleatic grounds)[22] that change is impossible, it follows that the senses are, after all, not to be trusted: 'it is evident, then, that we did not see correctly'.

Further statements denigrating the senses appear in Plato's middle period dialogues, although several of the views presented there are modified in his later works. Thus in the *Phaedo* – in the context of a discussion of the soul's immortality during the last hours of his life – Socrates associates the senses with the body and with the world of particulars, in contrast to reason, associated with the soul and with the Forms,[23] and he argues that it is only when the soul is 'separated' as far as possible from the body that it can grasp truth,[24] that we neither hear nor see anything exactly,[25] and that inquiry through the senses is full of deception.[26]

In *Republic* VII especially we find a series of comments on the role of perception in certain branches of science that are of cardinal importance for assessing Greek attitudes towards empirical research. In connection with astronomy, in particular, Plato makes a number of remarks that at first blush appear to amount to a total condemnation of observational methods. First at 529 cd he says that although the stars are 'the most beautiful and exact' of visible things, they 'fall far short' of the truth, which is to be grasped by reason and thought alone,[27] and he develops an analogy between the stars and geo-

---

[21] The examples Melissus cites are far from being all straightforward cases of what we should consider direct sense-perception. 'For the hot seems to us to become cold, and the cold hot, and the hard soft, and the soft hard, and the living to die and *to come to be from the not-living*, and all these things are altered and what they were and what they now are are in no way alike, but iron, being hard, seems to be rubbed away by contact with the finger, as do gold and stone and everything else that seems to be strong, and earth and stone seem to come to be from water.'

[22] Change implies the coming-to-be of what is not: see above, pp. 76f.

[23] Especially *Phd.* 79a ff. Contrast the reservations expressed concerning the beliefs of the 'friends of the Forms' at *Sph.* 248a ff. In the later dialogues Plato points out that in perception it is the *soul* that grasps sense-objects *through* the senses (e.g. *Tht.* 184bc, *Phlb.* 33c ff).

[24] See, e.g., *Phd.* 64c ff, 66d–67b, 81b. What hinders the soul, in its search for truth, is not just the sensations, but also the passions, of the body.

[25] See especially *Phd.* 65b: 'Do sight and hearing have any truth in them for men, or, as the poets are always telling us, do we neither hear nor see anything exactly?' (Where Melissus, Fr. 8, had used ὀρθῶς, 'correctly', Plato here uses ἀκριβές, 'exactly'.)

[26] *Phd.* 83a, cf. 65b and 79c. Cf. also the contrast, at *R.* 523b, between cases where perception does not provoke thought, and cases where it does, since it 'yields nothing sound': ὡς τῆς αἰσθήσεως οὐδὲν ὑγιὲς ποιούσης.

[27] What the true student of astronomy should aim to grasp is described in terms of 'speed

metrical diagrams. 'Anyone experienced in geometry who saw such diagrams would grant that they are most beautifully constructed, but think it absurd to examine them seriously as if one could find in them the truth concerning equals or doubles or any other ratio.'[28] He concludes that 'it is by using problems...as in geometry, that we shall study astronomy too, and we shall let the things in the heavens alone, if we are to participate in the true astronomy and so convert the natural intelligence in the soul from uselessness to a useful possession'.[29]

The interpretation of this text is disputed,[30] but one point seems essential. What Plato is concerned with, in *Republic* VII, is not astronomy as such, so much as what astronomy, among other studies, can contribute to the education of his philosopher-kings. The overall purpose of the propaedeutic studies is to train these Guardians in abstract thought, to make them cultivate reason rather than the senses.[31] In that context, Plato naturally emphasises the distinction between an observational, and an abstract, mathematical, astronomy and advocates the latter.[32] Even so, several of his remarks are exaggerated or ambiguous or both.[33] In the analogy between stars and geometrical diagrams, especially, there is an assimilation or confusion of two ideas that should have been kept distinct, the true and obvious point that we cannot observe the mathematically determined courses of the heavenly bodies as such,[34] and the highly

itself' and 'slowness itself' in 'true number'. But whether these expressions should be taken to refer to the absolute Forms themselves, or to Forms present in the movements of the heavenly bodies (cf. 'greatness in us' in *Phd.* 102 d f) is far from clear.

[28] *R.* 529 e 3ff. At 530 b 1–4 he puts it more strongly still, that it is absurd in every way to seek to grasp the truth of the visible, corporeal stars: on the syntax and interpretation of this sentence, see Vlastos forthcoming.

[29] *R.* 530 b 6ff.

[30] I have attempted an interpretation in my 1968. There is a helpful detailed analysis in Vlastos forthcoming : see also Heath 1913, pp. 135ff, Dicks 1970, pp. 103ff.

[31] Time and again in the passages that precede the discussion of astronomy, the criterion used to determine whether a study is suitable for the Guardians' education is: does it encourage abstract thought? See, e.g., *R.* 521 cd, 523 ab, 523 c, 525 b–e, 526 ab and cf., in our passage, 530 b 8f.

[32] It is clear from Glaucon's remarks at *R.* 529 c 4ff and 530 c 2f that Plato saw the astronomy he was advocating in the *Republic* as a radical departure from the usual modes of doing astronomy in his day.

[33] Thus at *R.* 529 b 7–c 1 he says that 'there is no knowledge at all' of sensible objects. Like other statements we have noted from *Phd.* (e.g. 65 b 1ff) and *R.* (523 b), this might stand for a variety of theses, ranging from the unobjectionable point that is made in *Tht.* and elsewhere, that knowledge cannot be identified with perception, to the more extreme position that knowledge can in no way be derived from, or applied to, the objects of sensation.

[34] The diagrams are necessarily imprecise: one does not get out a ruler to verify the length of the hypotenuse of a right-angled triangle whose shorter sides are 3 inches and 4 inches, cf. Crombie 1963, p. 187. From *R.* 530 a 7–b 4 it is clear that, unlike Aristotle, Plato did not assume the heavens to be completely unchanging.

contentious (and misguided) thesis that it serves no use at all to observe the heavenly bodies.[35] In advocating the abstract, theoretical study, Plato writes as if he thought it necessary not merely to distinguish it from observational astronomy, but to run the latter down.[36] Understandable as they may be in the discussion of the education of the Guardians, such remarks as 'we shall let the things in the heavens alone', taken out of context, were interpreted already in antiquity[37] – as they have been again in modern times – as a ban on observational methods as a whole.

These and other texts show that the idea of the untrustworthiness of the senses had powerful advocates in the ancient world.[38] Yet these passages represent only one side, even if at times the more articulate side, of the debate, and other positions were proposed and defended both in the Presocratic period and later. When Heraclitus castigated 'much learning', he had something to attack, even if his target included more than just what we should call empirical research. One of those he singled out, Xenophanes, had indeed insisted that 'the gods have not revealed all things to men from the beginning: but by seeking (ζητοῦντες) men find out better in time'.[39] Alcmaeon, too, suggests that even though men cannot obtain 'clear knowledge' about 'unseen things and mortal things', they can at least make inferences or conjectures.[40]

More important, after Parmenides' denial of the validity of the senses, both Empedocles and Anaxagoras reinstate them, even if both

---

[35] That this latter thesis does not correctly represent Plato's intention can be argued, for example, from the high value set on sight in such passages as *Ti.* 47 ab. Indeed in our passage itself he insists on using the stars at least as *diagrams* (*R.* 529 d 7ff). But what he nowhere points out, or even recognises, is that observations of the movements of the planets have – as geometrical diagrams do not – the status of evidence.

[36] Similar points can be made about the passage that follows on acoustics (*R.* 530 d ff) where again Plato advocates a mathematisation of the science, but again argues (with even less justification than in astronomy) against observational methods, categorising them at *R.* 531 a 1–3 as useless labour. Here too in his zeal to criticise the empiricists for not ascending to problems (cf. 531 c) Plato denies that empirical methods have any value at all for the purposes he has in mind.

[37] E.g. by Proclus, *Hyp.* Proem, 2.1ff.

[38] Apart from the philosophers we have already considered, several others are represented by the doxographers as having held that the senses are false: in Aetius IV 9.1 the list of those who did so includes Pythagoras, Empedocles, Xenophanes, Anaxagoras, Democritus and Protagoras, as well as Parmenides, Zeno, Melissus and Plato. But in several cases, as we shall see, our other evidence shows this to be a drastic oversimplification.

[39] Xenophanes, Fr. 18. In Fr. 34 (cf. Fr. 35) he expresses the limitations of human knowledge – 'seeming is wrought over all things' – though we should note that the first subject he mentions to illustrate this is religion: 'No man knows or ever will know the truth about the gods and about everything I speak of.'

[40] Alcmaeon Fr. 1: 'concerning unseen things and mortal things, the gods have clear knowledge, but as far as men may infer (τεκμαίρεσθαι) . . .'.

express certain reservations about them. Thus although in the obscure but suggestive Fr. 2 Empedocles describes the means we have of grasping things as 'narrow',[41] at the end of Fr. 3 he says:

Come now, consider by every means of grasping how each thing is clear, neither holding sight in greater trust than hearing, nor noisy hearing above what is made plain by your tongue, nor withhold trust from any of the other limbs, by whatever way there is a channel of understanding, but apprehend each thing in the way in which it is clear.

Again, although Anaxagoras[42] is reported to have said that 'we are unable to judge the truth because of the weakness' of the senses (Fr. 21), in Fr. 21a he advocates using the 'things that appear' as a 'vision' of 'the things that are obscure'. This dictum – ὄψις ἀδήλων τὰ φαινόμενα[43] – was, we are told, praised by Democritus[44] who, whilst he categorised the senses as 'bastard' knowledge as opposed to the 'legitimate' knowledge we have of such matters as atoms and the void themselves,[45] nevertheless acknowledged that the mind derives its data from the senses.[46]

This epistemological debate was far from confined to the philosophers and we have valuable supplementary evidence on attitudes towards the issue between reason and the senses in the medical writers especially. Several echo Anaxagoras' dictum,[47] and the theme of the importance of experience and research in medicine is a

---

[41] Fr. 2 goes on to emphasise the difficulties: 'thus these things are not to be seen nor heard by men, nor grasped by the understanding'.

[42] A famous fragment of Euripides (Fr. 910 N) on the happiness of a life devoted to the inquiry (ἱστορία) into the ageless order of nature has often been taken to refer to Anaxagoras. Whether or not that is correct, the fragment provides good evidence of a positive attitude towards that inquiry.

[43] On the importance of this methodological principle in early Greek thought, see Regenbogen 1931, H. Diller 1932, Lloyd 1966, pp. 338–41, 353–5.

[44] See Sextus, *M.* VII 140. Cf. Clement, *Strom.* I 15.69 (Fr. 299) where – if the quotation is authentic – Democritus refers to the travels he went on in his inquiries, ἱστορέων.

[45] Fr. 11: 'There are two kinds of knowledge, one legitimate, one bastard: to the bastard belong all these, sight, hearing, smell, taste, touch: but the other is legitimate and separate from these.' The latter, he goes on to suggest, operates on objects too fine for the senses to perceive. Elsewhere he puts it that the objects of sensation exist 'by convention' (νόμῳ) not 'in reality' (ἐτεῇ), and that 'in reality we understand nothing exactly, but as it shifts according to the disposition of the body and of the things that enter and press on it' (Fr. 9). These fragments indicate that despite such apparently unqualified statements as Fr. 7 ('we know nothing truly about anything', cf. Frr. 6, 8, 10, 117) Democritus maintains a modified, not an extreme, scepticism. At the same time the view that the senses yield only a 'bastard' form of knowledge shows that Aristotle's repeated statement that the atomists found truth in appearance (e.g. *GC* 315b9ff) must be understood as an interpretative comment based on Aristotle's own conception of the distinction between sensibles and intelligibles.

[46] Fr. 125: the senses address the mind: 'Wretched mind, taking your proofs from us, do you overthrow us? Our overthrow is your fall.'

[47] Eg. *VM* ch. 22, *CMG* I, 1 53.12f, *Flat.* ch. 3, *CMG* I, 1 93.5, *Vict.* I ch. 11, L VI 486.12f, cf. *de Arte* ch. 12, *CMG* I, 1 18.14ff. See also Herodotus II 33f.

recurrent one.[48] The author of *On Ancient Medicine*, in particular, writes of the tried and tested methods of discovery in medicine. In chapter 2 he refers to the 'way' (ὁδός) of medicine, 'through which many and excellent discoveries have been made over a long period, and by which the rest will be discovered, if anyone is clever enough to conduct his researches knowing the discoveries that have already been made and taking them as his starting-point'.[49] More important still, in chapter 1 he draws a distinction between medicine and other inquiries, for example those about 'things in the sky or things under the earth'. The former has no need of ὑποθέσεις, 'hypotheses' or 'postulates':[50] as for the latter, 'if anyone were to speak and declare the nature of these things, it would not be clear either to the speaker himself or to his audience whether what was said was true or not, since there is no criterion to which one should refer to obtain clear knowledge'.[51] Here, then, is a demand that physical theories, at least in certain areas, must be verifiable, at any rate according to the writer's ideas of verifiability. What he has in mind in medicine itself can be judged in part from a subsequent passage where he remarks that, in determining treatment, the doctor has no other measure to refer to than the feeling (αἴσθησις) of the body.[52]

Moreover if Parmenides and Melissus represent one extreme view in their total rejection of the senses, the opposite extreme view was also discussed, at least by the fourth century. The thesis that knowledge is perception[53] is one that Plato has Theaetetus propose in the dialogue named after him, where Socrates represents it as following from Protagoras' dictum that 'man is the measure of all things'.[54] Now how far Socrates' interpretation of that dictum should be taken to correspond to the views of the historical Protagoras is a vexed question.[55] Yet it is enough, for our present purposes, to observe that

---

[48] See, e.g., *Flat.* ch. 1, *CMG* I, 1 91.15, *Praec.* ch. 1, *CMG* I, 1 30.3ff, *Off.* ch. 1, L III 272.1–5. Similarly the historians discuss the problems, and stress the laboriousness, of historical research, e.g. Hdt. II 29, III 115, Th. I 1, 20 and 22.

[49] *VM* ch. 2, *CMG* I, 1 37.1ff, cf. also ch. 3, 37.20ff, ch. 8, 41.8f, ch. 20, 51.17ff.

[50] The writer has physical, not mathematical, assumptions in mind. But on the possibility that his opponents' methodology may not be uninfluenced by the mathematical use of 'hypotheses' (on which see above, pp. 111ff and 113ff), see Lloyd 1963.

[51] *VM* ch. 1, *CMG* I, 1 36.2ff, especially 15ff.   [52] *VM* ch. 9, *CMG* I, 1 41.20ff.

[53] Cf. Aristotle's remark, *de An.* 427b3, that some hold that all appearances are true: however his frequent attribution to the atomists of the view that truth lies in appearance is open to question (see above, p. 134 n. 45).

[54] *Tht.* 151e, 152a, cf. also Protagoras' 'defence', *Tht.* 166a ff, e.g. 166d.

[55] Other later testimonies (such as Aristotle, *Metaph.* 1062b12ff – cf. 1053a35ff – and Sextus, *P.* I 216ff, *M.* VII 60f) report Protagoras' view as being that the 'appearance' is the measure, but these may well be largely, if not wholly, dependent on Plato's interpretation. Guthrie 1969, pp. 181–92 provides a survey of views on the issue of the original meaning and application of Protagoras' dictum.

the *Theaetetus* itself indicates that – whether or not it had any pre-Platonic supporters – the identification of sense-perception and knowledge was at least discussed in the classical period.[56]

Finally while Aristotle shared many of Plato's epistemological doctrines, perception and experience are allotted a more positive, indeed from some points of view a basic, role in his theory of knowledge. Whilst agreeing with Plato that knowledge is of forms, he disagreed with him, in certain fundamental respects, on the nature of forms. Whereas in Plato they are transcendent, that is they can and do exist independently of particulars, in Aristotle forms do not, in fact,[57] exist in separation from the particulars of which they are the forms. It is the particulars themselves, each a concrete whole analysable in terms of form and matter, that are, in the doctrine of the *Categories*, what primarily exists.[58] Again, while Aristotle, like Plato, insisted on certainty as a condition of understanding in the fullest sense,[59] he drew a distinction between the method of proof and the method of discovery. In the latter, the starting-point is not 'what is better known absolutely', but 'what is better known to us': this will vary according to the subject, but he describes it as what is closer to perception and to the particulars (as opposed to the universal).[60] Thus while the Platonic elements in Aristotle's ontology and epistemology are considerable, the emphasis is, at certain points, very different, not only in the account he gives of the status of particulars, but also in his analysis of perception. This is now accommodated alongside reason as one of the cognitive faculties of the soul,[61] and its basic role comes out clearly in his account of how we apprehend the primary, immediate starting-points of demonstrations in the final chapter of the *Posterior Analytics*. We do so by a process he calls ἐπαγωγή, 'induction', the origin of which he traces back to perception: this is a faculty common to men and animals, but only a few animals have memory, and it is from memory that

---

[56] Cf. also below, p. 138 n. 65, on the roles of perception and the appearances in post-Aristotelian philosophies.

[57] ἔργῳ: they are, of course, separable λόγῳ or in thought. This rule applies to all sublunary objects: in the superlunary sphere, however, Aristotle talks of pure actualities, ἐνέργειαι.

[58] *Cat.* ch. 5, 2a11ff.

[59] Understanding is of what cannot be otherwise than it is, e.g. *APo.* 71b9ff.

[60] E.g. *APo.* 71b33ff, *Top.* 105a13ff, *Ph.* 189a4ff, *Metaph.* 1018b30ff, cf. *APr.* 68b35ff.

[61] As reason is of the intelligible forms, so perception is of perceptible forms. Whilst he recognises and analyses the possibility of error in perceiving, for example, the common sensibles (such as movement or size, e.g. *de An.* 428b23ff), and draws attention to the deceptiveness of appearances in his account of imagination (φαντασία) (*de An.* III 3, e.g. 428a11ff, 16ff, b2ff), he holds that perception of the special objects of the individual senses (e.g. colours, sounds) is 'infallible or least subject to falsehood' (*de An.* 428b18f).

experience, and from experience in turn that both 'art' and understanding in the strictest sense, come.⁶²

Aspects of his theory are picked up in certain recurrent themes in the body of his scientific work. He often employs a broad distinction between appeals to λόγοι, theoretical arguments, and appeals to the φαινόμενα, or to the 'facts' or the 'data', ἔργα or ὑπάρχοντα, or again to 'what happens', the γιγνόμενα or the συμβαίνοντα. What counts as such 'data' varies in different contexts, and – as has already been noted for φαινόμενα – these terms encompass much more than what we should call the results of observation. Thus to cite one characteristic instance, when he castigates the Pythagoreans for 'not seeking arguments and causes in relation to the φαινόμενα, but trying to drag the φαινόμενα into line with certain arguments and opinions of their own' (*Cael.* 293a25ff), the criticism is a general one: in introducing their doctrine of the 'counter-earth' the Pythagoreans violated 'what appears to be the case' both in the sense of the common opinions and in the sense of what is observed. But the inclusion of the latter is sometimes made clear by the addition of the specification κατὰ τὴν αἴσθησιν, what appears 'according to perception', as it is, for instance, in another text in the *De Caelo* (297b23ff) where he turns from theoretical arguments establishing the sphericity of the earth to consider such points as the shape of the earth's shadow in eclipses of the moon and the changes in the observed positions of the stars at different latitudes.⁶³

Similarly when, in the zoology, he discusses how to proceed in the study of animals and says that one must first view the φαινόμενα concerning each kind of animal and then proceed to state their causes,⁶⁴ the φαινόμενα here include a good deal more than what is directly observed. On the other hand the role of perception is once again made explicit in a famous text in the *De Generatione Animalium* (760b27ff) which contrasts λόγοι, arguments or theories, not just with φαινόμενα but also with αἴσθησις. Completing his account of how bees reproduce, he writes:

⁶² *APo.* 99b32ff (see also above, pp. 116f), cf. *Metaph.* 980a27ff; at *APo.* 100a16ff he states that while we perceive the individual (e.g. Callias), perception is of the universal (man), cf. *APo.* 81b6f, 87b28ff, 37ff, *de An.* 417b22f, *EN* 1142a25ff, 1147a25ff. On the role of experience in securing the starting-points of different kinds of inquiry, see, for example, *APr.* 46a4ff, 17ff.

⁶³ See below p. 206 and cf. also *Ph.* 253a32ff, *Cael.* 306a5ff, 16ff, *GC* 315a3ff, 325a13ff.

⁶⁴ *PA* 640a14f, cf. 639b8ff, referring to the parts of each animal, with the protreptic to the study of all the parts of every kind of animal in *PA* 15, e.g. 644b29ff, 645a6f, 21ff, where, however, he characteristically emphasises that the inquiry is directed to the formal and final causes especially, e.g. 645a23ff, 33ff, b13ff. Cf. also *HA* 491a9ff which describes the starting-point as the differentiae of animals and τὰ συμβεβηκότα πᾶσι – what is the case with all of them.

this then seems to be what happens with regard to the generation of bees, judging from theory (λόγος) and from what are thought to be the facts (τὰ συμβαίνειν δοκοῦντα) about them. But the facts (τὰ συμβαίνοντα) have not been sufficiently ascertained, and if they ever are ascertained, then we must trust perception (αἴσθησις) rather than theories (λόγοι), and theories, too, so long as what they show agrees with what appears to be the case (τὰ φαινόμενα).

This survey of the positions adopted during the fifth and fourth centuries on issues related to observation and research[65] has been summary, but it suffices to show that there is no simple orthodoxy on the question. Anti-empiricist views (including the denial of the validity of perception) are expressed, sometimes with an element of rhetorical exaggeration, by a number of influential writers, several of whom are more concerned with the development of the notion of deductive demonstration than with the analysis of the empirical foundations of knowledge.[66] But – obviously – other views are also expressed that are a good deal less hostile to, and some that positively recommend, observation and empirical research. To put the point negatively, it is not as if the practising scientists, including the natural philosophers, worked against a background of a consensus of opinion condemning, or even consistently devaluing, the use of the senses.

THE PRACTICE OF RESEARCH

With these points in mind we may now turn to our second and more substantial topic, the actual practice of Greek scientists in different fields of research. If the epistemological debate I have reviewed is well known, the level of comment on Greek scientific practice in the matter of the use of observation has sometimes been superficial. In particular the differences in performance between – and within – the main areas of natural scientific inquiry have often been ignored, although they are in some cases quite marked. While there are some

---

[65] From the late fourth century the debate continues and develops. Both the Stoics and the Epicureans took perception to be one – and the basic – source of knowledge. But against the 'dogmatists' of all kinds (including the Peripatetics) the Sceptics mounted arguments to undermine both perception and reason – and indeed any criterion that might be used as a foundation for knowledge: yet even the Sceptics referred to the φαινόμενον as a criterion for regulating conduct (though not as a basis for statements about reality), see Sextus, *P.* I 21ff. Once again, the debate is not confined to those whom we think of chiefly as philosophers. Thus the view that 'one must call the φαινόμενα primary, even if they are not so' is attributed to the Hellenistic biologist Herophilus in Anonymus Londinensis XXI 22f, and the contrasting views of the three main medical schools, the Dogmatists, the Empiricists and the Methodists, on the roles of reason on the one hand, and experience (ἐμπειρία, πεῖρα) and the appearances (φαινόμενα) on the other, are analysed at great length by Galen (e.g. *Sect. Intr.* and *Opt. Doctr.*) who himself advocates a method that combines both criteria.

[66] See above, ch. 2.

excellent examples of the practice of empirical research in early Greek science, there are also areas where the use of observation is very limited, and we must attempt to do justice to and explain these discrepancies. We may divide our study into four main sections, first the Presocratic natural philosophers, second the Hippocratic writers, third the development of empirical research in geography and astronomy, and finally its role in Aristotle. In the second and third sections, especially, we shall need to refer to material from later than the fourth century B.C. in order to put the earlier stages in the development of Greek science in perspective.

PRESOCRATIC NATURAL PHILOSOPHY

Anyone who studies the extant sources for the Presocratic natural philosophers is likely to be struck by the almost total dearth of references to anything that looks like a deliberate observation. Although well-known data from ordinary experience are used often enough, the occasions in either the pre- or the post-Parmenidean philosophers when empirical research appears to have been carried out on set purpose either to obtain new facts or to support or undermine hypotheses are rare. One consideration that immediately goes some way to explain this relates to the kinds of problems they investigated. As we have already remarked in another context,[67] many of the phenomena they were interested in are frightening or rare or both. The doxographers report theory after theory on such questions as the nature or causes of lightning, thunder, earthquakes, comets and the stars, where the opportunities for direct[68] empirical research are either restricted or non-existent. Here the philosophers generally proceeded by drawing on analogies with familiar objects, as Anaximenes, for instance, did when he attempted to support an explanation of lightning as being due to the wind cleaving a cloud by referring to the flash made by an oar in water.[69] Yet this procedure is in no way surprising, given that lightning is obviously not a phenomenon that the Presocratics – or anyone else in antiquity – could investigate directly, as in a laboratory. Another much discussed problem was what keeps the earth up, where the explanations proposed depend either on purely abstract argument,[70] or on simple analogies, such as Thales' reported view that the earth floats on

---

[67] See above, ch. 1, p. 32.
[68] That is of the phenomena themselves, as opposed to real or assumed analogues for the phenomena.
[69] Aetius III 3.2, DK 13 A 17, cf. Lloyd 1966, pp. 315–17.
[70] See above, ch. 2, pp. 67f, on Anaximander's theory.

water,[71] or the theory attributed to Anaximenes and others that it is supported, as flat objects are, by air.[72] It is precisely in these areas of speculation, concerning 'things in the sky or things under the earth', that – as the writer of *On Ancient Medicine* objected – 'there is no criterion to which one should refer to obtain clear knowledge'.[73]

In several of the areas of inquiry in which the earliest philosophers were interested, the lack of references to direct observation and research may be said to reflect the nature of the problems themselves. Yet they also attempted theories on other topics where this no longer holds, at least not to the same degree. In particular they put forward a great variety of doctrines concerning the fundamental constituents of material objects, and this provides something of a test case for the relation between observational data and theories in their work.

The physical theories in question include different kinds of monistic doctrines, according to which every type of material object is seen as a modification of a single basic element,[74] and of pluralistic ones, appealing to either a definite[75] or an indefinite[76] number of elementary substances – and we can supplement the fragments and reports for individual Presocratic philosophers with the information from a number of Hippocratic writers who also discuss the problem.[77] The extent to which the theories in question were either suggested, or supported, by empirical evidence is extremely limited. The same familiar 'data' – or what were assumed to be such – recur frequently and in connection with widely differing theories.

One instance of this is the group of observed or supposed changes

---

[71] Aristotle, *Metaph.* 983 b 21f, *Cael.* 294 a 28ff, cf. Lloyd 1966, pp. 306ff.

[72] Aristotle, *Cael.* 294 b 13ff (referring also to Anaxagoras and Democritus). Cf. Plato, *Phd.* 99 b, where an unnamed earlier philosopher is said to prop up the earth like a flat kneading-trough on air as its base. Cf. Hippolytus, *Haer.* 1 7.4 (DK 13 A 7) on Anaximenes' view of what holds the stars up, and cf. on Empedocles, Aristotle, *Cael.* 295 a 14ff, on which see Tigner 1974.

[73] *VM* ch. 1, *CMG* 1, 1 36.2ff, 15ff, see above, p. 135.

[74] Apart from physical theories such as those of Anaximenes and Diogenes of Apollonia, based on air, or that of Hippon on water, the atomists too explained the differences in material objects as being due ultimately to modifications in the shape, arrangement and position of the atoms, which do not themselves differ in substance, although in addition to the atoms they also postulated the void.

[75] As in theories based on two elements, such as that in Parmenides' Way of Seeming (light and night) or on four (usually earth, water, air and fire), versions of which appear in Empedocles, in Philistion (according to the report in Anon. Lond. xx 25ff) and in the Hippocratic treatise *Carn.* (ch. 2, L VIII 584.9ff) as well as in Plato, Aristotle and many later writers.

[76] As in Anaxagoras, see below, p. 141.

[77] Thus *Nat. Hom.* ch. 1 (L VI 32.3ff, cf. above, pp. 92f) refers to monistic physical doctrines based on air or fire or water or earth (which the writer later compares with the theories of the physicians who maintain that man consists of blood or bile or phlegm). Cf. further below, pp. 146ff, 149ff on the Hippocratic writers' own theories.

involving air, cloud, water, earth and stones. First our sources represent Anaximenes as suggesting that everything else is a modification of air, which becomes other things by a process of rarefaction and condensation. Simplicius, following Theophrastus in all probability, illustrates this by saying: 'Rarefying, it becomes fire, condensing, it becomes wind, then cloud, then condensing further water, then earth, then stones and the rest come from these.'[78] Hippolytus provides a similar list of changes[79] and there is little reason to doubt that some such examples represent the idea underlying Anaximenes' original theory correctly. But then at the other end of Presocratic speculation,[80] Anaxagoras cites similar 'data' in connection with his own quite different theory, that 'in everything there is a portion of everything'[81] and that apparent changes are to be explained in terms of changes in the proportions of the different substances in the objects we see.[82] Fragment 16 of Anaxagoras states that 'from the clouds water is separated off, and from water earth, and from earth stones are condensed by the cold'.[83] On the analogy of fragment 10, which puts the question 'How could hair come to be from not-hair?' and implies the answer that hair must be present in some form in the food we eat and the water we drink, Anaxagoras presumably held that water contains the earth and the stones that are separated off from it – and indeed every other kind of natural substance.

It is especially striking that when, in fragment 8,[84] Melissus mounts an argument against the senses on his opponents' own assumptions – where he would clearly avoid, so far as possible, any view that was controversial or not commonly agreed – he too says that 'earth and stones seem to come to be from water'. Nor is this particular set of examples confined to the Presocratic period, since (with one excep-

---

[78] *In Ph.* 24.29ff, DK 13 A 5.   [79] *Haer.* I 7.3, DK 13 A 7.

[80] Heraclitus, too, spoke of water coming to be from earth and of earth as the 'death' of water (Fr. 36) and in Fr. 31 refers cryptically to the 'changes of fire' – 'first sea, and of sea half is earth and half πρηστήρ' (perhaps thought of as a form of fire). Cf. also the testimonies collected in 'Fr.' 76, though the authenticity of these is doubtful since they appear, anachronistically, to attribute to Heraclitus a doctrine of four simple bodies.

[81] E.g. Frr. 6, 11 and 12 and cf. Fr. 4. Although mind is the one exception to this general principle, the reference of 'everything' is otherwise apparently quite unrestricted: it includes not only every kind of natural substance (gold, flesh, bark, etc.) but also the opposites 'the hot' 'the cold' 'the wet' 'the dry' and so on (Frr. 4, 8, 12, 15). Although many aspects of the interpretation of Anaxagoras' physical theory are disputed, this does not affect his use of the examples in Fr. 16.

[82] But, with the exception of mind (Frr. 11 and 12), everything always has a portion of everything else in it. As the end of Fr. 12 puts it, each single thing is and was most clearly those things it has most of – that is, as Simplicius explains (e.g. *In Ph.* 155.23ff), each thing acquires its character from what predominates in it. Thus what appears to us as gold has most gold in it, though it has a portion of everything else as well – and conversely every other object has a portion of gold.

[83] Cf. also Simplicius, *In Ph.* 460.13f, DK 59 A 45.   [84] See above, p. 131 and n. 21.

tion) they are to be found also in Plato,[85] in connection with his version of the four-element theory, in which earth, water, air and fire are represented as composed of two basic kinds of primary triangles.[86]

Here, then, the same assumed 'data' – interpreted quite differently – turn up in relation to divergent theories. First no attempt seems to have been made to scrutinise or check the 'data' themselves: it was just commonly supposed that not only clouds become water, but also 'water' 'earth', and 'earth' 'stones'.[87] Secondly, the data in question are – we should say – indifferent as between the theories in relation to which they are cited, the theories being sufficiently vague, or their empirical content sufficiently thin, that they are all equally well able to accommodate the phenomena. Now it is not as if the very same examples are represented or assumed to be *proof* of each of the theories concerned. At the same time they are not simply what was to be explained, for once incorporated into a given theory, they were no doubt held to *corroborate* it. Provided a theory could give some account of these and a few other similar familiar phenomena, this was taken as an adequate empirical basis for that theory. The doctrines in question were certainly competing with one another, but they were judged – one must suppose – not so much in terms of the empirical evidence that could be presented in their support, or the range of phenomena they could explain or predict, as first by their economy and consistency, and then by their ability to meet certain general philosophical difficulties relating to the nature, and indeed after Parmenides the possibility, of change and coming-to-be.

---

[85] See, e.g., *Ti.* 56c ff for the transformations of fire, air and water, and 60b ff on the formation of stones from a kind of earth. In Plato's theory, however, earth is composed of a different kind of primary triangle from the other three simple bodies, and so earth does not come to be from water or change into it. On Plato's physical theory, see most recently Vlastos 1975, ch. 3.

[86] Aristotle too speaks of transformations between the simple bodies (e.g. *GC* II 4) and now, against Plato, includes earth in these changes. For him stones are compounds of water and earth in which earth predominates (e.g. *Mete.* 383b20).

[87] For Empedocles, it is true, strictly speaking water (the element or as he calls it the root) does not become earth (the element). But given the vagueness with which 'water' 'earth' and so on were used, the issue between Empedocles, who denied changes between the four elements, and Plato and Aristotle, who asserted them (though as we have seen Plato held earth to be an exception), was not one that could be resolved by appeal to familiar phenomena – though that is not to say such phenomena were not invoked in the debate (as notably by Aristotle against Plato at *Cael.* 306a4f, see further below, p. 207). Where Plato and Aristotle would say that 'air' became 'water' (e.g. when rain fell) or 'water' 'air' (when water boiled), it was open to Empedocles to say that this 'air' was not the root, but either a form of water or a compound containing it. Our own notions of chemically pure elements are thus quite anachronistic, and each of the terms 'earth' 'water' and 'air' is applied to a wide range of substances (predominantly, though not exclusively, solid, liquid and gaseous respectively).

The element theories that the Presocratic philosophers proposed were not of a kind that could be decisively tested whether by new or by existing data,[88] and as a general rule they made little or no attempt to conduct observations – let alone systematically to extend the range of data under discussion – in this field of inquiry.[89] There are, however, a few exceptions to that rule, such as Anaximenes' use of the fact that breath exhaled from compressed lips feels cooler than from an open mouth to support a theory that the hot and the cold are to be identified with the rare and the dense respectively,[90] and Anaxagoras' proof of the corporeality of air by referring to the resistance it offers when trapped in inflated wine-skins or in the clepsydra.[91] To such examples can be added a handful of other instances of what may be deliberate observations made by the Presocratic philosophers in other contexts. One of the more notable is Xenophanes' reported use of the evidence of fossils to support his view that the relations between land and sea are subject to fluctuation and that what is now dry land was once covered by the sea. According to Hippolytus he cited as proof of this not only shells found inland and on mountains, but also the impressions of certain living organisms in the quarries of Syracuse, on Paros and on Malta.[92] We cannot know how much of this report to trust,[93] but if it is at least substantially correct, it is an interesting – if quite unusual – instance of ἱστορίη.[94]

---

[88] That Presocratic physical theories were not vulnerable to refutation by simple appeals to straightforward observations is a point that was successfully urged against Cornford's thesis (1952) by Vlastos (1955) 1970.

[89] We have to wait until Aristotle for the first attempt to collect and systematise information concerning, for example, which substances are combustible, which fusible, which soluble in water and other liquids and so on: see below, pp. 209f on *Mete.* IV.

[90] Plutarch, *de prim. frig.* 7.947F, DK 13 B 1.

[91] Aristotle, *Ph.* 213a24ff, DK 59 A 68, and cf. on the clepsydra, the account in the pseudo-Aristotelian *Problemata*, 914b9ff (A 69), which notes that Anaxagoras held air to be the cause. In the *Physics* passage Aristotle refers to others (unnamed) who try to disprove the void by this means, and this has sometimes been taken to include Empedocles, who certainly held that air is one of the four 'roots', and who cited the phenomena of the behaviour of air and water in a clepsydra, though in a different context – namely his theory of respiration – in Fr. 100. This was, for example, suggested by Burnet (1892) 1948, p. 229, but cf. the cautions expressed by Guthrie 1965, pp. 224f.

[92] *Haer.* I 14.5–6, DK 21 A 33. The living organisms appear to have included fish and a bayleaf, and seaweed if, as seems likely, we should read φυκῶν for the MSS (and DK) φωκῶν 'seals'. On the text, see Guthrie 1962, p. 387 and notes 2–4, who also remarks that similar observations can be paralleled in fifth-century authors, of shells on mountains in Herodotus (II 12) and of fossils in Xanthus of Lydia (Strabo I 3.4).

[93] It is clear from Fr. 8, at least, that Xenophanes had travelled extensively in the Greek world.

[94] This was one of Burnet's two examples of 'biological and palaeontological observations' (see above, p. 126 n. 2), the other being Anaximander's 'discoveries in marine biology', a reference to the group of testimonies (DK 12 A 30, A 10, A 11 (6)) concerning his ideas on the origin of living creatures. One of these texts, Plutarch, *Quaest.*

These are, to be sure, all isolated examples of observations. There is, however, one field in which more sustained empirical research appears to have been undertaken by philosophers before Plato, namely harmonics or acoustics. Admittedly many of the stories concerning the work of Pythagoras himself in this area must be discounted. Thus several of the experiments he is supposed to have conducted to discover the numerical relations of the main musical intervals, octave, fifth and fourth, must be rejected for the simple reason that they do not, in fact, yield the results reported.[95] That is not the case, however, of accounts relating to the measurements of lengths of pipe corresponding to different notes or to investigations of the mono-chord.[96] Although we have every reason to be cautious about Pythagoras' own involvement,[97] we have other evidence that tends to confirm that empirical investigations were carried out in this area at least by the early fourth century.

First, in a fragment preserved by Porphyry,[98] Archytas cites a

Conv. VIII 8.4.730E f, illustrates a theory on the origin of man with an allusion to γαλεοί (if the usual emendation is accepted). These are dog-fish, one species of which – as Aristotle pointed out in a famous description *HA* VI 10, 565b1ff – is remarkable for the placenta-like formation by which the young are attached to the female parent. But these evidences provide no firm basis for ascribing extensive, or indeed any, biological observations to Anaximander. The stimulus to his theorising was a *problem* (the origin of living creatures and of man) which was already the subject of mytho-logical accounts (e.g. the story of Pyrrha and Deucalion). The particular reference to dog-fish is Plutarch's (based ultimately on Aristotle, no doubt), and even if, as seems possible, Anaximander himself knew and referred to viviparous sea-animals of some sort, that does not necessarily imply that his knowledge came from personal obser-vation (cf. on Aristotle's sources of information, below, pp. 211ff).

95 This is true of three of the favourite stories in our sources, (1) that he made his discovery by weighing the hammers which produced different notes, (2) that he did so by attaching weights to strings and noticing that the weights gave the numerical relations of the concords (but in fact the pitch will vary with the square root of the weight, not with the weight itself), and (3) that he filled jars with water and discovered the con-cords by noting the relation beteeen the amounts of water in the jars and the sounds they gave when struck (but the concords would be revealed only if the columns of air in the jars, rather than the jars themselves, were vibrated). The main sources for these stories (1) (2) and (3) are as follows: Nicomachus, *Harm.* ch. 6, 245.19ff (1 and 2); Theon of Smyrna, 56.9ff (2 and 3); Gaudentius, *Harm.* ch. 11, 340.4ff (1 and 2); Censorinus, *de die nat.* ch. 10, 17.19ff (2); Iamblichus, *VP* 115ff, 66.12ff (1 and 2), and *In Nic.* ch. 6, 121.13ff (1); Macrobius, *Somn. Scip.* II 1.8ff, 583.28ff (1 and 2); Boethius, *Mus.* I chh. 10f, 196.18ff (all three); Chalcidius, *In Ti.* ch. 45, 94.13ff (2), The combination of the repeated clear references, in these sources, to the idea of *varying the conditions of the test*, and the inaccuracy of the results reported, is remarkable.
96 Of the sources mentioned in the last note, Nicomachus, Theon, Gaudentius, Censorinus, Iamblichus and Boethius all mention one or both of these methods, cf. also Diogenes Laertius VIII 12. For investigations on wind-instruments, see also below, p. 145, on Archytas Fr. 1, and cf. ps-Arist. *Problemata* XIX ch. 50, 922b35ff.
97 The Pythagoreans tended to ascribe their doctrines to him out of respect and to gain the prestige of his authority. On the other hand we should recall Heraclitus' attack on Pythagoras by name for his 'much learning', Fr. 40 and cf. Frr. 35 and 129, above, p. 130.
98 *In Harm.* 56.5ff, especially 57.2ff, 14ff, DK 47 B 1. A rather inflated assessment of the

variety of phenomena in an attempt to establish a theory relating the pitch of a note to its 'speed', and some of the examples may suggest first-hand observation. Thus he refers not only to the notes made by different lengths of pipe (which he interprets in terms of variations in the 'strength', and so of the speed, of the air at the holes it comes out of), but also to the variations in the pitch of the sound produced by a stick moved at different speeds, and to similar changes in the pitch of the ῥόμβοι or 'bull-roarers' swung in religious ceremonies.[99]

Secondly, in a passage in *Republic* VII which we have already had occasion to mention,[100]Plato too refers to empirical investigations in acoustics. Socrates introduces this discussion by agreeing with the view he attributes to the Pythagoreans that harmonics and astronomy are sister sciences (*R.* 530 d), but he then goes on to criticise the 'useless labour' of measuring audible sounds and concords against one another. Glaucon, in turn, speaks of those 'who lay their ears alongside' the strings, 'as if trying to catch a voice from next door: and some state that they can hear another note in between and that this is the smallest interval to use as a unit of measurement, while others contest that the sounds are the same, both parties preferring their ears to their minds', whereupon Socrates develops a comparison with the procedures used in the courts for examining witnesses by torture: 'they torture the strings and rack them on the pegs...'. Although the Pythagoreans are then distinguished from the more extreme type of empirical investigator,[101] they too are criticised for 'looking for numbers in these heard harmonies, and not ascending to problems'.

The evidence in Plato is all the more convincing, since he is here objecting to such methods of inquiry.[102] Nevertheless we must observe that, so far as the Pythagoreans themselves are concerned, they had a quite special motive for engaging in such investigations. Aristotle reports that they held that 'all things are numbers' or that

significance of this fragment for the role of experiment in Presocratic, particularly Pythagorean, philosophy is to be found in Senn 1929, pp. 271ff. Cf. also the obscure report from Aristoxenus, preserved in a scholium on Plato, *Phd.* 108d, that Hippasus constructed bronze disks of varying thickness to produce certain harmonies (cf. also Theon of Smyrna 57.7 on Pythagoras).

99 Archytas also develops several analogies to explain acoustical phenomena, (1) with water poured into vessels, to support a suggestion that some sounds may not be perceptible because of their extreme intensity, and (2) with missiles, where the force of the missile is compared with the intensity of the sound.

100 *R.* 530d–531c, cf. above, p. 133 n. 36.

101 See *R.* 531b7.

102 That is, for the purposes he has in mind in *R.* VII, where he is describing how to train the Guardians in abstract thought: see above, pp. 132f.

'things imitate numbers'.[103] a doctrine that was evidently applied in a wide variety of contexts, including not only acoustics and astronomy,[104] but also to such examples as justice (equated with the number four, the first square number) and marriage (equated with the number five, the union of male – identified with the number three – and female – two).[105] Clearly acoustical inquiries were part of a general search for support for this overall principle that 'things are numbers'. It would be quite unwise to assume that the earlier Pythagoreans drew any sharp distinction between, on the one hand, those acoustical investigations, and, on the other, reflections about the symbolic associations of certain numbers. Our evidence certainly suggests that what we might describe as empirical research was carried out in acoustics before Plato: but the context shows how rash it would be to ascribe to the Pythagoreans themselves any clear idea of the value of empirical research in general, let alone any recognition of the role and conditions of what we should call the experimental method. We must recognise, therefore, that in the one major case where we have evidence of philosophers engaging in empirical research before Plato, exceptional – and complex – motives were at work.

HIPPOCRATIC MEDICINE AND THE DEVELOPMENT OF DISSECTION

Our second and much richer source of information for the study of the use and development of empirical methods in early Greek science is the Hippocratic Corpus. Where the observational support for the philosophers' speculations is generally acknowledged to have been thin, the medical writers have often been represented as excellent practitioners, indeed sometimes as the founders, of the empirical method. Our task in this section is to assess those claims, and we may begin where a direct comparison is possible between some of the medical writers and the natural philosophers. Our extant Hippocratic treatises include many discussions of problems that are raised in natural philosophy, notably such questions as the fundamental constituents of physical objects in general or of the human body in

---

[103] E.g. *Metaph.* 985 b 23ff, 987 b 11 f, 27ff, 1090 a 20ff. On the problems of reconciling the statements that things are, and that they imitate, numbers, see, for example, Guthrie 1962, pp. 229ff.

[104] See further below, pp. 173f.

[105] See, e.g., Aristotle, *Metaph.* 985 b 29ff, 990 a 18ff, 1078 b 21ff, ps-Arist. *MM* 1182 a 11ff, Alexander, *In Metaph.* 38.10ff. At *Metaph.* 1093 a 13ff Aristotle is particularly scathing in his objections to those who saw a special significance in the number seven and who connected different instances of sevens together.

particular, and with the medical writers we have the advantage that we do not depend on quotations or interpretations, but can judge their arguments and evidence as they presented them.

As we have seen, *On Ancient Medicine* attacked the use of a method based on 'hypotheses' or 'postulates', where it is 'not clear either to the speaker himself or to his audience whether what was said was true or not',[106] a procedure which the writer evidently took to be typical of natural philosophy,[107] even if his own primary concern is to criticise the use of this 'new-fangled' way of inquiry in medicine itself.[108] But the question we must now ask is how far he implemented his principles in practice in his own accounts of such matters as the constituents of the body and the causes of diseases. The physiological and pathological theories he singles out for particular criticism are those based on the hot, the cold, the wet and the dry, and he pays particular attention to arguing that the hot and the cold are among the weakest 'powers' in the body.[109] Yet when he comes to present his own view concerning what the body is composed of and what causes diseases, the constituents he identifies turn out to be such things as the salty, the bitter and the sweet, the acidic, the astringent and the insipid.[110] It is true that this doctrine is more complex than those he rejects: he insists that there are many different components that have 'powers' of many different kinds, both in number and in strength.[111] Yet it is otherwise open – we should say – to very similar objections. The 'salty', the 'bitter' and so on are left vague and ill- or un-defined. Although the writer's ideas about isolating the operative factors in pathological conditions are admirable,[112] he does not, in practice, follow his analysis through to the point where he can show that the types of constituents he refers to are indeed the causes of particular complaints.[113] He sees that treatment must

---

[106] *VM* ch. 1, *CMG* 1, 1 36.15ff, cf. above, p. 135.

[107] See *VM* ch. 20, *CMG* 1, 1 51.6ff, where he refers to those 'who have written about nature' and (if our text is correct) mentions Empedocles by name in particular: the writer's argument at that point is that the correct way of discovering 'about nature' is from medicine.

[108] See especially *VM* ch. 13, *CMG* 1, 1 44.8, and cf. also ch. 1, *CMG* 1, 1 36.16 (where, however, the reading is disputed).

[109] *VM* chh. 1 and 13–17 especially, *CMG* 1, 1 36.2ff and 44.8ff.

[110] *VM* ch. 14, *CMG* 1, 1 45.26ff.   [111] *VM* ch. 14, *CMG* 1, 1 45.28ff.

[112] Especially *VM* ch. 19, *CMG* 1, 1 50.7ff, on which see above, ch. 1, p. 54.

[113] Thus in ch. 19 (*CMG* 1, 1 50.2ff) he observes that hoarseness, sore throats, erysipelas and pneumonia are accompanied by salty, watery and pungent discharges, and that when these discharges are 'concocted' the fever and the other pains cease, in an argument to show that the humours in question are instrumental in bringing about these conditions. But he obviously fails to counter the possibility that the discharges are merely the symptoms, not the causes, of both the onset, and the cessation, of the conditions. Cf. also a similar argument concerning the role of yellow bile at ch. 19, 50.14ff.

largely consist in the application of familiar substances and the control of diet.[114] Yet his own interpretation of the effective ingredients is, we might say, almost as arbitrary and dogmatic as that in terms of the hot, the cold, the wet and the dry.[115]

The difficulties the writer encountered, in attempting to bring evidence for his theories, emerge from two passages in particular. First in ch. 15, to support his views concerning the differences between the hot and astringent on the one hand, and the hot and insipid on the other, he says he 'knows their effects to be quite opposite' 'not only in man, but also in a leathern container or a wooden vessel'.[116] Then in ch. 24, having suggested that the powers of the various humours must be examined, he takes an example. What will the sweet humour change into, if it changes its own nature 'by itself', not by admixture with something else? Will it first become bitter, or salty, or astringent, or acidic? He answers his own question by saying 'the acidic, I think', and again goes on to refer to the possibility of studying such changes outside the body. 'If a man could light upon the truth by searching outside the body, he would always be able to select what is best.'[117]

The writer of *On Ancient Medicine* has, then, not just a general ideal, that theories should be testable, but also a particular method of approach in physiology. Given that it was not feasible to investigate what goes on inside the body directly, he has an alternative procedure to propose, that of studying the changes that take place in substances outside the body and drawing inferences – by analogy – concerning what happens inside it. Yet the gap between theory and practice is still wide. If he did indeed conduct investigations of the type he refers to in chh. 15 and 24, he does not report them.[118] Moreover, *even if he had carried them out*, we may wonder what they

---

[114] See chh. 13ff, especially ch. 15, *CMG* I, 1 46.22ff, and cf. ch. 5, 39.21ff.

[115] Nor is it clear precisely how he would define 'strong' and 'weak', although no doubt in practice there was a fair measure of agreement about what counted as 'strong' foods (e.g. ch. 6, *CMG* I, 1 39.27ff). One of his objections to 'the hot' 'the cold' 'the wet' and 'the dry' is that those who appeal to them are unable to point to them existing in a pure state (ch. 15, 46.20ff): but it was obviously open to his opponents to object that the same might also be said concerning his own 'salty' 'bitter' 'sweet' 'astringent' and 'insipid'. Again, he argues that if the hot is what causes pain, then relief should be procured by the cold (ch. 13, 44.18ff): but while he asserts that conditions are relieved when, for example, the salty is counteracted and concocted, how far that was always true might be questioned.

[116] *VM* ch. 15, *CMG* I, 1 47.5ff.

[117] *VM* ch. 24, *CMG* I, 1 55.4ff. Cf. also the method recommended in studying the effect of different structures in the body, ch. 22, 53.12ff, see below, pp. 158f.

[118] In ch. 15, the verb οἶδα, 47.6, may suggest past general experience, and in ch. 24 the optatives in the conditional sentence (55.12) might more naturally refer to hypothetical, than to an actual, investigations.

would have revealed. The writer's insistence on excluding arbitrary postulates from medicine, admirable as it is as an expression of the need to challenge assumptions, represents a quite impracticable ideal. In practice, the conceptual framework of his own theories is not much less purely speculative than that of those of his opponents, and this would clearly have been the major limiting factor to the usefulness of the type of empirical investigations he envisaged.

The point can be further illustrated by reference to other treatises that propose general physiological or pathological doctrines. In several cases, to be sure, the empirical support for such doctrines is negligible. This is true not only of such sophistic works as *On Breaths*,[119] but also, for instance, of the treatise *On Regimen*. The writer first states that a knowledge of the constitution of man is necessary for an understanding of dietetics,[120] and then advances the view that the elements of which the body is composed are fire (which is hot and dry) and water (cold and wet).[121] But although the bulk of the treatise contains a detailed account of, for example, the effects of different foods, which, while being largely theoretical and schematic, appears also to draw on experience and perhaps even the writer's own first-hand observations,[122] the general physiological theory of the opening chapters of book I is asserted as baldly and dogmatically as any doctrine attributed to the Presocratic philosophers.[123] Chapter 3, in fact, contains neither evidence nor argument to establish its dualist element theory.

Yet two other treatises make more serious attempts to offer evidence in support of general physiological doctrines. As we have already remarked, *On the Nature of Man* not only refutes monistic element theories whether proposed by philosophers or doctors, but also promises to prove its own four-humour doctrine.[124] The main

---

[119] Having promised to show that air is the cause of every kind of disease (ch. 5, *CMG* I, I 94.6f, cf. above, p. 88 n. 151), the writer goes through some of the more obvious pathological conditions first establishing that air plays *some* role, and then claiming that it is the cause. Thus in chh. 6–8 (94.8ff) he argues that there are two kinds of fever, epidemic and sporadic. Epidemic fevers are caused by the air we breathe (they are common because all men breathe the same air). Sporadic fevers come from food, but with the food we eat we also take in much air (as is 'clear' from belching, 95.6ff), and so air is responsible for these fevers too.

[120] *Vict.* I ch. 2, L VI 468.6ff.　　　　[121] *Vict.* I chh. 3 and 4, L VI 472.12ff, 474.8ff.

[122] See especially *Vict.* II chh. 39ff, L VI 534.17ff.

[123] *Vict.* I chh. 4 and 5 (L VI 474.8ff), which contain several almost verbatim quotations from Heraclitus, Parmenides and Anaxagoras, show that the writer is well aware of, and has been influenced by, the work of these philosophers.

[124] See above, pp. 93f. In addition to the vocabulary of 'demonstration' and 'necessity' (ἀποδείκνυμι, ἀποφαίνω, ἀνάγκη), discussed above, p. 103 n. 244, the writer makes extensive use of that of 'evidence': see, for example, τεκμήριον at ch. 1, L VI 32.13, ch. 2, 36.15, ch. 7, 46.11, and μαρτύριον at ch. 1, 32.13, ch. 6, 44.10, ch. 7, 50.9.

evidence the writer adduces relates to the effects of drugs, which he goes into in some detail. He first mentions that there are drugs that draw out phlegm, and others that purge bile and black bile,[125] and then points out, against the monists, that when a man dies from excessive purgation he does not vomit just one single humour, but when, for example, he has 'drunk a drug which withdraws bile, he first vomits bile, but then also phlegm, and then in addition to these, under the constraint of necessity, also black bile, and they end up vomiting pure blood and that is how they die'.[126] Finally he argues that the proportions of the humours vary in the body according to the seasons,[127] and for this too he suggests 'a most clear testimony', namely that 'if you will give the same man the same drug four times in the year, his vomit will be most phlegmatic in winter, most liquid in spring, most bilious in summer and blackest in autumn'.[128] While the evidence cited reveals the presence of certain substances in the body clearly enough, the writer asserts that they must be congenital merely on the grounds that they can always be found in the body and the drugs always have the same effect.[129] Moreover, apart from the doubtfulness of that inference, the writer assumes, what he needs to prove, that the humours in question are the *sole* constituents of the body.

Then a second treatise that marshals evidence in connection with a doctrine concerning the constitution of the human body is *On Fleshes*. After putting forward a version of the four-element theory in ch. 2,[130] the writer gives an account of the formation of the parts of the body in which 'the glutinous' (associated with the cold) and 'the fatty' (associated with the hot) play the chief role.[131] As 'clear evidences' (τεκμήρια...σαφέα) of the distinction between these two he proposes a simple test, that of cooking the different parts of the body, when, he says, it will be found that the 'glutinous' and 'sinewy' parts do not cook easily, while the 'fatty' parts do.[132]

---

[125] *Nat. Hom.* ch. 5, L vi 42.8ff.
[126] *Nat. Hom.* ch. 6, L vi 44.11ff, cf. 15ff.
[127] *Nat. Hom.* ch. 7, L vi 46.17ff, 48.10ff.
[128] *Nat. Hom.* ch. 7, L vi 50.9ff.
[129] *Nat. Hom.* ch. 5, L vi 42.18ff. It is true that the first reference to the effect of drugs, in that chapter, is made primarily to establish the *differences* between the humours (42.8ff): yet no other proof is offered to support the assertion that man *is* blood, phlegm, yellow bile and black bile, 'both by convention and according to nature', let alone the claim that this had been *shown*.
[130] *Carn.* ch. 2, L viii 584.9ff.
[131] *Carn.* chh. 3ff, especially ch. 4, L viii 588.14ff.
[132] *Carn.* ch. 4, L viii 588.25ff. In later chapters the author conducts simple tests with blood, showing that it does not coagulate when shaken, though it does when it is allowed to rest and goes cold (ch. 8, 594.14ff: this is taken to support a theory that the

As in *On the Nature of Man*, the bid to muster evidential support for a general physiological theory is apparent. Even though the question of the fundamental constituents of the body was both highly obscure and much disputed, both treatises have suggestions to make about how aspects of the problem may be investigated empirically. Now whether, or how far, the tests described were actually carried out is again (as with *On Ancient Medicine*) problematic.[133] Yet even if they had been conducted systematically, they would have fallen far short of establishing the general theories in connection with which they were suggested. In both cases the results of the tests were interpreted solely in terms of the author's own assumptions – when what was needed was for those assumptions to be *tested against others*. The function of these tests was to *corroborate* the theories in question, not to provide data to *decide between* competing theories.

Where the medical writers attempted to conduct empirical research in relation to such problems as the fundamental constituents of the body, their efforts met with little success, although their comparative ineffectiveness should not lead us to underestimate the significance of the fact that the efforts were made at all. The drawback, in this field of inquiry, was that their investigations were not open-ended, but designed specifically to provide support for theories that appear to have been adopted usually on the basis of general, often philosophical, considerations and arguments. It is, however, rather in relation to other aspects of their work that the Hippocratic writers have been celebrated for their championship of empirical methods, particularly in connection with clinical observations and prognosis.[134]

First we have two extended accounts, in *Prognosis* and the first book of the *Epidemics*, that show that these authors, at least, held that prognosis should be based on a very thorough examination of the patient. In *Prognosis*, which is particularly concerned with 'acute'

liver is formed from the blood 'when the cold defeats the warm', 594.12ff), and that if the 'skin' is removed from coagulating blood, another one is formed (ch. 9, 596.9ff, this being taken to show that the skin of the body itself is formed from the blood under the effects of the cold and the winds). This writer too appeals to visible effects outside the body to explain processes within it.

133 In *Nat. Hom.* ch. 5, L vi 42.10ff, the protasis of the conditional referring to the first test is introduced by ἤν, with the verb in the subjunctive (the main verb is present indicative): in *Carn.* ch. 4, L viii 590.1ff, the protasis is introduced with εἰ with the optative, though the main verbs are, again, present indicatives.

134 Where a modern doctor concentrates his attention on diagnosis, the ancient practitioner focused, rather, on 'prognosis', but this covered the past and present, as well as the future, of the disease. *Prog.* ch. 1, L ii 110.2ff, shows that apart from predicting the outcome of the disease, the doctor aimed to inform his patients also on their present and past condition.

diseases, that is those accompanied by high fever, such as pneumonia, the writer gives detailed instructions about how the doctor should proceed. First he should examine the patient's face, for example the colour and texture of the skin, and especially the eyes, where he should study whether 'they avoid the glare of light, or weep involuntarily', whether 'the whites are livid', whether the eyes 'wander, or project, or are deeply sunken', and so on.[135] He should also inquire[136] how the patient slept, about his bowels and his appetite: he should take into account the patient's posture, his breathing and the temperature[137] of the head, hands and feet, and separate chapters[138] are devoted to how to interpret the signs to be found in the patient's stools, urine, vomit and sputum.[139] Thus he has this to say on the second of these:

> urine is best when there is a white, smooth, even deposit in it the whole time up to the crisis of the disease...Sediment like barley-meal in the urine is bad...Thin white sediment is a very bad sign, and it is even worse if it resembles bran...So long as the urine is thin and yellowish-red, the disease is not ripened...When a patient continues to pass thin raw urine for a long time and the other signs indicate recovery, the formation of an abscess should be expected in the parts below the diaphragm. When grease forms patterns like cobwebs on the surface of the urine, this constitutes a warning, for it is a sign of wasting...[140]

Our second, more summary, account of the factors the doctor should consider in prognosis is in *Epidemics* I ch. 10. This begins by noting that 'the nature of man in general and of each individual, and the characteristics of each disease' should be learned and then proceeds:

> Then we must consider the patient, what food is given to him and who gives it... the conditions of climate and locality both in general and in particular, the

---

[135] *Prog.* ch. 2, L II 112.12ff, the famous description of the so-called Hippocratic 'facies'.
[136] *Prog.* chh. 3ff, L II 118.7ff.
[137] Judged, no doubt, simply by touch, *Prog.* ch. 9, L II 132.6ff. Cf. *Nat. Hom.* ch. 5, L VI 42.3ff, and ch. 7, 46.11ff, *VM* ch. 18, *CMG* I, 1 49.10f, *de Arte* ch. 9, *CMG* I, 1 15.9ff, *Epid.* III case 15 of the second series, L III 142.11f (cf. cases 6 and 8 of the first series and 5 of the second, L III 50.9, 56.9, 118.11).
[138] *Prog.* chh. 11–14, L II 134.13–146.15.
[139] The most notable absentee from the list of things the doctor should consider is the pulse. Although the phenomena of pulsation, throbbing and palpitation are referred to by Hippocratic writers, the value of the pulse in diagnosis was not appreciated until after the date of most of the Hippocratic treatises. The first person to restrict the pulse to a distinct group of vessels, the arteries, and to recognise that it can be used as an indicator of disease, was probably Praxagoras of Cos, working around 300 B.C. Auscultation, applying the ear to the chest, though not referred to in *Prog.*, is clearly described elsewhere in the Hippocratic Corpus, e.g. *Morb.* II ch. 61, L VII 94.16f, *Morb.* III ch. 16, L VII 152.21f.
[140] *Prog.* ch. 12, L II 138.15ff (trans. Chadwick–Mann 1978). It is notable that the writer ends: 'You must not be deceived if these appearances result merely from a diseased condition of the bladder, for they may then indicate not a disease of the whole body, but merely of that organ' (142.12ff).

patient's customs, mode of life, pursuits and age. Then we must consider his speech, his mannerisms, his silences, his thoughts, his habits of sleep or wakefulness and his dreams, their nature and time.[141]Next we must note whether he plucks his hair, scratches or weeps. We must observe his paroxysms, his stools, urine, sputum and vomit. We look for any change in the state of the malady, how often such changes occur and their nature, and the particular changes which induce death or a crisis. Observe, too, sweating, shivering, chill, cough, sneezing, hiccough, the kind of breathing, belching, wind, whether silent or noisy, haemorrhages and haemorrhoids. We must determine the significance of all these signs.[142]

These accounts of how to conduct a clinical examination display a remarkable appreciation of the variety of points to be considered, and an acute sense of the need for thoroughness and attention to detail. Moreover the principles they set out were not just idealised recommendations, but – sometimes, at least – closely followed in practice. Apart from the so-called 'constitutions' (general descriptions of the outbreaks of diseases concentrating on such matters as the climatic conditions), the first and third books of the *Epidemics* contain a total of forty-two individual case-histories. These are certainly not the first extant clinical reports in the history of medicine: they are anticipated by more than a thousand years in Egypt by the famous series of surgical cases in the Edwin Smith papyrus.[143] But whereas the Edwin Smith papyrus limits itself to brief notes under the five headings of title, examination, diagnosis, treatment and explanations of terms, the Hippocratic treatises engage in sustained and often much more detailed reports in which the progress of particular patients is recorded, generally day by day, over quite long periods. The entries under each day vary from a single remark to an elaborate description of some nine or ten lines, and in some cases occasional observations continue to be set down up to the 120th day from the onset of the illness. Thus, to illustrate from a single example, after briefer comments on the second and third days' symptoms, case 3 of the first series in *Epidemics* III proceeds:

Fourth day: vomited small quantities of yellow bilious matter and, after a while, a small quantity of rust-coloured material. There was a small haemorrhage of pure blood from the left side of the nose; stools and urine as before; sweating about the head and shoulders; spleen enlarged; pain in the region of the thigh: a rather

---

[141] What we should describe as psychological factors, such as, for example, despondency, are also referred to, for instance, in case 6 of the first series, and cases 2, 11 and 15 of the second, in *Epid.* III, L III 52.8f, 112.11f, 134.2ff, 142.6ff.

[142] *Epid.* I, section 3, ch. 10, L II 668.14ff (trans. Chadwick–Mann, 1978, ch. 23).

[143] This dates from around 1600 B.C., but incorporates material from a much earlier period, see Breasted 1930. The Edwin Smith papyrus is exceptional, so far as our extant evidence for the medical practice in *any* ancient Near Eastern civilisation is concerned, in being almost entirely free from magic and superstition.

flabby distension of the right hypochondrium; did not sleep at night; slight delirium.[144]

In this case observations continue daily till the twenty-first day, and further occasional entries are made up to the fortieth day when – exceptionally[145] – this patient reached a crisis and recovered.

These case-histories are undoubtedly one of the star examples of detailed observations in early Greek science. They were evidently carried out with great care and thoroughness, and they contain few interpretative comments and no explicit overall theory of disease. Yet the terms used in the descriptions are, of course, 'theory-laden', and they reveal certain assumptions concerning the nature and causes of diseases. Thus although these treatises do not propose any schematic doctrine of humours, such as we have in *On the Nature of Man,* they often refer – as our example does – to the 'bilious' or 'phlegmatic' matter in the patients' discharges.[146]

The primary aim of these case-histories is clearly to provide as exact a record as possible of the cases investigated. But we can and should be more precise than this as to the writers' motives.[147] First, the object of assembling the collections of case-histories was probably less to facilitate diagnosis, than to provide information which would help the doctor to predict the outcomes of diseases, especially whether the patient would die or recover.[148] Secondly, and more importantly, the writers have, as is by now well known, a particular motive for carrying out and recording their observations *daily* – over and above the laudable desire to be thorough in their work – namely that they adhere to the common Greek medical theory that the course of acute diseases is determined by what were called 'critical days', when marked changes take place in the patient's symptoms. Establishing the periodicity of the disease was crucial in diagnosing it as a 'quartan' a 'tertian' a 'semi-tertian' or even an 'irregular' fever, for instance. Moreover several case-histories draw attention to the fact that the pains or exacerbations occurred on the *even* days.[149]

---

[144] *Epid.* III, L III 38.7–44.8, especially 40.7ff.

[145] Of the 42 cases described, 25, or nearly 60%, end in death.

[146] But we may contrast the less 'theory-laden' terms, such as the reference to 'rust-coloured' material, ἰώδεα, in the discharges at L III 40.8.

[147] I say 'writers' because although *Epid.* I and III as we have them form a continuous whole, the observations they record may well be the work of several different practitioners.

[148] Similarly in *Prog.* the writer concentrates much of his attention on whether the 'signs' he identifies are favourable or unfavourable.

[149] E.g. *Epid.* I case 1, L II 684.9, *Epid.* III cases 3, 10 and 12 of the second series, L III 116.12f, 132.4f, 136.13. The symbolic associations that odd and even might have for some Greeks are clear from their inclusion in the Pythagorean Table of Opposites,

A concern to determine the pattern in the times of crisis and relapse is apparent in the generalisations in the third constitution in *Epidemics* I,[150] and in the final chapter of that constitution we find a detailed table of critical periods for diseases that had crises on even, and those that had crises on odd, days.[151]

It would evidently be quite mistaken to represent what motivated these writers as simply some general idea of the value of careful empirical investigations. At the same time these texts remain good evidence of a capacity for carrying out sustained and meticulous observations when there were particular reasons for so doing, that is when there was a general theory – in this case the doctrine of critical days – to prompt them. The data are collected and set out systematically and with care. Moreover although some of the conclusions in the constitutions take the form of sweeping generalisations concerning the periodicities of diseases,[152] many are explicitly qualified: the writers state what happened 'for the most part' or 'in the majority of cases' and exceptions are noted.[153] It is not the case that the writers conducted their observations merely to confirm *rules that they had already formulated in detail*. Rather those detailed rules are, in the main, generalisations which they arrived at on the basis of their particular observations[154] including, no doubt, many others besides those recorded in the case-histories as we have them.[155]

The case-histories in *Epidemics* I and III provide clear testimony of the systematic observations that were undertaken, in certain circum-

---

reported by Aristotle, *Metaph.* 986a22ff, where 'odd' is correlated with right, male, light and good, 'even' with left, female, darkness and evil.

[150] E.g. 'There was, however, a change in the periods at which the crisis occurred, it taking place usually on the fifth day from the beginning of the illness. A remission of four days would be followed by a relapse with the crisis on the fifth day...Most of those who behaved in this way were children, but it happened occasionally in adults. In some cases a crisis occurred on the eleventh day, a relapse on the fourteenth and the final crisis on the twentieth. But if shivering fits supervened about the twentieth day, the crisis took place on the fortieth' (*Epid.* 1 ch. 9, L II 666.11ff, ch. 22 Chadwick-Mann).   [151] *Epid.* I ch. 12, L II 678.5ff (ch. 26 Chadwick–Mann).

[152] See especially *Epid.* I ch. 12, L II 678.5ff (ch. 26 Chadwick–Mann).

[153] As, for example, at *Epid.* I ch. 9, L II 666.11ff, 668.7ff, quoted above, n. 150. Earlier in the same chapter, 660.6ff, the remark is made that the crises were sometimes similar, sometimes dissimilar, and this is followed by a reference to the cases of two brothers who fell ill at the same time, but had their crises on different days; cf. also *Epid.* I ch. 11, L II 674.14–676.10, III chh. 4 and 9, L III 70.14–76.4, 90.1ff, and *Prog.* ch. 20, L II 168.16ff, where the author observes that critical periods cannot be calculated in whole days any more than the solar year or the lunar month can: cf. also *Morb.* I ch. 16, L VI 168.23–170.8.

[154] Cf. Deichgräber 1933a, pp. 20f.

[155] Thus the constitutions refer by name to several individual patients for whom there are no corresponding case-histories in either *Epid.* I or III. On the disparities between the accounts in the constitutions and those in the case-histories, see Deichgräber 1933a, pp. 11ff.

stances, by some Hippocratic physicians. Yet if we turn to other subjects that they were interested in, the application of empirical techniques of research was, in certain respects, severely limited. One topic that offers a good opportunity to test this is the use of dissection.[156] The origin of this method has sometimes been traced back to Alcmaeon, that is to some time about the middle of the fifth century, but this is quite dubious.[157] Much of the account in our chief source, Chalcidius, must be taken to relate not to Alcmaeon, but to another much later anatomist whom he mentions, namely Herophilus, one of the foremost Alexandrian biologists who undoubtedly dissected, and may even have vivisected, not just animals, but also human subjects.[158] The most we can infer from Chalcidius (assuming that his report has some basis in fact) is that Alcmaeon excised, rather than dissected, the eye.[159] Our other sources for him, notably the account of his theory of the senses in Theophrastus, yield no evidence that justifies the conclusion that he dissected, and some grounds for believing that he did not.[160]

But while the method should not be thought to have originated as early as Alcmaeon, we have good evidence, nevertheless, that the idea of opening animals to conduct certain investigations had occurred to certain individuals in some contexts by the late fifth or early fourth century. First there is the text in *On the Sacred Disease* which we have already discussed, where the author mentions the possibility of carrying out a post-mortem dissection on the brain of certain animals – he specifies goats – in order to establish the causes

---

[156] What follows is largely based on Lloyd 1975a, to which I may refer for a detailed analysis of the evidence.

[157] Apart from Lloyd 1975a, see also Mansfeld 1975. On the date of Alcmaeon's work (more likely to be around 450 than early fifth or late sixth century), see Lloyd 1975a, p. 114.

[158] Cf. further below, pp. 165f.

[159] Of the three distinct investigations, (1) cutting out the eyeball, (2) dissecting the eye itself, and (3) cutting open the skull to investigate the structures communicating with the eye within the skull itself, Alcmaeon is much more likely to have undertaken (1) than either (2) or (3), that is his investigation was probably confined to the orbit, to reveal the structures leading off from the back of the eye towards the brain: see Lloyd 1975a, pp. 118ff on Chalcidius, *In Ti.* ch. 246, 256.16–257.15.

[160] Theophrastus, *Sens.* 25f, DK 24 A 5. This indicates that Alcmaeon was interested in the elements in the eye (fire and water) but does not suggest that he had any clear idea of its internal structure. It also gives details of his theory of pores or channels linking the senses with the brain, which he held to be the seat of consciousness. Yet *if* he conducted empirical investigations to support his theory, it is far more probable that he did so by using a probe, than by carrying out dissections in the strict sense. Indeed if he had explored the cranial cavity by dissection, it is surprising that he continued to hold, for example, that there are channels leading directly from the ear to the brain. On this and our other evidence (especially Aristotle, *HA* 492a14ff, *GA* 752b25ff, Aetius v 16.3, 17.3 and 24.1, and Censorinus, *de die nat.* ch. 5, 10.7ff), see Lloyd 1975a, pp. 121–8.

of the sacred disease.[161] Then Aristotle, who often himself refers to the results of dissections, whether carried out by himself or by his associates,[162] provides clear testimony to his predecessors' use of the method. At *HA* 511 b 13ff, especially, he criticises earlier work on the blood-vascular system and remarks on the difficulties of conducting observations: 'Those who have examined dead bodies[163] by dissection have not observed the principal sources of the blood-vessels, while those who have examined very emaciated living men have inferred the sources of the blood-vessels from what could then be seen externally.'[164] This shows both that dissection had been used by those who had attempted to describe the blood-vessels before Aristotle, and that it was not their only method, since some investigators apparently relied on observations of emaciated human subjects.

On the authority of these evidences we may accept that dissections were sometimes performed in the period before Aristotle. Yet the occasions when we can definitely confirm the use of the method for the purposes of a scientific inquiry in fifth- or fourth-century texts are rare,[165] despite the fact that many Hippocratic authors pronounce on anatomical and physiological questions to which dissection was, or might be thought to be, relevant. Thus we have already noted that the author of *On the Sacred Disease* maintains some highly fanciful notions on the courses of the 'veins' in the body, notions which were evidently neither based on, nor checked by, dissection.[166] Of our other main pre-Aristotelian accounts of the blood-vascular system, the three that Aristotle himself quotes and attributes to Syennesis of Cyprus, Diogenes of Apollonia and Polybus[167] bear out

---

[161] On *Morb. Sacr.* ch. 11 paras. 3–5 (G) L VI 382.6ff and e.g. Hdt. IV 58, see above, p. 24 and n. 79.   [162] See below, pp. 162ff.

[163] The context makes clear that he is talking of animals, not men.

[164] *HA* 511 b 20ff, cf. 496 a 11 and b 4ff.

[165] The infrequency of dissection for the purposes of research is all the more striking in that, in another context, divination by the inspection of entrails or haruspicy, animals were regularly opened and their parts examined, even if that examination was usually confined to an inspection of *surface* features of, for example, the liver (see e.g. Cicero, *Div.* II 12.28ff, cf. Plato, *Ti.* 71 a ff). But although we find occasional references to anatomical data learned from sacrificial victims in our 'scientific' authors (e.g. Aristotle, *HA* 496 b 24ff, *PA* 667 b 1ff, and cf. *Carn.* ch. 8 which refers to blood taken from sacrificial animals), the contrast in the context and aims of divination, and those of anatomical studies, were no doubt sufficiently marked to act as an effective barrier to communication. Cf. Rufus, *Onom.* 158.5ff, who notes the irrelevance of certain of the terms used in haruspicy from the point of view of medicine.

[166] See above, ch. 1, pp. 20–4.

[167] *HA* III chh. 2f, 511 b 23–513 a 7. The report of Syennesis corresponds to part of ch. 8 of the composite treatise *Oss.*, that of Polybus to the longer accounts in *Oss.* ch. 9 and *Nat. Hom.* ch. 11. On the development of notions of the blood-vascular system, see especially Harris 1973.

his general criticisms: such dissections as they carried out must have been quite cursory or superficial. An over-fondness for bilateral symmetry is a feature of all three accounts, and to judge from the references to blood-letting in the theories of Diogenes and of Polybus, their views seem to have been partly based on (and no doubt served in turn to justify) current practices in venesection.[168] It is true that some of the surgical treatises, such as *Fractures* and *Joints*, show a fair knowledge of surface anatomy and of osteology, but this was, in all probability, gained through the actual treatment of fractures, dislocations and wounds. There is nothing in those works to suggest that their authors attempted to increase their anatomical knowledge by dissecting animals or men.[169]

In other treatises the theories put forward concerning the inter-relations of the main organs in the body are bizarre. *On Regimen* I chh. 9f[170] has a doctrine of three 'circuits' in the body corresponding to the three circuits of the heavenly bodies. *On Diseases* IV ch. 39 imagines that the four main sources of the humours in the body, heart, head, spleen and liver, all communicate directly with the stomach.[171] Even where we have a reference, as in *On the Nature of the Child*, to the possibility of investigating the development of an embryo chick by incubating a batch of twenty or more hen's eggs and opening one each day, the author does not *in fact* set out the results of the inquiry he proposes in any detail but confines himself to a single observation concerning the membranes extending from the umbilicus.[172] When such treatises as *On the Art* and *On Ancient Medicine* discuss how to proceed to gain insight into the internal functionings of the body, neither author mentions dissection. *On the Art* suggests that nature may be made to yield information by the administration of certain foods and drinks, for example,[173] and the writer of *On Ancient Medicine*, having noted the difficulty of obtaining knowledge about the internal structures of the body, recommends his usual procedure of studying objects outside it.[174] The vagueness of his

---

[168] E.g. *HA* 512a30f (Diogenes), b17ff, 24ff (Polybus).

[169] *Art.* chh. 1 and 46 (L IV 80.1ff, 196.19ff) mentions surgical interventions that involve dissection, but does so merely as 'thought experiments' and with no intention of using such procedures in the treatments he is discussing.

[170] L VI 482.13ff, especially 486.3ff.

[171] He illustrates what he believes happens inside the body by referring to what he takes to be an analogue to it, namely the behaviour of water in a system of intercommunicating vessels, L VII 556.17ff.

[172] *Nat. Puer.* ch. 29, L VII 530.10ff. When Aristotle undertakes a similar investigation at *HA* 561a4ff, his account of his observations is much more detailed.

[173] *De Arte* ch. 12, *CMG* I, 1 18.3ff.

[174] *VM* ch. 22, *CMG* I, 1 53.1ff, 12ff: cf. above, p. 148.

classification of internal structures can be judged from the fact that the bladder, skull and womb are said to be similar, namely 'broad and tapering', like a cupping-glass.

But if the general impression given by most fifth- and early fourth-century Hippocratic works[175] is that their authors' knowledge of internal anatomy was extremely limited,[176] two treatises provide something of an exception. Both *On the Places in Man* and *On Fleshes* give quite detailed and in parts, at least, fairly accurate accounts of the sense-organs and the blood-vessels. Thus *On the Places in Man* distinguishes three membranes in the eye, a thick outer one, a thinner middle one and a third inner one 'that guards the moist element': the writer further identifies certain communications between the eye and the brain, and distinguishes two membranes of the brain itself, a thick outer one and a thin one that touches the brain.[177] In *On Fleshes* there is an account of the blood-vessels in which two main vessels communicating with the heart are clearly distinguished, the ἀρτηρίη (corresponding to the aorta and its branches) and the hollow φλέψ (vena cava),[178] and the organs of hearing, smell and sight are described with some care in separate chapters.[179]

We should certainly not rule out the possibility that these accounts were based, in part, upon dissection. Yet even here we cannot establish this for certain. It is striking that while both authors refer to the evidence to be gained from studying lesions,[180] neither does to any dissection. *On Fleshes* ch. 19 is particularly remarkable in that

---

[175] The one Hippocratic treatise that refers extensively to dissection is *Cord.*, which gives a brief but quite detailed account of the anatomy of the heart, including unmistakable references to two of the valves of the heart, where both the structure and function are correctly understood (chh. 10 and 12, L IX 86.13ff, 90.11ff). Yet first Abel 1958, and then Lonie 1973, have shown that this treatise is much later than most of the Hippocratic collection, being roughly contemporary with the work of Herophilus and Erasistratus (early third century B.C.). *Cord.* is evidence for dissection, but for its use in the third, not the fifth or fourth, century.

[176] A similar conclusion applies also to the natural philosophers. Diogenes' account of the blood-vessels has already been mentioned (p. 158). Empedocles' theories of vision and respiration (Frr. 84 and 100) are based on analogies and are quite unlikely to have been confirmed by dissection: though they include references to the membranes of the eye, for instance, those references are quite vague and imprecise. Although Democritus evidently had wide biological interests, and Theophrastus' report of his account of the eye suggests he distinguished between its 'outer membrane' and its inner parts (*Sens.* 49ff), none of our testimonies can be said to show conclusively that he had dissected (cf. Lloyd 1975a, pp. 131f). On Anaxagoras, see above, p. 24 n. 79 and below, p. 161, n. 185.   [177] *Loc. Hom.* ch. 2, L VI 278.14ff: cf. ch. 3, 280.10ff on the blood-vessels.

[178] *Carn.* chh. 5f, L VIII 590. 5ff.

[179] *Carn.* chh. 15–17, L VIII 602.19ff (see Lloyd 1975a, p. 136). Elsewhere, however, as we have noted (above, p. 150), *Carn.* proposes highly speculative physiological theories.

[180] *Loc. Hom.* ch. 2, L VI 280.5f notes that in an injury of the eye when the middle membrane is broken, it 'comes out like a bladder'. *Carn.* ch. 17, L VIII 606.10ff says 'we have often seen glutinous moisture oozing from an eye that has ruptured'.

the author mentions what indeed he claims he has often seen, namely an aborted human foetus. Yet although he describes what this is like 'if you put (it) in water and inspect it', there is no suggestion that the foetus might be dissected. His account is confined to surface features and there is no reference to any internal organ at all.[181]

The situation our analysis reveals poses an obvious question and one that will now take us beyond the Hippocratic writers themselves. Evidently it was not the case that, once dissection had begun to be used in particular contexts, the method came fairly rapidly to be applied generally to a wide range of anatomical topics, and we may well ask why this was so, while recognising that such suggestions as we can put forward on such an issue must be conjectural. First it is clear from the way in which Aristotle sets out to justify the investigation of the internal parts of animals[182] that some Greeks[183] felt a certain squeamishness on the matter. Indeed Aristotle himself writes that 'it is not possible to look at the constituent parts of human beings, such as blood, flesh, bones, blood-vessels and the like, without considerable distaste'.[184] Yet the importance of this factor should not be exaggerated. There can be no doubt that the general public felt a certain revulsion from the study of internal anatomy: nor is there anything exceptional about the ancients here, since the same holds true today. But how far that provides an answer to our question is hard to judge. After all, Aristotle himself was successful in overcoming his own expressed distaste so far as animals are concerned, and we have no reason to believe that as a general rule the medical practitioners (at least) would have been more inhibited than Aristotle in this regard.

A more fundamental factor concerns the assumptions that directed dissection and the problems it was brought to bear on. It is all too

---

[181] The foetus, he says (L VIII 610.6ff), 'comes out like flesh: if you put this flesh in water and inspect it, you will find that it has all the members, and the places for the eyes, and the ears and the limbs; the fingers of the hands, the legs, the feet and the toes, the genitals and all the rest of the body are clearly visible'. Cf. a similar procedure suggested at *HA* 583b14ff by Aristotle, whose account also refers only to external features: see below, p. 163 n. 194. A third text (*Nat. Puer.* ch. 13, L VII 488.22ff) purporting to describe aborted 'seed' six days after intercourse, including its membranes and what the author says looked to him like an umbilicus, is largely fanciful.

[182] *PA* I chh. 1 and 5, see below, pp. 163f.

[183] Although Aristotle probably has Platonists in particular in mind in the protreptic to research in *PA* I, the points he makes have a general application.

[184] *PA* 645a28ff. Although Aristotle refers to the parts of human beings in this text, his point is a general one. There were, no doubt, other, religious factors inhibiting the dissection of humans – though for all the ancients' respect for the dead, there are plenty of cases recorded both in the classical period and later where the corpses of enemies were desecrated. But we are concerned, at this point, purely with the dissection of animals.

easy for us to underestimate the difficulties the earliest researchers faced. For a dissection to be carried out successfully requires not only patience, attention to detail and practical skill, but also and more importantly a clear conception of what to look for. To advance beyond a merely superficial account of, say, the contents of the skull or the heart depends, above all, on an idea of what there is to find. In fact in two of the references to the possibility of opening the skull that we have considered[185] the aim was the strictly limited one of determining that an abnormal condition had a natural cause. But even if the investigators in question had attempted a more general examination, it is doubtful how much they would have seen, that is, how much of what they saw they would have understood. Again, after dissections had begun to be made on the heart, it was some time before the valves, for instance, came to be recognised *as such*, as we can document by a comparison between Aristotle's accounts and that in the treatise *On the Heart*.[186] In the earliest history of anatomy experience in dissection undoubtedly contributed to the development of both anatomical and physiological doctrines: but practice itself – the nature of the dissections that were carried out – always depended on, in the sense that it was prompted and guided by, theory.

Before the method was accepted as a more or less routine procedure of investigation, the immediate stimulus to undertake a dissection had to come from a problem – a phenomenon to be explained or a controversy to be resolved. One of the keys to the slow development (as *we* see it) of the use of dissection lies, indeed, in the nature of the problems the Greeks were interested in and the way those problems were formulated. Two contrasting examples illustrate this very clearly. One of the topics that both philosophers and medical writers tackled was sensation in general and the functioning of each of the special senses. Yet so far as sight was concerned, many of the earliest Greek investigators were primarily interested in the *elementary constituents* of the eye,[187] rather than in its structure. It was commonly

---

[185] *Morb. Sacr.* ch. 11, L VI 382.6ff, and Plutarch, *Pericles* ch. 6, mentioned above, p. 24 and n. 79. Even though the latter story, of Anaxagoras having the skull of a one-horned ram opened, may be apocryphal, it may be taken to illustrate the type of *exceptional* occasion when recourse might be had to dissection.

[186] Aristotle's principal discussions of the anatomy of the heart are in *HA* I 17, *HA* III 3, *PA* III 4 and *Somn. Vig.* 458a15ff. But while he mentions what he calls sinews (νεῦρα) in the heart at *HA* 496a13, 515a28ff, *PA* 666b13f, he does not identify these as valves. On the account in *Cord.*, where at least two valves are clearly described as such, L IX 86.13ff, 90.11ff, cf. above p. 159, n. 175.

[187] This is true of Alcmaeon (Theophrastus, *Sens.* 25f, above, p. 156 n. 160) and cf. Aristotle's account of his predecessors (especially Empedocles, Democritus and Plato), and his own view, in *Sens.* 2, 437a19ff.

assumed that vision was to be explained in terms of some interaction (such as that of 'like to like' or between unlikes)[188] between fundamental physical elements such as earth, water, air or fire. Yet with the possible exception of 'water' none of these elements could be directly observed in the eye. Some theorists simply assumed, while others inferred,[189] their presence in the eye, but the common overriding concern was to bring their explanations of sight into line with their general physical and psychological doctrines. In that connection the accounts that were given of the functioning of the eye were often not capable in practice, indeed in some cases not even in principle,[190] of direct verification.

The contrast with the controversy concerning the seat of sensation in general is illuminating. Here too there were plenty of theorists who were content with purely speculative accounts. Yet in this case there was a clearer opportunity for advocates of the two main views – that the common sensorium is the heart, or that it is the brain – to support their ideas by direct reference to visible structures in the body. Some early investigators did just that: yet for the purposes of that controversy all that was *necessary* was to show *some* connection between sense-organ and common sensorium, and in several cases it appears that that was indeed where their demonstrations ended.[191] Just as with the valves of the heart, the discovery of the nerves *as such* had to wait for the right questions to be pressed, and in both cases this did not happen until after Aristotle.[192]

The way the problems were formulated is relevant also to the scope of the use of dissection on the second main topic where the method was employed, namely the blood-vascular system. Here too some of the early accounts were far from being undertaken for purely descriptive purposes, since venesection was clearly a practical concern for the physicians. For those for whom blood-letting was an important therapeutic procedure the question was to find links between surface blood-vessels and deep structures in the body. However,

---

[188] See Theophrastus, *Sens.* 1ff, where he lists Parmenides, Empedocles and Plato as having held that sensation comes about by the like, and the followers of Anaxagoras and Heraclitus as having held that it comes about by the opposite.

[189] As Alcmaeon inferred the presence of fire from what happens when the eye is struck, Theophrastus, *Sens.* 26, DK 24 A 5.

[190] Thus for the atomists, the ultimate explanation of sensation lies in the interaction of atoms that differ in shape, size and position but are not, in principle, observable. Even in Empedocles' analogy of the lantern (Fr. 84), the pores in the eye and in the lantern itself are below the level of what can be directly observed.

[191] This appears true not only of Alcmaeon (above, p. 156 n. 160) but also of Aristotle, who asserted that the sense-organs are connected with the heart, e.g. *Juv.* 469a 10ff; see Lloyd 1978a, pp. 222ff.

[192] See further below, p. 165.

although this generated interest in the subject of the paths of the blood-vessels, it may also have acted as an inhibiting factor, in so far as current practices in venesection may themselves have been taken as evidence enough for vessels linking different parts of the body.

Yet if a powerful specific motive was a necessary, it was not a sufficient, condition for recourse to dissection. Although on several topics where we might have expected dissection, we find that the problems were framed in such a way that the method was incapable of resolving the issue, there are many other occasions when opportunities to dissect were not taken.[193] It was not until Aristotle that the method came to be applied fairly generally, even though it was still confined to animals,[194] and much of his work can still be criticised as superficial.[195] Yet Aristotle offers something that is missing from the extant remains of any earlier writer, namely a statement setting out what is to be gained from an inquiry into the animal kingdom which provides a rationale for the *method in general*. In *PA* I 5 he says that animals are inferior to the heavenly bodies as objects of study in that the latter are unchanging, while the former belong to the world of change. But animals have the advantage in that 'we have much better means of obtaining information' about them and 'anyone who is willing to take sufficient trouble can learn a great deal concerning each of their kinds [i.e. animals and plants]'.[196] The investigation is directed to the four types of causes, especially form and finality, and involves much else besides dissection. But his

[193] Thus one of the controversies that goes back to Alcmaeon and seems positively to invite the use of dissection was the question of which part of the embryo is formed first. Censorinus (*de die nat.* ch. 6, 10.9ff, cf. Aet. v 17.1–6) cites the theories of Empedocles, Hippon, Anaxagoras, Diogenes of Apollonia, Democritus, Aristotle and Epicurus on this. Yet there is no evidence that any of these, with the sole exception of Aristotle, tried to determine the answer with the help of dissection – even though, as we have noted (p. 158), an investigation of hen's eggs is proposed in another context in *Nat. Puer.*

[194] *HA* 494 b 21ff shows that the possibility of dissecting a human body did not occur to Aristotle: 'The inner parts of man are for the most part unknown, and so we must refer to the parts of other animals which those of man resemble, and examine them.' (Cf. also *HA* 511 b 13ff, 513 a 12ff.) Some commentators, however, have suggested that he may have dissected a human embryo (Ogle 1882, p. 149, Shaw 1972, pp. 366ff). Yet despite his considerable knowledge of mammalian embryos, this seems doubtful. One striking passage that tells against the suggestion is *HA* 583 b 14ff, where he records what happens when a male human embryo, aborted on the fortieth day from conception, is put into water. It holds together, he says, in a sort of membrane, and if this membrane is pulled to bits, the embryo is revealed inside. On this occasion, at least, he appears not to have proceeded to a dissection – any more than the author of *Carn.* did, see above, pp. 159f – since his subsequent brief remarks are confined to surface anatomical points.

[195] I have argued in my 1978a, p. 216ff, that this is true in particular in connection with his psychological doctrines.

[196] *PA* I 5, 644 b 22ff, especially b 28ff.

protreptic not only justifies, but obliges him to undertake, the study of every kind of animal,[197] including their internal as well as their external parts, and their various vital functions.

With Aristotle, dissection can be said to become, for the first time, an integral part of a general programme of research with well defined aims. The exact extent to which he and his co-workers used the method is impossible to judge, but apart from the many more or less detailed reports that give information that could only have been obtained from dissections in his zoological works, he refers to a separate treatise (now lost) on *Dissections*.[198] In several contexts he mentions the difficulties of carrying out dissections and the mistakes that may arise from carelessness.[199] At *HA* 496 a 9ff, for instance, when he notes that 'in all animals...the apex of the heart points forwards', he adds: 'although this may very likely escape notice because of a change of position while they are being dissected'. Again in his account of the male generative organs of viviparous land-animals in general he remarks that the membrane now known as the *tunica vaginalis* must be cut to reveal the relation between the ducts it encloses: 'The ducts that bend back again and those that lie alongside the testicle are enclosed in one and the same membrane, so that they appear to be one duct, unless the membrane is cut open' (*HA* 510 a 21ff). The accuracy of his descriptions of some of the internal organs of a wide variety of animals (the first time any such systematic study had been attempted) is testimony to the effective use made of the method by him and by those who worked with him in the Lyceum.[200]

Yet despite his successes, Aristotle's dissections are, in certain respects, still primitive and crude compared with those of his successors. Although as a first approximation his descriptions of the principal anatomical features of the main groups of animals he identified are admirable, his account of certain organs are quite vague and obscure. To turn from his descriptions of, for example, the eye,[201]

---

[197] See especially *PA* 645 a 6f and 21–3.

[198] This appears to have contained, and may even have consisted in, anatomical diagrams: see e.g. *HA* 497 a 32, 525 a 8f, 566 a 14f, *GA* 746 a 14f, *PA* 684 b 4f.

[199] *PA* 676 b 33f is one passage that draws attention to the dangers of generalising concerning the whole species or group on the basis of observations of one or a few specimens.

[200] But see further below, pp. 213ff on certain limitations to his inquiry.

[201] The three main parts he usually identifies in the eye (e.g. *HA* 491 b 20ff) are the pupil, the 'so-called black' (i.e. iris) and the white (i.e. the visible white surrounding to the iris). There are references to a membranous coat at *de An.* 420 a 14ff, *Sens.* 438 b 2, *GA* 780 a 26ff and 781 a 20, but no attempt is made to identify its separate components. On the problems of interpretation posed by his references to communications leading from the eye to the brain (e.g. *HA* 495 a 11ff, *Sens.* 438 b 13f), see Lloyd 1978a, pp. 219f.

the brain,[202] and even the heart,[203] to the work of Herophilus and Erasistratus – difficult as this is to reconstruct[204] – is to enter a new world. Again, where Aristotle, like all earlier investigators, had been content with only a very general account of the transmission of movement and sensation, the Hellenistic biologists began to draw fundamental distinctions both between the different kinds of nerve (sensory and motor) and between the nerves proper and other tissues that had also been called νεῦρον.[205]

In part the achievements of Herophilus and Erasistratus are attributable to their having dissected – they may even have vivi-sected – humans as well as animal subjects.[206] Yet that is certainly not the only, and it may not even have been the most important, consideration. What marks out their work is, quite simply, the comparative detail and precision of their descriptions, and the way in which the inquiry is pressed home, whatever the subject under dissection. At this stage in the history of anatomy major advances could be, and no doubt were, made even without recourse to human dissection, as we can illustrate by reference to the approximately contemporary treatise *On the Heart*. Although we have it on Galen's authority[207] that it was Erasistratus who discovered the valves of the heart – that is, he was the first to recognise their true function – our first extant account of the semi-lunar valves at the base of the aorta and the pulmonary artery is, as we have noted, in *On the Heart*. But while this author refers clearly to both dissections and vivisec-tions, his investigations were evidently carried out on animal sub-

[202] Thus he asserts that the brain itself is bloodless and devoid of 'veins' (e.g. *HA* 495a4ff, 514a18ff, *PA* 652a35ff, but contrast *PA* 652b27ff and *Sens.* 444a10ff), and that the back of the skull is empty (*HA* 491a34ff, 494b33ff, *PA* 656b12ff). Although at *HA* 494b29ff he distinguishes two membranes round the brain, at *PA* 652b30 and *GA* 744a10 he speaks of a single membrane.

[203] One of the chief puzzles relates to the idea that the heart has three chambers, a doctrine that persists in all four of his main accounts (see above, p. 161 n. 186) despite their other divergences: see most recently Shaw 1972, pp. 355ff, Harris 1973, pp. 121ff. It is particularly striking that the notion that the central chamber is the ἀρχή for the other two persists even when his views on the identity of the three cavities changed (see Lloyd 1978a, pp. 227f, and cf. Byl 1968, pp. 467ff).

[204] None of the works of the major Hellenistic biologists has survived: we rely on the fortunately often extended quotations in such writers as Rufus, Celsus and, especially, Galen.

[205] See Rufus, *Anat.* 184.15–185.7. Galen, while recognising the importance of Hero-philus' work on the nerves, criticises it nevertheless on the grounds of incompleteness, K VIII 212.13ff, cf. VII 605.7ff: Erasistratus' interest in the problem of the origin of the nerves is clear, for example, from K XVIII A 86.11ff. The classic account of the discovery of the nerves is Solmsen 1961.

[206] Celsus, *De Medicina*, Proem 23ff is our chief evidence. Although the veracity of his testimony has often been impugned, there seems no good reason to reject it.

[207] See Galen K V 548.8ff, especially 549.5ff, cf. *Nat. Fac.* II 1, *Scr. Min.* III 156.24ff Helmreich, K II 77.4ff, K V 166.10ff.

jects.[208] In fact the number and extent of human dissections undertaken by Herophilus and Erasistratus may well not have been very great.[209] Yet Herophilus was responsible for, among other things, the first clear description of the four membranes in the eye,[210] for the identification of the ovaries,[211] and an impressive list of other anatomical and physiological discoveries,[212] while Galen credits Erasistratus with what was probably the first clear account of the several ventricles of the brain.[213]

Evidently the opportunities presented by the use of dissection only began to be fully exploited in the Hellenistic period. Once they were, and thanks to the fact that they were, dramatic progress was made not only in descriptive anatomy, but also in physiology, in, for example, the investigations of the blood-vascular, the nervous and the digestive systems. In each of these three areas, several of the principal problems that occupied later physiologists not just in antiquity but again in the early Renaissance and indeed in some cases down to Harvey and beyond, were first formulated either by the Hellenistic biologists or as a direct result of their work.[214] Yet although dissection continued, its use remained controversial, not only among the lay public but also in the medical profession. Celsus reveals that one of the three chief medical sects of the late Hellenistic period, the Empiricists, rejected both vivisection and dissection as irrelevant and superfluous for medical practice,[215] and texts in Rufus

---

[208] E.g. *Cord.* ch. 2, L ix 80.13ff describes a vivisection on a pig to show (as the writer claims) that drink goes to the lungs, cf. also ch. 8, 86.5f. Though the dead creature is not specified at ch. 10, 88.3ff, it was no doubt an animal: cf. the clear reference to the dissection of a slaughtered animal at ch. 11, 90.5ff.

[209] Although we have good evidence of observation of human subjects in, for example, Herophilus' naming of the duodenum from its length in man (twelve finger's breadths, see below, n. 212), in his comparisons between human and animal livers (Galen, K ii 570.10–571.4) and in Erasistratus' between human and animal brains (*UP* i 488.14ff Helmreich, K iii 673.9ff, K v 603.9ff), those very comparisons suggest, what we would in any case have expected, that much of their work was done with animals.

[210] See Rufus, *Onom.* 154.1ff, *Anat.* 170.9ff, Celsus, *De Medicina* vii 7.13.

[211] See Galen K iv 596.4ff, which gives an extended quotation from the third book of Herophilus' work *On Dissections*.

[212] E.g. of the duodenum (Galen K viii 396.6f), the torcular Herophili (*UP* ii 19.6f Helmreich, K iii 708.14f), the calamus scriptorius (K ii 731.6ff) and the prostate glands (*UP* ii 321.8ff Helmreich, K iv 190.2ff).

[213] See Galen K v 602.18ff, especially 604.6ff, and cf. *UP* i 488.14ff Helmreich, K iii 673.9ff. Another text in Galen (K ii 649.5ff) indicates that Erasistratus may well have been the first to describe the vasa chylifera in the mesentery.

[214] This applies, for example, to the question of the movements of blood – and, as some maintained, of air – in the heart and arterial and venous systems, to the tracing of nerves controlling the various vital functions, and to the debate on how far the processes of digestion could be explained in purely mechanical terms.

[215] *De Medicina* Proem 27ff. The Empiricists argued that the inquiry about obscure causes and natural actions is superfluous because nature cannot be comprehended. Rejecting the use of reasoning partly on the grounds that in theorising it is always

and in Galen show that human dissection, if it did not quite die out completely, was severely restricted.[216] Nevertheless, as Galen himself shows, the method continued in some quarters at least to be practised extensively on animals, indeed sometimes with great skill and sophistication. Much as he owed to his distinguished predecessors, Galen's own experimental dissections and vivisections in connection with the digestive and nervous systems[217] are the high-water-mark of the use of the method in the ancient world.

The history of dissection is one in which an empirical technique was eventually applied with great success to a wide range of problems in anatomy and physiology. Yet it also clearly shows how hard won those successes were. The slow development – as we see it – of the method cannot be put down simply to a reluctance to engage in empirical research, or even to a squeamishness about looking at some unpleasant objects. Inhibitions about opening human bodies were an obstacle, but generally a minor one compared with some other assumptions that were made concerning the use of the method. The fruitful exploitation of dissection depended on a complex interaction of theories and observations. Initially applied – before Aristotle – only to a few quite specific questions, the method had first to be recognised, as it was by Aristotle himself, to be one of quite general value and applicability. But that was only the first step. The next was to use it as a far more open-ended tool of research,[218] under-

possible to argue either side of a question, they held that practical experience of treatments is the sole source of medical knowledge. There is no need to inquire how we breathe, but only what relieves laboured breathing, no need to find out what moves the blood-vessels, only what the various types of movement signify (para. 39). Dissection is superfluous: and vivisection should be rejected on the further grounds that it is cruel (paras. 40ff).

216 Rufus, who was active around A.D. 100, indicates that he worked with animal subjects though he contrasts this with earlier anatomical demonstrations on humans, which still clearly represented the ideal, if an impracticable one (*Onom.* 134.10ff). Galen shows that human subjects were still used at Alexandria in his day (second century A.D.) in the teaching of osteology (*De Anat. Admin.* 1 2, K 11 220.14ff), though otherwise the opportunities for human dissection were rare, and one passage (*Mixt.* 11 ch. 6, 77.13f Helmreich, K 1 632.5ff) rules out human vivisection. He writes of observations of a corpse from a grave exposed in a river flood, of the skeleton of a robber who had been killed and whose body had been left unburied, and of the dissection of the body of a dead German enemy (see *De Anat. Admin.* 1 2 and 111 5, K 11 221.4ff, 14ff, 385.5ff), though most of his own work was evidently done on animals.

217 See especially *Nat. Fac.* 111 ch. 4 (*Scr. Min.* 111 213.11ff Helmreich, K 11 155.6ff) and *De Anat. Admin.* 1x 13f (Duckworth 1962, pp. 20ff).

218 It is no mere coincidence that an extended quotation from Erasistratus in Galen (*Consuet.* ch. 1, *Scr. Min.* 11 16.5ff Müller) provides us with an eloquent statement of the need for determination and persistence in research: 'Those who are completely unused to inquiry are, in the first exercise of their mind, blinded and dazed and straight-way leave off the inquiry from mental fatigue and an incapacity that is no less than that of those who enter races without being used to them. But the man who is used to

taking dissections not merely to substantiate a view on an ante-
cedently conceived problem, but in the recognition that the problems
themselves might need to be redefined in the light of what the
preliminary explorations suggested. Dissection was always guided by
theories and assumptions: but once one of those assumptions was a
realisation that the problems themselves might be more complex than
anticipated, the investigator was more open to the unexpected. Used
in such a way, dissection could and did generate what were effectively
new problems, which in turn – as we see most notably in the history
of work on the nervous system – generated new programmes of
research.

The study of dissection has taken us far away from the Hippocratic
physicians, but this digression helps to set their achievements in the
domain of empirical research into perspective. The claim that they
were, or were among, the founders of the empirical method in Greek
science can be upheld only if certain all-important reservations and
qualifications are added. It is true in respect of the meticulous clinical
observations that some of the doctors undertook, but there were
particular reasons for their sustained inquiries in that area, and their
successes there were not matched in other fields. Several writers,
especially the author of *On Ancient Medicine*, made important contributions to the methodological debate, criticising the use of arbitrary
assumptions, insisting on the need for theories to be testable, on the
importance of established methods of discovery in medicine and on
the use of what is apparent as the vision of the obscure. Yet the gap
between stated ideal and actual practice was often wide. It is true
that alongside some theorists who make little attempt to collect
detailed evidence, there were others who exercised considerable
ingenuity in seeking empirical support for their general pathological
and physiological doctrines. Yet the claims they made – that their
evidence clearly demonstrated their views – were generally excessive.
Such empirical data as they appealed to – whether observations or
simple tests – were regularly used to support their theories, not to
decide between them and their rivals. While the ambition to marshal
evidence is clear – and this distinguishes some of the medical writers
from many of the philosophers – most of the fundamental issues they
tackled were not to be resolved by the straightforward means they
brought to bear. Meanwhile the limitations of their performance as

inquiry tries every possible loophole as he conducts his mental search and turns in
every direction and so far from giving up the inquiry in the space of a day, does not
cease his search throughout his whole life. Directing his attention to one idea after
another that is germane to what is being investigated, he presses on until he arrives at
his goal' (17.11ff).

practitioners of the empirical approach appear clearly in the field of anatomy, where the fifth- and fourth-century treatises take only the first very hesitant steps in the use of dissection.

ASTRONOMY

The next main areas where we can study the use and development of empirical methods are geography and astronomy. On the former we can be brief. The obtaining and recording of information about the known part of the inhabited world[219] is one of the fields where ἱστορία was practised at a very early stage. The first Greek map[220] is that attributed to Anaximander, though we cannot reconstruct its contents, and to judge from Herodotus' comments on early map-makers,[221] it is likely to have been highly schematic. But it was another Milesian map-maker, Hecataeus, who founded this branch of ἱστορία, initiating the tradition that continued through Herodotus, the writer of the treatise *On Airs Waters Places* and Eudoxus (who wrote a lost γῆς περίοδος or 'circuit of the earth') down to Hipparchus, Posidonius, Strabo and Ptolemy. This work was important both as providing the very first example of sustained research, and for one of its results – a greatly increased knowledge of other peoples.[222] Yet the aim of early geographical accounts was largely descriptive, and so of far less importance than astronomy in the development of scientific theory. The Greeks themselves distinguished such inquiries, called 'chorography', from the more strictly scientific mathematical geography,[223] which included such topics as the determination of the size of the earth and especially – from the Hellenistic period at least – the problems of projection or cartography. Geography understood in the latter sense was essentially a branch of applied mathematics and the use it made, or needed to make, of empirical data was limited.

Astronomy itself, however, like medicine, provides an excellent opportunity to study the relation of observation and theory in Greek science. Admittedly our sources, especially for the earlier periods,

---

[219] Descriptions of foreign countries usually included accounts not only of the main geographical features such as rivers and mountains, but also of the flora and fauna and of the customs of the inhabitants.

[220] This had been anticipated by earlier ancient Near Eastern, particularly Babylonian, maps, though most of those known are local sketch maps, not attempts to relate all the parts of the known world in a single whole, as was the case with the maps referred to in Herodotus ɪᴠ 36.

[221] Herodotus ɪᴠ 36 criticises earlier map-makers on the grounds that they make the world symmetrical, with Oceanus running round the earth as if drawn with a compass, and Asia made equal to Europe. Cf. also the map referred to in Hdt. ᴠ 49.

[222] See below, ch. 4, pp. 236ff.

[223] A clear distinction between these two is drawn, for example, in the opening section of Ptolemy's *Geography*.

leave much to be desired: for the Presocratic philosophers we rely often, as usual, on second-hand reports, many of which are evidently overenthusiastic in their attributions of the discoveries of astronomical facts and theories to individual thinkers.[224] Even so, we can gain a general idea of the level of astronomical knowledge in the sixth and fifth centuries. Thus whereas Anaximander's account of the relations between the heavenly bodies apparently ignored the difference between planets and stars,[225] that distinction was grasped not long after him, perhaps by Anaximenes,[226] but certainly by some of his successors.[227] Various authors are credited with knowing – or themselves indicate that they know – that the moon shines with reflected light,[228] that the Morning Star and the Evening Star are one and the same body,[229] and that the interposition of the moon and of the earth causes eclipses of the sun and moon respectively.[230] Although

[224] Thus much sophisticated astronomical knowledge – for example of the equinoxes and solstices – is ascribed to Thales, partly perhaps because of Herodotus' report of his foretelling an eclipse of the sun. But even in that instance whereas Herodotus himself says merely that he foretold this eclipse to within a year (1 74), later writers, such as Clement (*Strom.* 1 14.65, DK 11 A 5) and Diogenes Laertius (1 23, DK A 1) drop that qualification: yet at no stage were precise predictions of solar eclipses within the competence of ancient astronomers. It is unlikely, too, that Thales had a very accurate idea of the sizes and distances of the sun and moon when his successor, Anaximander, apparently gave a quite schematic and in parts grossly aberrant, account (see next note).

[225] When we piece together the admittedly fragmentary reports of his astronomy, it appears that he pictured the heavenly bodies as disposed in three concentric circles, the outermost containing the sun, the middle the moon and the innermost the stars (see Hip. *Haer.* 1 6.4–5, DK 12 A 11, Aet. 11 20.1, 21.1, 25.1, DK A 21–2). Since there is no separate ring, or rings, for the planets, it is presumed that he included these with the stars. It is true that this reconstruction is conjectural, but it is one that has the preponderance of such evidence as we have in its favour.

[226] As Heath 1913, pp. 42f, suggested on the basis of a corrupt passage in Aet. 11 14.3–4, DK 13 A 14, interpreted as distinguishing between the stars fixed like nails in the crystalline, and the planets like 'fiery leaves'. But on the difficulties in this view, see, for example, Guthrie 1962, p. 135, Dicks 1970, p. 47.

[227] Thus Aristotle, *Mete.* 342 b 27ff, reports that Anaxagoras and Democritus both held that comets are due to the conjunctions of planets: by the time we come to the Philolaic system, at the end of the fifth or beginning of the fourth century, the five planets known in the ancient world are assigned to separate circles, see below, pp. 173f.

[228] This is stated, admittedly rather obscurely, in the fragments of Parmenides (Fr. 14), Empedocles (Frr. 43, 45) and Anaxagoras (Fr. 18). Knowledge of this is also ascribed to Thales (Aet. 11 28.5, DK 11 A 17b) and to Anaximenes (Theon of Smyra, 198.19–199.2, DK 13 A 16) though we may be sceptical. Our secondary sources also report, however, that Anaximander, Xenophanes and Antiphon all held that the moon shines by its own light (Aet. 11 28.1 and 4, DK 12 A 22, 87 B 27).

[229] This is ascribed alternatively to Pythagoras and to Parmenides, e.g. D.L. vIII 14 and ix 23, DK 28 A 1.

[230] It appears from Empedocles Fr. 42 (though the text is corrupt and the interpretation problematic) that he may have known the true cause of solar eclipses. Hippolytus, *Haer.* 1 8.9, DK 59 A 42, reports that Anaxagoras held both that solar eclipses are caused by the interposition of the moon and that the moon is eclipsed both by the earth and by other bodies under the moon (a similar idea recurs in the Pythagorean system described by Aristotle, who indicates that it was used to explain the greater

our sources disagree about who precisely 'discovered' these facts (that is who among the Greeks first recognised them), we may take it that some at least of those who engaged in astronomical speculation in the late fifth and early fourth centuries were well aware of them, even though many ordinary people, and indeed some educated writers, remained extremely vague or ignorant concerning some elementary astronomical data. Thus apart from the evidence of popular fears concerning eclipses provided by the famous account in Thucydides (VII 50) of Nicias delaying the retreat of the Athenians from Syracuse because of the reaction among the soldiers to an eclipse of the moon, we find the author of *On Breaths* maintaining that the wind is responsible for the movements of the sun, moon and stars.[231]

Yet knowledge of the points we have mentioned does not require elaborate, or even very sustained, astronomical observations – certainly not much more sustained than those that were ordinarily carried out for practical purposes. The morning and evening risings and settings of some of the major constellations were closely noted and used to mark the seasons from an early period, as is clear from Hesiod's *Works* especially.[232] For evidence of more systematic observations undertaken by Greek astronomers we have to turn to two areas particularly, the determination of the solar year and planetary theory.

Although none of the writings of Meton and Euctemon has survived, we have good evidence concerning some of their work around the year 430 B.C.[233] Their most notable achievement was the nineteen-year cycle – now named after Meton[234] – that established a correlation between the solar year and the lunar month. Nineteen solar years were equated with 235 lunar months, 110 of them 'hollow' months of twenty-nine days, the remaining 125 'full' ones of thirty, a total of 6,940 days. This gives a mean lunar month of

---

frequency of lunar eclipses – that is as observed from any given point on the earth's surface, *Cael.* 293 b 23ff). Yet other explanations of eclipses continued to be put forward: thus the idea that eclipses of the moon are due to the tilting of its bowl is ascribed to Antiphon (as well as to Heraclitus and Alcmaeon) by Aetius II 29.3 (DK 87 B 28).

[231]　*Flat.* ch. 3, *CMG* I, 1 93.9ff. Cf. Hdt. II 24 (which talks of the sun being driven from its course by storms) and Hippolytus, *Haer.* I 8.9 (who ascribes to Anaxagoras the view that the air is responsible for the sun's and the moon's 'turnings').

[232]　It was not only astronomical phenomena that were used to mark the seasons, but also, for example, the movements of birds, see, e.g., *Op.* 564ff.

[233]　The observation recorded by Ptolemy at *Syntaxis* III 1, ii 205.15ff, is of 432 B.C., and this is the presumed starting date for the Metonic cycle.

[234]　The same cycle was also introduced some time in the fifth century in Babylonia, but it is not certainly attested there before Meton, and may have been either an independent development or even a case of Greek influence on Babylonia (see Neugebauer 1957, p. 140, 1975, I pp. 354f and II pp. 622ff, but cf., e.g., Toomer 1974, p. 339).

$29\frac{25}{47}$ days and a mean solar year of $365\frac{5}{19}$ days.[235] Now it is clear that this cycle was in part – probably in large part – the product of computation and extrapolation. But we can be certain it had some observational basis, even if we cannot determine how extensive this was. Ptolemy, who reports in the *Phaseis* that Meton and Euctemon made observations at several different places in the Greek world,[236] cites a particular observation of the summer solstice of 432 B.C. in the *Syntaxis*.[237] In the second-century B.C. astronomical papyrus known as the *Ars Eudoxi* we find attributed to Euctemon figures for the lengths of the seasons of the year determined by the solstices and equinoxes.[238] Finally both Meton and Euctemon are often quoted in the parapegmata literature – almanacs that set out astronomical and, more especially, meteorological data for each day of the month – and they may even have drawn up such an almanac themselves.[239]

This constitutes our first good evidence for some fairly sustained Greek astronomical observations. But the context in which they occur is clearly significant. Although the determination of the solar year is of fundamental importance for theoretical astronomy in the construction of models of heavenly motion, there was an additional purely practical motive for work of the type that Meton and Euctemon undertook. The chaotic character of the civil calendars of Athens and other Greek city-states – a subject with which Aristophanes makes a good deal of play[240] – is well known.[241] Unlike the Egyptians, who adopted a notional year of 365 days,[242] the Greek city-states relied on luni-solar calendars with a thirteenth month intercalated in some

---

[235] This gives an error of 30′ 11″ for the mean tropic year, and of not quite 1′ 54″ for the mean lunar month, according to Heath 1913, p. 294.

[236] *Phaseis* II 67.2ff. Ptolemy specifies Athens, the Cyclades, Macedonia and Thrace.

[237] *Syntaxis* III 1, ii 205.15ff.

[238] Starting from the summer solstice, the figures are 90, 90, 92 and 93 days (the first three are given in the papyrus, the fourth may be inferred from them): the errors, compared with a modern determination for the period, range from 1.23 to 2.01 days (Heath 1913, p. 215).

[239] On the history of parapegmata, see especially Rehm 1941. Our two main examples, Geminus, *Isagoge* and Ptolemy's *Phaseis*, are both late, but both cite their authorities by name (they include Democritus, as well as Euctemon and Meton, from the fifth century). The conservative nature of this literature can be judged by the fact that data about the risings and settings of constellations were copied out for use in places of quite different latitudes from those where the original observations were made. We can only guess at how the meteorological data – which were sometimes quite specific, for example about which wind blows on a particular day – were interpreted.

[240] See, e.g., *Nu.* 615ff: cf. Thucydides' dissatisfaction with time reckoning by archonships (v 20).

[241] The chief discussions are Meritt 1928, Pritchett and Neugebauer 1947, Pritchett 1957, van der Waerden 1960, Meritt 1961, Pritchett and van der Waerden 1961, Pritchett 1964 and 1970.

[242] This was the year that was eventually adopted as standard for computational purposes by Greek astronomers such as Ptolemy.

years to keep the calendar more or less in step with the seasons. Yet the question of when to intercalate a month was left to the magistrates to resolve. Their decisions were evidently quite arbitrary, and indeed they continued to be so even after Meton and Euctemon had provided a reasonable basis on which a stable luni-solar calendar could have been drawn up. Nevertheless for our purposes one conclusion is clear, that it was in part at least the problems that arose in regulating the calendar that provided the stimulus for Meton's and Euctemon's observations. Once again, then, as with the case-histories in the *Epidemics*, we find that some of the best early sustained observational work has a quite specific motivation: here too it would be rash to infer that the individuals concerned necessarily recognised the value of empirical research of this kind in astronomy *in general*.

The first comprehensive solution to the problems posed by planetary motion was Eudoxus' theory of concentric spheres, later adopted and modified by Callippus and Aristotle. But two earlier contributions must first be discussed, the so-called Philolaic system and the astronomical passages in Plato. In several texts Aristotle attributes to certain unnamed Pythagoreans[243] a theory in which no less than ten heavenly bodies are distinguished, including the five visible planets. As we noted earlier, Aristotle is highly critical of the way in which – as he thinks – the Pythagoreans did not 'seek arguments and causes in relation to the φαινόμενα' but tried to 'drag the φαινόμενα into line with certain arguments and opinions of their own',[244] and elsewhere he objects that the doctrine of the counter-earth in particular sprang from the Pythagoreans' desire to make the number of the moving heavenly bodies total ten, the perfect number.[245] Although the reconstruction of their doctrines is extremely problematic – and it may be that the original system itself was not fully consistent[246] – this was the first Greek astronomical theory for which we have good early evidence in which each of the planets, sun and

[243] The attribution to Philolaus is made by Aetius: we have no means of verifying this or of dating the system more precisely than to the late fifth or early fourth century.

[244] *Cael.* 293a25ff, see above, p. 137. Aristotle there also suggests that the Pythagoreans shifted the earth from the centre of the universe and gave that place to fire out of considerations of value: the most honourable place befits the most honourable thing and fire is more honourable than earth. This passage thereby incidentally confirms that the idea that the earth is one of the planets had been suggested before Aristotle, though not in connection with a heliocentric hypothesis.

[245] *Metaph.* 986a8ff. The ten heavenly bodies are the fixed stars (counted as a single sphere), the five planets, sun, moon, earth and counter-earth (on this version of the system the central fire is a separate body at rest in the centre of the universe, although another view of the central fire is also reported in our sources, that it is within the earth, the earth itself being in the centre).

[246] See the discussions in Heath 1913, ch. 12, van der Waerden 1951 and Dicks 1970, ch. 4, especially.

moon is assigned to a separate ring or sphere. This was the essential preliminary to any attempt to investigate and explain their individual motions, although we have no reliable means of telling how far the Pythagoreans themselves,[247] or any other thinker before Plato,[248] progressed in any such inquiry.

With Plato himself, however, we are already on much firmer ground. In the myth of Er in the *Republic* he distinguishes between two main movements, which in the *Timaeus* he calls the circles of the Same and of the Other, the westerly movement of the Same being responsible for the phenomena we should ascribe to the daily rotation of the earth on its axis, and the easterly movements of the Other accounting for the movements of the planets, sun and moon along the ecliptic.[249] Moreover Plato clearly refers to the different speeds of revolution of the various heavenly bodies, and although he nowhere gives specific values for their periodicities, in both the *Republic* and the *Timaeus* he gives the order of their relative speeds:[250] the moon is the fastest, then the sun, Mercury and Venus (these three are thus given the same mean easterly motion), then Mars, Jupiter and Saturn in that order. He also assigns different breadths to the 'whorls' in the 'spindle of necessity' in the *Republic*, and this may correspond to the different distances separating the bodies in question, although once again he gives no definite figures but contents himself with a statement of their order.[251] He draws back from a detailed account, putting it, in the *Timaeus*, that to attempt such a discussion without being able to consult visible models is useless labour;[252] and *Timaeus* 39cd provides clear evidence of a general lack of information, among his contemporaries, about the periods of the planets. 'Except for a few',[253] he says, 'men have not grasped' their periods: 'they do not name them nor do they investi-

---

[247] The late sources who ascribe elaborate systems concerning the harmonies of the heavenly spheres to the Pythagoreans are generally untrustworthy.

[248] Thus Democritus, who is reported to have written a work on the planets, is said by Lucretius (v 621ff) to have held that the speeds of the heavenly bodies decrease with their distance from the outermost heavens, though there is no evidence that he gave definite values to their various periodicities.

[249] *R.* 617a, cf. *Ti.* 36b ff. The *Timaeus* mentions, however, what the *Republic* ignores, namely the obliquity of the ecliptic (*Ti.* 36c7).

[250] *R.* 617ab, *Ti.* 36d, 38de. *Ti.* 38e–39b contrasts the speeds of the planets, sun and moon relative to one another, with their apparent absolute speeds (i.e. in the direction of the motion of the fixed stars).

[251] *R.* 616de. *Ti.* 36b–d however refers to the 'double and triple' intervals, giving the series 1, 2, 3, 4, 8, 9, 27, and this may correspond to Plato's view of the *ratios* between these distances, though the point is disputed (Heath 1913, p. 164).

[252] *Ti.* 38de, 40cd. Whether he had some kind of orrery in mind, or more simply a celestial sphere, is disputed: see, for example, Cornford 1937, pp. 74ff, Dicks 1970, pp. 120, 137 and n. 193.

[253] Eudoxus, would no doubt, have been counted as one such exception.

gate and measure them in relation to one another by means of numbers.' Yet Plato's own excursions into astronomy allow us to infer that he had access to some quite complex information concerning the main celestial motions, and in particular he provides us with our first definite extant reference to the phenomenon that was thereafter to remain at the centre of ancient astronomical theory, namely the retrogradations of the planets.[254]

The general structure of the theory that enabled Eudoxus to suggest a resolution to that problem is not in doubt,[255] but the two main questions that concern us are first the extent of the observational data on which his model was based, and secondly the author or authors of those observations. Simplicius gives us definite figures for the two main periodicities, that is the sidereal and the synodic periods, and with the exception of the synodic period of Mars, they are tolerably accurate.[256] But neither Simplicius nor any other source provides us with precise information on one point that is essential for a reconstruction of Eudoxus' theory of the planets, namely the angles of inclination of the fourth sphere.[257] This lacuna in our evidence means that we do not know to what extent Eudoxus' model was indeed a fully quantitative one. Schiaparelli was the first in modern times to attempt a detailed reconstitution of the model, but his interpretation – like most of those that have followed it[258] – depends on supplying figures for which we have no ancient authority. Moreover the figures that have usually been taken are those that are most favourable to the theory's success. Thus interpreters have generally assumed that Eudoxus had accurate knowledge of the maximum lengths of the retrograde arc of each of the planets, and they then take the true modern determinations of those arcs as the basis of their reconstructions of the relationships between the third

---

[254] *Ti.* 40c. Plato comes back to the problem of the movements of the planets in a puzzling passage in the *Laws*, 822a, where he insists that each of the heavenly bodies moves with a *single* circular movement: either he is referring to the resultant motion of (e.g.) the movements of the Same and the Other, or he believed that a simple model can be given. On general grounds it may be thought likely that the *Laws* was written at a time when Plato was aware of Eudoxus' theory of concentric spheres: but 822a can hardly be taken as an allusion to it.

[255] See, for example, Heath 1913, ch. 16. Neugebauer 1957, pp. 153f, Maula 1974, and Neugebauer 1975, II pp. 675ff.

[256] *In Cael.* 495.26–9, 496.6–9.

[257] The third sphere has its poles on the ecliptic and rotates in an easterly direction in the synodic period of the planet. We are told simply that the poles of this sphere are the same for Mercury and Venus, but different for each of the other planets. The fourth sphere, which carries the planet itself, rotates on an axis inclined to that of the third, and in the same period but in the opposite direction.

[258] See Schiaparelli 1877, and cf. Dreyer 1906, Heath 1913: but contrast the fresh attempt at an analysis of possible reconstructions in Maula 1974.

and fourth spheres. Even so the theory breaks down for Mars and Venus, where the evidence given by Simplicius already fails to provide for retrogradation at all.[259]

While it is conceivable that Eudoxus gave *all* the necessary parameters for his model, this is far from certain. That he gave *some* definite parameters is suggested not only by Simplicius, but also by the modifications later introduced by Callippus.[260] But the shortcomings of the theory are such that we cannot rule out – it may even be considered more likely[261] – that it was not, in fact, fully determined, and that once he had solved the problems of the stations and retrogradations of the planets qualitatively and geometrically, he did not proceed to provide complete parameters for the individual planets.

The question of who was responsible for such observational data as were available to Eudoxus takes us immediately to a major problem that we have yet to broach, namely that of the transmission of Babylonian astronomical records to Greece. Although many ancient commentators exaggerate both the antiquity, and the extent and accuracy, of Babylonian observations,[262] astronomical cuneiform tablets confirm that some detailed records of a limited number of phenomena had begun to be kept from at least the first half of the second millennium B.C.[263] It must be stressed that the extant tablets are often based as much on computation as on observation: the results have been schematised to conform to predetermined regularities.[264] Yet there can be no doubt that quite extensive observations were undertaken. At the same time it appears very unlikely that the Greeks had access to such data in the fifth century, for if they had, it is hard to explain their hesitant grasp of such basic

---

[259] See, for example, Heath 1913, p. 211 and cf. most recently, Maula 1974, pp. 73ff.

[260] Aristotle tells us (*Metaph.* 1073b32ff) that Callippus introduced extra spheres not only for the sun and moon (the former presumably to take account of the inequalities of the seasons, known already to Meton and Euctemon, but ignored by Eudoxus in his model), but also for three of the planets, Mars, Venus and Mercury: here Callippus' purpose may well have been to try to meet some of the difficulties Eudoxus' model encountered in attempting to explain their retrogradations.

[261] Cf. Neugebauer 1972, p. 248.

[262] See, e.g., Cicero, *Div.* II 46.97, Simplicius, *In Cael.* 117.24ff, 481.12ff, 506.8ff. Contrast, however, Ptolemy, who complained of the comparative lack of reliable information concerning the planets, *Syntaxis* IX 2, iii 208.12ff.

[263] One such early record relates to the appearances and disappearances of the planet Venus in the reign of Ammisaduqa, *c.* 1600 B.C. For an authoritative assessment of early Babylonian astronomy, see Neugebauer 1957 and 1975, I pp. 347ff.

[264] Thus Neugebauer 1955, II p. 281 commented on 'the minute role played by direct observation in the computation of the ephemerides. The real foundation of the theory is (*a*) relations between periods, obtainable from mere counting, and (*b*) some fixed arithmetical schemes (for corrections dependent on the zodiac).'

astronomical data as the identity of the planet Venus as both Morning and Evening Star.[265]

By the mid fourth century, however, we begin to have direct references to Babylonian and other Eastern astronomy in Greek writings. First the author of the *Epinomis* writes that the Egyptians and Syrians initiated observations of the planets, which thereafter became available to the Greeks,[266] and then Aristotle reports both that the occultations of stars by planets had been observed by the Egyptians and Babylonians 'who have watched the stars for very many years past and to whom we owe many trustworthy grounds for belief about each one of them',[267] and that the Egyptians had observed the conjunctions of planets, and of planets with fixed stars.[268] The references in all three texts to *Egypt* are odd since our direct evidence for Egyptian astronomy contains no hint that they had carried out observations of planetary movements before the Hellenistic period. This may, no doubt, be due simply to the fragmentary nature of our sources: or it is possible that knowledge of Babylonian work arrived in Greece in part via Egypt and that this was the source of some confusion, in Greek writers, on the origin of the data they were interested in.[269] In any event, so far as the Babylonians themselves are concerned, we cannot doubt that the Greeks could have learned a great deal – and certainly in the Hellenistic period did learn a great deal – from them.[270]

Yet the question of the precise extent of Eudoxus' debts to eastern

---

[265] Although the doxographers claimed that Anaximander 'discovered' the gnomon (evidences in DK 12 A 1, 2 and 4) we hear from Herodotus (II 109) that this and the polos came to Greece from Babylonia. Yet we should distinguish between the use of a simple astronomical sighting instrument, and items of astronomical knowledge contained in texts that would normally not be at all readily intelligible to the Greeks. Even in such an apparently promising case as the origin of the constellations, the differences between Greek and Babylonian representations are as great as their similarities, and independent development cannot be ruled out (see Dicks 1970, pp. 164f, though contrast van der Waerden 1954, p. 84).

[266] *Epin.* 986 e f. If, as is likely, this work was not by Plato himself, it was probably by a pupil or close associate.

[267] *Cael.* 292 a 7ff, cf. also 270 b 13ff.

[268] *Mete.* 343 b 9ff.

[269] This is not as implausible as might at first appear. Egypt remained a province of the Persian empire during much of the fifth and fourth centuries, intermittently revolting from Persian rule, only again to be subjugated to it, and the sea voyage to Egypt was appreciably easier for the Greeks than the overland journey to Babylonia. Moreover there is some direct evidence that Babylonian eclipse- and lunar-omina arrived in Egypt in the reign of Darius, see R. A. Parker 1959.

[270] This is not to deny the fundamental differences in the problems that interested the original Babylonian astronomers on the one hand, and the Greeks on the other. The former were not, while the latter undoubtedly were, chiefly concerned with constructing geometrical models to explain the movements of the heavenly bodies. See further below, ch. 4, p. 230.

astronomical observations remains open. We know that he wrote a
'circuit of the earth' and our secondary sources refer to his travels,
for example that he visited Egypt,[271] and indicate that he had some
knowledge of Chaldaean astronomy/astrology.[272] In view of this,
and of Aristotle's testimony, it seems likely that *some* information –
including perhaps data relating to the major periodicities of the
planets – would have been available to Eudoxus.[273] But it would
certainly be rash to suppose that he had access to the extensive
astronomical records to which Ptolemy refers, some at least of which
were evidently available to Hipparchus.[274] Although a reference to
Saturn as 'the star of the sun' (its Babylonian name) in Simplicius'
report of Eudoxus' figure for its sidereal period has been taken as a
possible indication of the Babylonian origin of some of his data,[275]
it would appear from another passage in Simplicius that records of
Babylonian observations only began to reach Greece in considerable
quantities after the conquest of Alexander – that is, too late for
Eudoxus.[276] As we have said, Eudoxus seems to have worked with at
most only very general data concerning the retrogradations of the
planets. Even if he knew the maximum value for the retrograde arc
for each planet, his model cannot accommodate variations in either
its length or shape, which are quite noticeable in such a case as that
of Mars. Precise observations from which the retrograde arcs could
be plotted in detail were, then, either not available to him, or were
ignored by his theory.

Similar reservations must apply not only to such data as Eudoxus
obtained from Babylonia, but also to his own observational work, the
general nature of which can be inferred from our secondary sources.
First it is clear that, like Meton and Euctemon, he engaged in
investigations of the parapegma type, making astronomical and
meteorological observations for different days of the month, for like
them he is often mentioned in the parapegma literature. Secondly
the numerous fragments of his *Phainomena* and *Enoptron* ('Mirror')
that are preserved by Hipparchus[277] show that those works not only

---

[271] Plutarch, *De Is. et Osir.* 353c, 354de, Strabo XVII 1.29–30, D.L. VIII 87. Although in
general reports in late writers about the eastern travels of Greek wise men should be
treated with caution, the circumstantial details of these (for example the letter of
introduction to Nectanebis from Agesilaus) suggest they have some foundation in fact.

[272] E.g. D.L. Proem I 8 and Cicero, *Div.* II 42.87.

[273] Yet the order of the planets in Babylonian astronomy differs from that in Eudoxus –
which rules out the possibility that he derived his ideas on their general arrangement
from that source: see, for example, Dicks 1970, p. 175.

[274] See below, pp. 180 and 185.

[275] *In Cael.* 495.28–9, cf. Dicks 1970, p. 167.   [276] *In Cael.* 506.10ff.

[277] In his *Commentary on Aratus* (I 2.1f and frequently elsewhere) Hipparchus tells us that
Aratus' poem was based on Eudoxus' *Phainomena*.

contained a detailed description of the heavens but also set out a good deal of information about, for example, which constellations rise and set together. At the same time he lacked the division of the celestial globe into 360 degrees,[278] and identified and located individual stars generally quite imprecisely, with reference to the conventional constellation figures.[279] We may conclude that he carried out some, perhaps comparatively quite extensive, observations of his own.[280] But what remains quite doubtful is whether he undertook any detailed investigations of the courses of the planets. Any suggestion to that effect may not be excluded, but is rendered unlikely, by the already mentioned fact that his theory was incapable of accommodating the variations in the lengths and shapes of the retrograde arcs.[281]

At the risk of labouring the point we may refer to Aristotle for supplementary evidence concerning the type of observation carried out in the fourth century. With the exception of some isolated earlier references to particular eclipses, Aristotle provides us with our first extant original records[282] of Greek astronomical observations. Passages in the *De Caelo* and the *Meteorologica* especially refer to what, he says, 'we have seen', where the first person plural shows that it is a question of contemporary Greek observations, not Babylonian records, even if not necessarily observations made by Aristotle personally. Thus he reports an occultation of Mars by the moon,[283] describes the course of a comet in the archonship of Nicomachus (341/0 B.C.),[284] and the tail appearing on one of the stars in the Dog,[285] and mentions the conjunction of Jupiter with one of the stars in the

---

[278] Although by the fourth century the idea of the celestial sphere was well established, it was not until later that the division into 360° was brought into Greek astronomy from Babylonia. In the absence of the notion of degrees, proportions or ratios were used to describe, for instance, the length of arc of the summer tropic visible above the horizon at a given latitude (i.e. the maximum length of daylight).

[279] For example, 'beneath the tail of the Little Bear lie the feet of Cepheus, making an equilateral triangle with the tip of the tail' (Hipparchus I 2.11). Apart from the question of the precision of these descriptions, their correctness is often criticised by Hipparchus (e.g. II 2.37, 47, II 3.2f) who also notes certain discrepancies between the *Phainomena* and the *Enoptron* (II 3.29f).

[280] We can supplement the information from the parapegma literature and the fragments of the *Phainomena* and *Enoptron* with occasional reports of other observations, such as that of Canopus recorded in Strabo (II 5.14, cf. also XVII 1.30). Maula and others have recently attempted to reconstruct, on the basis of remains found at Cnidus, an astronomical instrument which may represent, or at least be derived from, Eudoxus' Arachne or Spider, mentioned at Vitruvius IX 8.1: see Maula 1977 and cf. Maula 1975–6.

[281] Yet Sosigenes, quoted by Simplicius, *In Cael.* 504.17ff, suggested that Eudoxus' theory failed to 'save' some of the 'phenomena' that were known at the time, and one clear example of this which we have noted is that his model takes no account of the inequality of the seasons, which had been discovered by Euctemon and Meton.

[282] As opposed, that is, to observations reported in *later* writers.

[283] *Cael.* 292a3ff.   [284] *Mete.* 345a1ff.   [285] *Mete.* 343b11ff.

Twins.[286] It is true that Aristotle himself disclaims expert knowledge in astronomical matters,[287] though he is clearly well-informed on the subject. Moreover in these texts he is not concerned with the problems of constructing a planetary model, so much as (in most cases) with certain physical questions. Yet the vagueness of his references is still striking. He has no satisfactory dating system: indeed on no occasion does he date the observation he records more precisely than to the archon-year and the month.[288] Nor does he have – or at least he does not use – a system of celestial coordinates, whether equatorial or ecliptic, by reference to which objects or events could be located precisely.[289]

As with dissection, it will help to put the fourth-century investigations into perspective if we refer to some later work, particularly to that of two of the greatest Greek astronomers, Hipparchus (second century B.C.) and Ptolemy (second century A.D.).[290] Three preliminary points are fundamental. First it is apparent that extensive Babylonian 'data' were available to, and used by, Hipparchus[291] as well as Ptolemy. Secondly, in addition to practical considerations such as the regulation of the calendar, a further powerful motive for undertaking astronomical observations came to be of increasing importance in Greece from the fourth century B.C. onwards, namely the belief in astrology, notably – eventually – in the form known as genethlialogy, the casting of horoscopes.[292]

---

[286] *Mete.* 343 b 30ff.

[287] As is well known, Aristotle often merely uses what the 'mathematicians' say on astronomical matters, or defers to them for a more exact account: see e.g. *Metaph.* 1073 b 10ff, 1074 a 14ff, *Cael.* 291 a 29ff, b 8ff, 297 a 2ff, 298 a 15ff.

[288] We may contrast Ptolemy's use of the first year of Nabonassar's reign (747 B.C.) as epoch, and of the conventional Egyptian calendar of 365 days.

[289] In *PA* I 5 where Aristotle contrasts the study of the stars with that of plants and animals, he remarks on how little is clear to perception concerning the heavenly bodies, *PA* 644 b 24ff.

[290] It was Apollonius of Perga, working at the beginning of the third century B.C., who was responsible for the model of epicycles and eccentrics that was used by both Hipparchus and Ptolemy. We have, however, no means of determining the extent of the observational basis of Apollonius' theory.

[291] As is clear, for example, from *Syntaxis* IV 2 and 9, ii 270.19ff, 332.14ff, and cf., e.g., Aaboe 1955–6 and Aaboe and Price 1964.

[292] Cicero's report, *Div.* II 42.87, that Eudoxus held that Chaldaean astrological predictions are untrustworthy, suggests that Babylonian astrological lore had already begun to penetrate Greece in the fourth century B.C., though at first this is more likely to have related to general beliefs about heavenly omens than to genethlialogy. To judge from our extant evidence, Babylonian horoscopes were rare before about 200 B.C. (Sachs 1952), nor should we underestimate the Greeks' own role in turning the 'art' into a universal system (see Neugebauer 1975, II pp. 613ff). As regards Hipparchus and Ptolemy themselves, it appears from Pliny (*Nat.* II 24.95), for example, that Hipparchus believed in astrology in some form, and Ptolemy wrote a treatise in four books (the *Tetrabiblos*) on it: in this he was careful to distinguish between on the one hand predictions concerning the movements of the heavenly bodies, and on the other pre-

Thirdly we can trace certain improvements in astronomical instruments, again beginning in the fourth century B.C. First there are mechanical models or orreries, designed to represent the movements of the sun, moon and planets.[293] Secondly there are developments in sighting instruments.[294] Thus apart from the gnomon Hipparchus was evidently familiar with the equinoctial (or equatorial) armillary – a ring mounted in the plane of the equator from which the equinoxes could be determined.[295] Moreover Hipparchus improved (and may even have invented) the instrument Ptolemy calls the 'four-cubit rod dioptra', a horizontal bar with fixed backsight and movable foresight used to measure, for example, the angular diameter of the sun.[296] In addition to these instruments Ptolemy himself refers to four others in the *Syntaxis*, (1) the meridional armillary, (2) the plinth or quadrant (these two were used to measure the midday altitude of the sun at the solstices, from which the obliquity of the ecliptic can be calculated),[297] (3) the parallactic

dictions concerning their influence on human affairs, but while the former are more certain than the latter, he claimed that the latter too are possible, even if both difficult and conjectural (*Tetrabiblos* I chh. 1–2). The classic work on ancient astrology is Bouché-Leclerc 1899, but see also, e.g., Cumont 1912, Boll and Bezold (1917) 1931, Capelle 1925, Gundel and Gundel 1966 and H. G. Gundel 1968.

[293] Apart from Plato, *Ti.* 40cd, on which see above, p. 174 n. 252, Archimedes is recorded as having constructed an orrery (Cicero, *Rep.* I 14.21–2, *Tusc.* I 25.63: Archimedes' lost work *On Sphere-Making* may have dealt with this), and so too is Posidonius (Cicero, *N.D.* II 34–35.88). The first-century B.C. anti-Kythera instrument has now been shown to be not a planetarium, but a calendrical computer, in which a sophisticated system incorporating differential gearing is used to show the sidereal motions of the sun and moon consistently with the phases of the latter: see Price 1974, who points out the implications of this instrument for our understanding of the technological capabilities of the ancient world.

[294] There are brief surveys of ancient astronomical instruments in Dicks 1953–4 and Price 1957, and cf. Aaboe and Price 1964.

[295] This is clear from passages from Hipparchus' *On the Precession of the Tropical and Equinoctial Points* quoted by Ptolemy, *Syntaxis* III 1, ii 194.23ff, 196.8ff, in which Hipparchus records several observations of the equinoxes made on the ring in the 'so-called Square Hall' at Alexandria. After recording Hipparchus' view that an error of up to a quarter of a day might arise 'in observation and calculation' in determining the exact time of the solstices, Ptolemy himself notes that a deviation of a mere 6 minutes of arc from the equatorial plane in the setting of the instrument generates an error of 6 hours in determining the time of the equinox, and he adds that the equatorial armillaries in the Palaestra at Alexandria were, in his own day, unreliable (*Syntaxis* III 1, ii 197.4ff). Bruin and Bruin (1976) have recently reconstructed such an instrument to investigate the systematic errors that may arise in its use.

[296] See *Syntaxis* V 14, ii 417.1ff (where Ptolemy says that he himself constructed the instrument described by Hipparchus). We may compare the method that Archimedes used to determine the angular diameter of the sun in the *Sand-Reckoner* II 222.11ff.

[297] The meridional armillary and the plinth are described one after another in *Syntaxis* I 12, ii 64.12ff and 66.5ff, and although Ptolemy says 'we shall construct' the former, the fact that he states that the observations may be made more conveniently on the latter (66.5ff) suggests that he actually used the plinth. The possibility of systematic errors of various kinds in the use of this instrument to measure the zenith distance of the sun has been discussed by Britton 1969.

ruler,[298] and (4) – the most complex of his astronomical instruments – the armillary astrolabe: this consisted in a nest of concentric rings representing the main circles of the heavens, and it had the great advantage that, once the instrument was set with reference to a known point (e.g. the moon or a bright star), the ecliptic coordinates of a heavenly body could be measured directly rather than determined from complicated observations of its position in relation to the zenith and horizon.[299] Finally although Ptolemy does not mention the plane astrolabe in the *Syntaxis*, it has been thought likely that it was known to him, if not indeed also to Hipparchus: Ptolemy at least appears to refer to it as the 'horoscopic instrument' in the *Planisphaerium*, a work which deals with the problems of stereographic projection on which both the plane astrolabe and the anaphoric clock are based.[300]

Ptolemy is at some pains to give quite full details concerning both the construction and the use of most of the astronomical instruments mentioned in the *Syntaxis*.[301] Thus he sometimes (though far from invariably) specifies their minimum dimensions, and he issues warnings concerning particular sources of inaccuracy in their use. Nevertheless the limitations of the instruments at his disposal are evident. First, although the elementary laws of optics were well known,[302]

[298] This was essentially a vertical rod at the top of which was hinged a second rod of equal length fitted with sights (an alidade): the distance from the back of the alidade to the lower end of the vertical rod was measured by a thin lath, and this measurement then gave the chord of the angle between the alidade and the vertical (and so the angle itself could be obtained, using a Table of Chords such as that in *Syntaxis* i 11). Ptolemy sets out the details of the construction of the instrument at *Syntaxis* v 12, ii 403.9ff, insisting that the two main rods should be 'not less than four cubits long' and thick enough to be rigid. Though he refers to the instrument only in the context of determining the moon's parallax, it could be used to measure the zenith distance of any star. Price 1957, p. 589, notes that it was 'perhaps the most serviceable of Ptolemy's instruments and the only one used in similar form by subsequent astronomers', including Copernicus, and Dicks 1954, p. 81, estimates that it 'probably gave results to an accuracy of 5".

[299] See *Syntaxis* v 1, ii 351.5ff where Ptolemy says he constructed this instrument and reports the discrepancies he found between the actual, and the predicted, positions of the moon by using it. See also, e.g., vii 4, iii 35.11ff, in connection with his star catalogue.

[300] *Planisphaerium* ch. 14, ii 249.19ff. See Neugebauer 1949 and cf. Drachmann 1953–4 who believes that the anaphoric clock preceded the plane astrolabe rather than the other way about. The plane astrolabe was also probably the subject of a work by Theon of Alexandria on the 'little astrolabe', the contents of which are preserved in the treatise by Severus Sebokht, and we have a book by Philoponus on the instrument (see Hase 1839).

[301] In many cases Ptolemy's own descriptions can be supplemented by those in Theon of Alexandria, Pappus and Proclus. Proclus especially shows, for a Platonist, a remarkable interest in, and knowledge of, astronomical instruments: see *Hyp.* ch. 3, 42.5–54.12, ch. 4, 128.6–130.26, ch. 6, 198.15–212.6, and cf. 72.20ff, 110.3ff, 120.15ff.

[302] Although atmospheric refraction is discussed at some length in Ptolemy's *Optics* v 23ff, the chief context in which its effects are mentioned in the *Syntaxis* is that of the

neither he nor any other Greek astronomer tried to develop sighting instruments that magnified the visible object. Secondly, the technological problems involved in constructing large examples of complex instruments such as the armillary astrolabe in metal were formidable, and, so far as we know, the Greeks attempted nothing on the scale of the massive bronze armillary spheres eventually built by the Chinese.[303] Thirdly, for time-keeping at night the Greeks relied principally on various versions of the constant-flow waterclock (invented by Ctesibius) and it is significant that none of the specific observations recorded by Ptolemy is timed more precisely than to one sixth of an equinoctial hour.[304]

After these preliminaries, we may confront the two interlocking questions that chiefly concern us, namely first the extent of the observations that Hipparchus and Ptolemy carried out, and secondly the relation between observation and theory in their work. On the first topic we may begin with the information provided by the star catalogues they made. Although Hipparchus' catalogue has not survived, we have some information concerning the number of stars he identified in most of the main constellations from which it has been estimated that it contained some 850 stars.[305] Ptolemy's own catalogue, in *Syntaxis* VII and VIII, includes over 1,020 stars, giving their longitudes and latitudes in degrees and fractions of a degree[306] and also their magnitudes. Although the identification of a few stars is in doubt, a comparison between Ptolemy's values and those calculated for the year A.D. 100 by the most recent editors of the catalogue[307]

apparent size of the heavenly bodies observed near the horizon (I 3, ii 11.20ff, 13.3ff, but cf. also IX 2, iii 209.16, 210.5ff) and corrections for refraction are not made in the body of the work.

[303] See Needham, Ling, Price 1960 and J. Needham 1954–, III pp. 342ff, with the table, pp. 344ff, giving the dimensions and dates of the principal instruments.

[304] Cf. Dicks 1954, p. 84, who concludes: 'it seems that the ancient astronomers could tell the time at night to an accuracy of within 10 minutes'. A much lower figure (one minute) was given by Schjellerup 1881, p. 39, based on a study of the constellations whose risings were used to mark the hours, but this is unduly flattering. Cf. also Fotheringham 1915 and 1923.

[305] See Boll 1901, who points out that figures such as 1,080 for the stars within the recognised constellations alone (in Anonymus edited by Maass 1898, p. 128) are – like the guess of a total of 1,600 stars in Pliny, *Nat.* II 41.110 – quite untrustworthy, and who attempts to work out the extent of Hipparchus' catalogue on the basis of the numbers of stars given for the main constellations in the list ascribed to him in the fourteenth-century astrological MSS, Cod. Parisinus 2506 (reproduced by Boll 1901, pp. 186f). It should, however, be stressed that that list is incomplete and the margin of error consequently large (Boll himself gives upper and lower limits of 761 and 881 stars).

[306] Only seven simple fractions are used, $\frac{1}{6}$, $\frac{1}{4}$, $\frac{1}{3}$, $\frac{1}{2}$, $\frac{2}{3}$, $\frac{3}{4}$, $\frac{5}{6}$, i.e. 10', 15', 20', 30', 40', 45' and 50'.

[307] Peters and Knobel 1915 (cf. the corrections in Moesgaard 1976). The actual date of Ptolemy's catalogue is A.D. 138, which increases the errors in longitude given by Peters and Knobel (where these are underestimates) by 32'.

shows that the mean error in longitude is about 51′ and in latitude about 26′. In the great majority of cases, however, the errors in longitude are in the same direction, that is they are too small, and it has been suggested that the reason for this is that – so far from working from independent observations – Ptolemy has simply taken over Hipparchus' catalogue and corrected it for precession, using his own figure for precession which underestimates it by some 13″ per year.[308]

The difficulties of determining just how Ptolemy's catalogue was drawn up are formidable, but from one point of view the question is not an essential one. Whoever is responsible for it, the catalogue as it stands is excellent testimony to some remarkably sustained and accurate observational work in Greek astronomy, and the more Ptolemy's part in this is downgraded, the more impressive Hipparchus' contribution must be considered.[309] Yet the hypothesis that *all* that Ptolemy has done is to copy Hipparchus with an adjustment for precession is hard to sustain. (1) If the estimate of the contents of Hipparchus' catalogue is sound, Ptolemy has evidently added some 170 stars that did not appear in it. (2) Where a direct comparison is possible between Hipparchan longitudes (obtained mostly from the data in his *Commentary on Aratus*)[310] and Ptolemy's figures, it emerges that the differences are not constant, as we would have expected them to be if Ptolemy had simply added a figure to Hipparchus' values.[311] If so, then it follows that while here as elsewhere Ptolemy may well have used Hipparchus' data as a starting-point, he did more than merely copy them with a simple numerical adjustment. It seems that we should conclude that Ptolemy's own account of how he set about observing the fixed stars, using the armillary astrolabe and trying to obtain the positions of as many stars as possible up to those of the sixth magnitude,[312] is not purely fictitious, although, to be sure, we cannot verify in detail just how many observations he carried out.[313]

---

308 See especially Delambre 1817, 1 p. 183, 11 p. 264, Tannery 1893, p. 270, Newton 1977, pp. 237ff, but cf. Boll 1901, pp. 194–5, Dreyer 1916–17 and 1917–18, Vogt 1925 and Pedersen 1974, pp. 252ff. The idea that Ptolemy had plagiarised an earlier astronomer – namely Menelaus – was already suggested by Arabic astronomers, see Björnbo 1901, Dreyer 1916–17, pp. 533ff and Vogt 1925, pp. 37f.

309 Ptolemy himself repeatedly expresses the highest admiration for Hipparchus, and there is no doubt that one of Hipparchus' major contributions was the discovery of precession.

310 These data relate to equatorial coordinates (right ascension and declination) or to mixed equatorial and ecliptic ones (the so-called polar longitude) far more often than to ecliptic coordinates and show that, as Neugebauer put it (1975, 1 p. 277), 'at Hipparchus' time a definite system of spherical coordinates for stellar positions did not yet exist'.

311 See Vogt 1925, and cf. Pedersen 1974, pp. 255ff, Neugebauer 1975, 1 pp. 283f.

312 See *Syntaxis* VII 4, iii 35.11ff.

313 Various suggestions have been made on the sources of systematic errors in Ptolemy's

This takes us already to our second and more complex problem, that of the relation between observation and theory, and again we must concentrate on the evidence supplied by the *Syntaxis*.[314] Ptolemy himself is explicit about the sources of much of his information. He not only tells us in general terms that he has astronomical data from the Babylonians going back to Nabonassar,[315] but he also cites several precisely observed lunar eclipses from the eighth, seventh and sixth centuries B.C.[316] He is evidently able to draw on some quite detailed information from previous Greek astronomers, even though he expresses his doubts about the reliability of some of their observations, especially those made before Hipparchus,[317] and the *Syntaxis* includes exact reports of some 35 observations that he tells us that he carried out himself in Alexandria.[318] Nevertheless the actual number of observations deployed to resolve complex problems in his theories of the sun, moon and planets is quite small, and in particular the authenticity of those he claims to have made himself has been called in question. Many years ago Delambre[319] suggested that what Ptolemy reports as his own observations are merely computations from the tables he sets out, and one recent commentator has concluded:

All of his own observations that Ptolemy uses in the *Syntaxis* are fraudulent, so far as we can test them. Many of the observations that he attributes to other astronomers are also frauds that he has committed...Thus Ptolemy is not the greatest astronomer of antiquity, but he is something still more unusual: he is the most successful fraud in the history of science.[320]

Many aspects of this longstanding controversy are obscure and,

catalogue: see Drayson 1867–8, Peters and Knobel 1915, p. 8, Dreyer 1916–17, pp. 536ff, 1917–18, p. 346, Vogt 1925, Czwalina 1956–8 and Neugebauer 1975, I pp. 280ff. Ptolemy himself says that he used certain bright stars as reference points, whose positions had in turn been obtained with reference to the moon. The whole catalogue would therefore be affected by the errors in the solar theory and the coordinate system (see below, p. 186 and n. 322).

[314] While Ptolemy takes over Hipparchus' solar theory in all essentials, and uses his lunar model as the starting-point of his own discussion, he reports that Hipparchus did not attempt a detailed solution to the problems of planetary motion, but contented himself with 'arranging the observations to make them more useful' – notably by determining the fundamental periods – and with showing how the phenomena conflicted with current theories, *Syntaxis* IX 2 and 3, iii 210.8ff and 213.16ff.

[315] See *Syntaxis* III 7, ii 254.11ff.

[316] These are conveniently collected in Appendix A in Pedersen 1974, pp. 408ff.

[317] See, for example, *Syntaxis* VII 1 and 3 and especially IX 2 (iii 3.1ff, 18.14ff, 209.5ff, 17ff), and contrast the remarks on Hipparchus' own observations in VII 1 and IX 2 (iii 3.8ff, 210.8ff).

[318] These are set out in Pedersen 1974, pp. 461ff. They range from A.D. 127 (or possibly 125) to 141. Ptolemy is also able to draw on observations made at Alexandria by his contemporary Theon (often identified with Theon of Smyrna).

[319] Delambre 1817, I pp. xxv ff, II pp. 250ff.

[320] Newton 1977, pp. 378–9, cf. also Newton 1973, 1974a and 1974b.

despite much recent work,[321] we are still far from being able to give definitive answers to all the questions it raises. In particular certain errors in the fundamental coordinate system and the basic parameters of the solar model[322] (the obliquity of the ecliptic, the length of the tropical year, the rate of precession and the eccentricity of the solar orbit) infect all his astronomical workings in such a way that it becomes extremely difficult to diagnose the precise sources of mismatches between Ptolemy's results and those obtained on the basis of modern calculations.[323] Nevertheless some points are not in dispute. First the fewness of the actual observations cited in his detailed accounts of the movements of the planets in books IX to XI is obvious and remarkable. Thus for Mercury he uses some 17 precise observations, for Venus 11, and for Mars, Jupiter and Saturn a mere 5 each. In each case he cites the absolute minimum – or very close to the minimum – number of observations that are necessary to determine the parameters of what are, even for the superior planets, complex models.[324]

How he proceeds can be illustrated by considering the example of Venus. First, even before he turns to the detailed discussion of each of the planets in turn, he has set out a table giving their two main periodicities, that is the 'movement in longitude' (interpreted as the

[321] See, for example, Pannekoek 1955, Czwalina 1956–8 and 1959, van der Waerden 1958, Petersen and Schmidt 1967–8, Petersen 1969 and Gingerich forthcoming.

[322] In *Syntaxis* III 1 Ptolemy claims to confirm the parameters of Hipparchus' solar model with observations of solstices and equinoxes of his own which turn out to manifest a systematic error of about a day (see Rome 1937, 1938, Petersen and Schmidt 1967–8, Bruin and Bruin 1976). The match between the results he gives and the figures predicted by Hipparchus' model is sufficiently close to suggest that Ptolemy has been influenced by, and may even simply have worked back from, the latter – that is he may have selected his cases to tally closely with the value of the tropical year that he takes over from Hipparchus ($365 + \frac{1}{4} - \frac{1}{300}$ day) and that he is convinced on independent grounds to be, as he puts it, the 'nearest approximation possible' (ii 208.13f). But in judging this set of results we should bear in mind first that Ptolemy particularly emphasises the difficulty of observation in this context, noting that there may be an error of up to a quarter of a day in both solstice and equinox observations (*Syntaxis* III 1, ii 197.1ff, cf. also 194.12ff, 195.1ff – quoting Hipparchus – 202.14ff and cf. above, p. 181 n. 295), and secondly that what Ptolemy presents as his results here cannot *all* be simple observations since they include a solstice calculated as occurring two hours after midnight (ii 205.21ff: equally the 'most accurately observed' Hipparchan autumn equinox he takes as a point of comparison at ii 204.1ff is one calculated to have occurred at midnight).

[323] Thus Neugebauer (1975, I p. 107) concluded: 'In all ancient astronomy direct measurements and theoretical considerations are so inextricably intertwined that every correction at one point affects in the most complex fashion countless other data, not to mention the ever present numerical inaccuracies and arbitrary roundings which repeatedly have the same order of magnitude as the effects under consideration.'

[324] The best brief exposition of Ptolemy's planetary models is that in Neugebauer 1957, Appendix A, pp. 191ff: cf. also Pedersen 1974, chh. 9 and 10, Neugebauer 1975, I pp. 145ff.

movement of the epicycle's centre on the eccentric deferent) and the 'movement in anomaly' (interpreted as the movement of the planet itself on its epicycle). In *Syntaxis* IX 3, iii 213.16ff, he says he will set out the periodic returns 'as calculated by Hipparchus' with corrections that he obtained himself for which he will give some justification in due course, but there is no mention of the data from which the original figures were derived: no doubt in most cases they go back to Babylonian parameters, and it should be remarked that they are independent of any particular geometrical model.[325]

His general preliminaries to planetary theory incorporate four further points of importance. (1) He decides to ignore the latitudinal movements of the planets (deviations north and south of the ecliptic) in his exposition of their longitudinal movements:[326] this is a problem that he returns to in *Syntaxis* XIII. (2) While he observes that the epicyclic and the eccentric models are both able to account for some of the phenomena of planetary motion[327] (that is their first, or zodiacal, anomaly, which in practice he will explain by using an eccentric deferent), he says that the anomaly in respect of the sun[328] possesses a property that is incompatible with an eccentric model, namely that the time from the greatest to the mean movement is always longer than the time from the mean to the least. This can, however, be explained on the hypothesis of an epicycle moving in the same sense as the eccentric deferent, and he therefore uses a combination of both models for each planet. But while he asserts that this property is always found,[329] he does not give the data themselves, although the point is clearly fundamental for his choice of model. (3) He notes that the apogees of the eccentric circles are fixed in respect of the tropic and equinoctial points, not in respect of the fixed stars: in other words precession has to be taken into account.[330] Finally (4) he introduces the notion of an equant: the epicycle's centre is carried round on an eccentric circle, but its motion is uniform not with respect to either the centre of the eccentric or with respect to the earth, but with respect to a point on a line joining these two centres and at the same distance from the eccentric centre as the

---

[325] Ptolemy points out, IX 3, iii 214.2ff, that approximate or uncorrected figures are adequate for the exposition of his model.

[326] See *Syntaxis* IX 2 and 6, iii 211.24ff, 254.3ff.

[327] In *Syntaxis* III 3 Ptolemy, following Apollonius in all probability, demonstrates the equivalence of the two models in his exposition of the theory of the sun.

[328] It is the anomaly in respect of the sun that produces the phenomena of stations and retrogradations.

[329] *Syntaxis* IX 5, iii 250.15ff.

[330] *Syntaxis* IX 5, iii 252.2ff and 11ff.

earth.[331] Here, unlike the data underlying point (2), he will cite some empirical observations in the course of his exposition of each planet in turn.

His discussion of Venus in book x proceeds step by step as follows.[332] In the first chapter he cites two pairs of observations in order to determine the planet's apsidal line – the diameter of the eccentric circle through the apogee and the perigee. Both pairs of observations are of the planet at equal greatest elongations on opposite sides of the sun (that is first as an evening, then as a morning, star),[333] and using a theorem previously established (ix 6), he obtains the apsidal line by simply bisecting the angle between the two observed positions of the planet.[334] Chapter 2 then cites a further pair of observations made

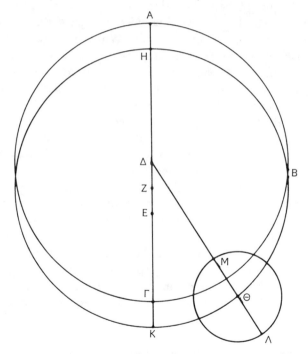

Diagram 1

[331] *Syntaxis* ix 5, iii 252.7ff and 18ff: in diagram 1 (from *Syntaxis* ix 6) E is the earth, AHΔZEΓK the apsidal line, Z the centre of the eccentric circle HBK: Θ, the centre of the epicycle, moves round the circumference of this circle, but its movement is uniform not with respect to Z, but with respect to Δ (the equant), that is the angle AΔ̂Θ increases/ decreases uniformly.

[332] Cf. Stahlman 1953, pp. 480ff, Pedersen 1974, pp. 298ff, Neugebauer 1975, I pp. 152ff.

[333] For the two inferior planets, which have maximum elongations from the sun, the centre of the epicycle can be identified with the position of the mean sun.

[334] In diagram 2 (from *Syntaxis* ix 6) Λ and M are two positions of the planet at greatest elongation (BÂZ and ΔM̂Z are right angles), Z is the earth, E the centre of the ec-

when (*a*) the sun was on the apsidal diameter that has just been determined, and (*b*) the planet was at greatest elongation from it. This immediately reveals which end of the diameter is perigee and which apogee. Ptolemy also asserts[335] (though he does not substantiate) that the sum of the greatest elongations on opposite sides of the sun is never less than the sum obtained when the sun is at apogee nor greater than the sum when the sun is at perigee – which confirms that a unique apsidal line, with a single perigee,[336] has been determined. Moreover the same observations enable him to obtain both the magnitude of the epicycle (its radius being expressed as a proportion of the radius of the eccentric circle) and the eccentricity (the distance from the earth to the centre of the eccentric circle, again expressed in terms of the radius of that circle).[337] The next step

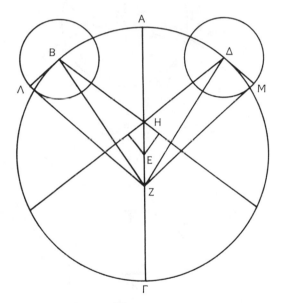

Diagram 2

centric and H the equant. In *Syntaxis* IX 6 Ptolemy first proves that when $A\hat{H}B = A\hat{H}\Delta$, then $H\hat{B}Z$ is equal to $H\hat{\Delta}Z$ and $B\hat{Z}\Lambda$ is also equal to $\Delta\hat{Z}M$ (that is the two greatest elongations are equal). In X 1 he uses this theorem in reverse: when $B\hat{Z}\Lambda$ and $\Delta\hat{Z}M$ are equal, the apsidal line AHEZΓ can be found by bisecting the angle $\Lambda\hat{Z}M$ (Λ and M being the two given positions of the planet).

335 *Syntaxis* X 2, iii 300.19ff.
336 This is true for all the planets except Mercury, where Ptolemy finds a double perigee and adapts his usual model: the centre of the eccentric circle itself moves in a circle round a point on the line between the earth and the apogee.
337 In diagram 3 (from *Syntaxis* X 2), where Δ is the centre of the eccentric circle and E the earth, the planet, at Z and H, is at maximum elongation (so $A\hat{Z}E$ and $E\hat{H}\Gamma$ are right angles) and ZA = HΓ (the radius of the epicycle is assumed to remain constant). By

is to determine the point round which the epicycle's centre moves uniformly. For this he uses two further observations made when the mean sun was a quadrant's distance from the apogee and the planet was at greatest elongation once as a morning and once as an evening star. Since the difference between the two elongations is twice the zodiacal anomaly, the zodiacal anomaly itself can be obtained,[338] and the result that Ptolemy reaches is that the centre of uniform motion (the equant) is twice as far from the earth as from the centre of the eccentric circle. The last main step is to determine the planet's

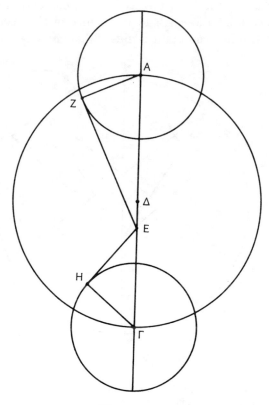

Diagram 3

simple trigonometry AZ can be found in terms AE (and that is AΔ+ΔE) and HΓ in terms of EΓ (and that is ΔΓ−ΓE), and so both AZ and ΔE in terms of the radius AΔ.

[338] In diagram 4 (from *Syntaxis* x 3), Z and H are the two observed positions of the planet, B the earth, Δ the equant, Θ the centre of the eccentric circle and E the position of the centre of the epicycle (which can be identified with that of the mean sun, here at a quadrant's distance from the apogee: so AÂE is a right angle). BE is first found from the angle EB̂H (which is half the sum of the two elongations, ZB̂H) and EH (the previously determined radius of the epicycle). But the difference between the two elongations is

mean motion in anomaly. For this he uses two more observations[339] separated by a considerable time. The earlier *rough* approximation had been that 5 cycles of anomaly correspond to 8 Egyptian years (of 365 days):[340] using this and taking the difference between the observed longitudes of the planet on the two occasions he can arrive at a more *exact* figure for the mean daily motion in anomaly by calculating the total distance travelled in degrees and dividing by the time taken.

Several features of this exposition call for comment. First he works with a geometrical model the main features of which are taken for granted. That the problem was to find a combination of *circular* movements to yield the resultant complex motion of the planet had been common ground to astronomers ever since Plato. It would be a mistake to explain this purely in terms of a quasi-religious respect for

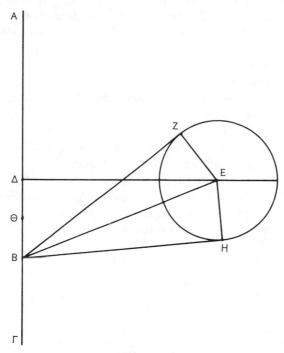

Diagram 4

double the zodiacal anomaly, $\hat{AEB}$: so $\hat{AEB}$ can be found: and from BE and $\hat{AEB}$, ∆B can be obtained.

[339] A third observation is cited merely to indicate that the planet was not at maximum elongation on the first of these two occasions.

[340] See *Syntaxis* x 4, iii 314.15ff, cf. ix 3, iii 215.5ff.

the perfection of the circle. The geometry of the circle is, after all, far simpler than that of any other curvilinear figure. Moreover the fact that the motion of the sphere of the fixed stars is circular could be used to justify the applicability of the circle to the problems of heavenly motions. The success of either an eccentric or an epicyclic model in giving a very close approximation to the movement of the sun meant that it was natural to use these models as a starting-point elsewhere, and although, as we have seen, Ptolemy does not specify all the grounds for his particular combination of eccentric and epicyclic hypotheses, some combination of the two had the advantage that the account given of the second irregularity in the motions of the planets was both consistent with and analogous to the account given of the first. The main point at which he diverges from all previous astronomers is in the introduction of the equant, which, since it broke the rule that motion along the circumference should be uniform with respect to the centre of the circle, was considered one of the major scandals of the Ptolemaic system down to Copernicus.[341] But this is, as we have seen, one point where, in his account of Venus at least,[342] Ptolemy claims that the observed data themselves necessitate the modification.

Yet while the geometry of the exposition of the model is both simple and elegant, the confrontation between theory and empirical data is kept to a minimum. We are given enough specific observations[343] to determine the parameters, but very little more than the absolute minimum. In particular he does not check his results by submitting them to further rigorous and extensive tests. Moreover several aspects of his workings are remarkable. It is clear that he makes two types of adjustments, first discounting minor discrepancies in the observational data that he reports, and secondly making rounding adjustments in the course of his calculations. Thus in the

---

[341] But contrast Neugebauer's judgement (1975, 1 p. 155) that the equant is 'probably Ptolemy's most important discovery in the theory of planetary motion'.

[342] The argument for the equant is clearest in the case of Venus (though even here no justification is offered for the assumption that it is located on the apsidal line). For the outer planets Ptolemy's procedure is very different. The equant is not arrived at as a conclusion from empirical data, but Ptolemy first finds the equant circle and then bisects the eccentricity of the equant to find the centre of the deferent. Pedersen 1974, pp. 277ff, conjectures that this was the first step in an iterative procedure of adjustment, once Ptolemy had decided that the equant could not be the centre of the deferent, but the *Syntaxis* gives no hint of the process by which he arrived at his conclusion. (If he had already worked out the model for Venus, he may, of course, have been influenced by it: but we do not know the order in which he tackled the planets.)

[343] On the difficulty of correctly determining the maximum elongations of the planet and on the inaccuracy of Ptolemy's reported positions, see especially Czwalina 1959, Neugebauer 1975, 1 p. 153 n. 1 and p. 158, and cf. Newton 1977, p. 307.

argument in *Syntaxis* x 1 that establishes that the apsidal line runs through the points on the zodiac defined as 25° within the Bull and 25° within the Scorpion, the first pair of observations gives the former result precisely, but the second yields a figure of 24° 58′ in the Scorpion: nor can the explanation of this be that we should make an allowance for precession, for (*a*) the figures still do not come out exactly, and (*b*) no such allowance was made in evaluating the first pair of observations. We should bear in mind his warnings concerning the difficulties of exact observation, and the way his own observations are presented here clearly reveals their imprecision. Thus in the fourth observation in *Syntaxis* x 1 (iii 298.11ff) he states Venus' position as 'nearly two-thirds of a full moon' from a particular fixed star. In this chapter, as often though not invariably elsewhere, he signals the approximate nature of his results by the use of the term ἔγγιστα – 'very nearly'[344] – though he does not specify what he takes to be the margin of error to which they are subject.

In his mathematical manipulations rounding operations occur frequently not only (1) in conversions from arcs to chords or vice versa, using the table of chords in *Syntaxis* 1 11, and (2) in the process of renorming that is a recurrent feature of his working with ratios between lengths and with chords and arcs (where we should use sines and cosines), but also in other contexts. Thus in the working in *Syntaxis* x 2, the chord of arc 94° 40′ is given as 'very nearly 88p13', where the table yields 88p14^i 12^{ii} 30^{iii}. The values he arrives at for the distance from the earth (*a*) to the centre of the eccentric circle, and (*b*) to the equant, are 'very nearly 1p15' (*Syntaxis* x 2, iii 302.17) and 'very nearly 2p30' (*Syntaxis* x 3, iii 305.17), but taking his own data, neither figure is correct to within a sixtieth,[345] and the neatness of his result – with one distance precisely twice the other – is factitious. He tells us in *Syntaxis* x 4 that he obtained the correction to the mean daily motion of the planet in anomaly from the observations he cites, putting it that 'just about that many degrees of surplus' are obtained from the tables for the planet that he had set out in ix 4. But if we divide the total number of degrees travelled (expressed sexagesimally as 25,35,38° 25′) by the total number of

344 *Syntaxis* x 1, iii 299.2.

345 This is also true of Ptolemy's figure for the radius of the epicycle (43p10), at least if we take his own figure of 1p15 for the eccentricity (from which we would obtain a figure of 43p9^{i}20^{ii}37^{iii} for the radius of the epicycle). But this is a case where two roundings cancel one another out: with the more accurate figure of 1p16 for the eccentricity, the result Ptolemy gives for the radius of the epicycle is correct to a sixtieth. Cf. Neugebauer's computation, from Ptolemy's data, of an eccentricity of 1p16^{i}48^{ii} and an epicycle radius of 43p10^{i}48^{ii} (Neugebauer 1975, I p. 154).

days (41,30,52;0) the figure we obtain is 0; 36^i 59^{ii} 25^{ii} 49^{iv} 8^v 51^{vi}, as against the figure given in the table[346] of 0; 36^i 59^{ii} 25^{iii} 53^{iv} 11^v 28^{vi}.

In each of the cases we have considered so far Ptolemy's adjustments are *transparent*. Even though we cannot say we fully understand the approximation and interpolation procedures he used, we can determine the relationship between the data he reports and the results he obtains. This does not tell us that the data are genuine, but if they were invented, they were not invented to correspond exactly and in all cases with his stated conclusions. But the charge of fudging sometimes takes the form not that he invented the data (reporting observations that he never undertook) but that he selected them to conform (exactly or approximately) to a preconceived theory. It has been argued that this is the case, for example, with his determination of precession, where the comparison between Ptolemy's discussion and the modern analysis done by Newton (1974*b*)[347] is revealing.

In *Syntaxis* VII 2 and 3 Ptolemy gives data concerning the declinations of eighteen stars observed by Timocharis, Hipparchus and himself and then cites six of these where he says that the differences in declination between Hipparchus' time and his own correspond to a difference of 2$\frac{2}{3}$° in longitude (that is a rate of precession of about 36″ a year). In his analysis Newton first asks how accurate Ptolemy's reported declinations are[348] and then compares the figures for the differences in declination for the stars Ptolemy uses with those for the stars he does not use. Here Newton cites the formula by which the longitude of a star can be obtained from its declination and its latitude, together with the obliquity of the ecliptic,[349] and he produces a table setting out his results. However he first discards four of the stars that Ptolemy mentioned on the grounds that the changes in declination are so small that a reliable value for pre-

---

[346] One of the surprising features of this table is that the figure for the yearly anomaly differs by 6^{v}5^{vi} from that obtained by multiplying the daily anomaly by 365. If Ptolemy worked back from an already existing table, using the yearly figure for the 409 whole years between the two observations cited, and the daily figure for the remaining days, he would reach a figure for the total number of degrees of anomaly of 25,35,38° 27^i 47^{ii} 51^{iii} 16^{iv} 28^v 11^{vi} as against 25,35,38° 25^i derived from his observations. Cf. other examples of imprecise mathematical manipulations mentioned by Neugebauer 1975, I pp. 91 and 197.

[347] Compare also Pannekoek 1955, and two earlier workings, Dreyer 1917–18, pp. 347f and Delambre 1817, II p. 254.

[348] They are all correct to within half a degree: but the error is on the same side for the six stars he uses, on either side for the other twelve.

[349] $\Delta(\sin \delta) = \Delta(\sin \lambda) \cos \beta \sin \epsilon$, where Δ denotes a change, δ is the declination, λ the longitude, β the latitude and ϵ the obliquity of the ecliptic.

cession cannot be inferred from them, and he then shows that in the remaining fourteen stars there is a marked discrepancy between the figure for precession for the six that Ptolemy uses and the figure for the other eight. Indeed he concludes that the *only* six stars that yield the conclusion that Ptolemy reached are the six he actually takes – which cannot then have been the result of a random selection from the original set of eighteen: the other eight in Newton's table give a figure of $52''.8$ per year for precession which is close to the correct value.[350]

Now if Ptolemy proceeded in the way that Newton does, the fraudulence is manifest. Yet his workings may well have differed from Newton's in two ways. First there is no reason to think that the four stars that Newton discards were discarded by Ptolemy.[351] Secondly and far more importantly, Ptolemy does not tell us how he proceeded in converting his declinations into longitudes.[352] The modern formula enables exact calculations to be made, as set out in Newton's table. But Ptolemy's way of referring to his results tells strongly against his using any precise ancient equivalent. Apart from the fact that he sometimes qualifies his conclusions by 'nearly' or 'almost',[353] roundings are in evidence at two points, (1) on one occasion he rounds the figure for the shift in declination,[354] and (2) more significantly still he never refers to the longitudes of the stars in question precisely. The shift in declination of $1\frac{1}{12}°$ for the middle star of the Pleiads is, he says, nearly the same as the shift in longitude of $2\frac{2}{3}°$ that we get for 'stars at the end of the Ram'. Capella has a shift in declination that corresponds to one of $2\frac{2}{3}°$ in longitude for 'stars in the middle of the Bull' (but Capella's position is given as $25°$ within the Bull in *Syntaxis* VII 5). Again the star in the western shoulder of Orion has a shift in declination that corresponds to one of $2\frac{2}{3}°$ in longitude for 'stars two-thirds within the Bull' (and this star's position is given as $24°$ within the Bull in *Syntaxis* VIII 1). It looks as if Ptolemy is here *working back* from the assumed figure of

350 Given by Pedersen 1974, p. 423 as $49''.86$, per year.
351 If, like Newton, he thought these four could yield no reliable value for precession (because the shift in declination is so small), there is no reason for him to have cited them in the first place: however we do not know what conclusions he reached in their case any more than we do for the other eight stars in his original list which he does not consider further.
352 The problem of finding the declinations and right ascensions of points on the ecliptic is discussed in *Syntaxis* I 14 and 16 and the general case of transformation from ecliptic to equatorial coordinates considered in *Syntaxis* VIII 5.
353 *Syntaxis* VII 3, iii 23.23, 24.13 and 25.10.
354 At *Syntaxis* VII 3, iii 24.12f he takes the difference between a declination of $1\frac{4}{5}°$ and one of $2\frac{1}{2}°$ (i.e. $42'$) to be 'very nearly' $\frac{2}{3}°$ ($40'$).

$2\frac{2}{3}°$ for precession and – it may be – consulting tables for conversions for *quite broad intervals* of longitude.[355]

That his conversions were rough and ready is further suggested by the way in which he turns from these data to a second group of data in the later part of VII 3 from which he says the question under discussion will be 'even clearer'.[356] Here he cites figures that he represents as the values for the longitudes themselves that follow from observations of the occultations of stars by the moon that were carried out by Timocharis, Agrippa of Bithynia and Menelaus,[357] and these give a range of rates of precession, although, with the exception of one figure for a period that Ptolemy may well have thought too short to yield a reliable value,[358] they all come close to his ultimate conclusion of about 36″ per year.[359]

If Ptolemy's workings with the declination data were as approximate as his descriptions of their results suggest, then those data would have been much less useful to him than they might have been. It is possible that he had at his disposal a table of precise results like that provided by Newton, and that he then fraudulently misrepresented them – although on this hypothesis we would have to suppose him not only fraudulent, but also very foolish in citing evidence from which his deceit could be deduced.[360] But a more likely story is that his workings were approximate and that the limits of tolerance within which he interpreted them were such that he felt justified in the conclusion that he usually (though not, it is true, invariably) expresses in the form of a rate of precession of '*very nearly*' 1° in 100 years. This was a figure that had some authority from Hipparchus (though it was probably his lower limit for precession) and it had the further advantage of mathematical convenience: but there is no reason to think that Ptolemy did not

---

[355] Cf. Manitius' notes and reference to the table of obliquities (*Syntaxis* I 15) in his commentary (1963, II pp. 20ff) and Pannekoek's table, 1955, p. 62.

[356] III 25.13ff: cf. the way the use of the plinth is introduced in the investigation of the obliquity of the ecliptic, above, p. 181 n. 297.

[357] The trouble here is that the longitudes are only obtained from the occultations by using Ptolemy's own lunar theory, which underestimates the mean lunar motion in longitude. Fotheringham and Longbottom 1914–15, for example, already took this to be the explanation for the low values for precession in these cases.

[358] 10′ in 12 years, III 30.15ff.

[359] They vary from 35″37‴ to 36″59‴ per year.

[360] Neither Delambre nor Newton provides a motive for Ptolemy's deliberate fraud, and it is hard to see why he should have preferred a figure that he appreciated to be widely inaccurate: it is not the case that a better approximation to precession would have threatened any part of his astronomical theories. Arguing against Delambre, Dreyer 1917–18, p. 347, put it that Delambre's own investigations furnished proof of the *bona fides* of Ptolemy since the observations of declinations do *not* agree with Ptolemy's preconceived notion of 36″ per year.

suppose that it was corroborated by the evidence he had at his disposal.[361] The magnitude of the error – more than 13″ per year – is no doubt surprising by *our standards of accuracy*, but it cannot be emphasised too strongly how dangerous it is to transfer those standards back to the ancient world.

Some evidence from outside the field of astronomy itself helps to throw further light on Ptolemy's attitudes to the relation between theory and observation. This comes from the *Optics*, where we have to rely on a Latin translation of a lost Arabic version of Ptolemy's work.[362] Book v, which deals with the problem of refraction, describes in chh. 6ff detailed experiments to determine the refraction from air to water, from air to glass, and from water to glass, and Ptolemy sets out the results of these in the form of tables giving – in degrees and half degrees – the angles of refraction for angles of incidence at 10° intervals from 10° to 80°.[363] In all three cases the first results are introduced with the qualification 'nearly',[364] but the interesting feature is that they all tally exactly with a general law taking the form $r = ai - bi^2$, where r is the angle of refraction, i the angle of incidence, and a and b are constants for the media concerned.

Both the similarity and the difference between this set of data from the *Optics* and the astronomical examples we have considered should be remarked. The similarity is that Ptolemy signals the *approximate* nature of the observational results he reports. But the more important difference is that the *Optics* – unlike the cases in the *Syntaxis* we have discussed – presents results *that have already been adjusted to tally perfectly* with the underlying general theory.[365] In this instance, it is clear that the observations have been interpreted before Ptolemy records them.

If we now turn back to his planetary models, it is obvious that we do not have enough evidence to reach confident conclusions concerning some aspects of his procedures. We may presume that he had other data besides those he quotes for each of the planets, and that he has therefore been selective in what he cites. But we cannot be sure about the principles on which his selections were made. His approximation and interpolation techniques are not those that we would use

---

361 This may and probably did include other data besides the observations specifically mentioned in vii 2 and 3.

362 See Lejeune 1956.

363 In Ptolemy's terminology the incident ray is the ray from the eye to the refracting surface, the refracted one from the surface to the object, the angles in both cases being measured to the perpendicular to the refracting surface.

364 See Lejeune 1956, p. 229.5, p. 234.2, p. 236.9.

365 The formula for the relationship between the angle of incidence and the angle of refraction is not, however, expressed anywhere in the *Optics*.

today, but how far he is prepared to adjust his data or to ignore conflicting evidence – how far he systematically biases what he records in favour of preconceived conclusions – is in large part a matter of guesswork. What *is* abundantly clear from the *Syntaxis* itself, however, is that he does not submit his results to rigorous and extensive controls.

He uses a general model – and in many cases specific parameters – from earlier astronomy, and there can be little doubt that as a whole he sought to *confirm* earlier results as far as possible, particularly those of Hipparchus. At the same time he is prepared to modify the current theory at certain points – to obtain a better fit with such evidence as he had at his disposal. His introduction of the equant is an example of this, and so too are the extra circles he employed in his models of Mercury and of the moon: in the last case he specifically says that this modification was a response to new data he had obtained concerning the moon's positions in the first and third quarters.[366]

The deductive nature of his exposition is clear.[367] It is an exercise in geometrical demonstration and that is where its great strength lies.[368] But it is not simply and solely a piece of pure mathematics. Unlike Aristarchus' *On the Sizes and Distances of the Sun and Moon* – at least if we take it that Aristarchus allowed himself quite arbitrary hypotheses[369] – Ptolemy attempted to provide a comprehensive and true account of the heavenly motions. Although, following Plato and Aristotle, ancient commentators often distinguished sharply between the mathematical, and the physical, aspects of astronomical inquiry,[370] there can be no doubt, in Ptolemy's case, that his aim was not simply to offer a mathematical model from which the positions of the

---

[366] Having set out Hipparchus' lunar theory, based on eclipse data – i.e. when the moon is at syzygy with the sun – in *Syntaxis* IV, Ptolemy remarks in *Syntaxis* V 1 and 2 on the discrepancies between this theory and certain data that he obtained with the astrolabe concerning the moon's positions in the first and third quarters. On the superiority of his final lunar model, in V 5, to its predecessors, see Petersen 1969 and Gingerich forthcoming.

[367] Pace Kattsoff's remarks (1947–8, pp. 21f) concerning the primary consideration to be given to the observations.

[368] One should not be misled by the inaccuracies of some of the parameters into underestimating the flexibility of the astronomical model he works with.

[369] See above, p. 121 and n. 328.

[370] Ancient and modern commentators alike often speak vaguely of 'saving the phenomena' as the aim of ancient astronomical theory, but by itself this slogan leaves open the answers to two vital questions: (1) the exactness of the fit expected between phenomena and theory if the latter is to be deemed to save the former, and (2) the status of the theoretical entities themselves (often contrasted as ὄντα with the mere appearances, φαινόμενα). I have argued elsewhere (1978*b*) that it is mistaken to hold, as Duhem and others have done, that ancient astronomers were, in general, interested purely in the mathematics of their problems to the exclusion of any concern with the truth of their accounts or with questions relating to the underlying physical realities.

heavenly bodies could be predicted. One of his other astronomical treatises, the *Planetary Hypotheses*, confirms that he hoped for a true physical account, indeed one that covered not just the kinematics, but also the dynamics, of heavenly movement.[371] In book II ch. 6 (II 117.17ff) of that work the planets, sun and moon are imagined as carried on tambourines or segments of spheres, and in II ch. 7 the dynamics of their movement is interpreted in vitalist terms, each planet being said to possess a vital force and to impart motion to the bodies it is connected with. In the *Syntaxis*, too, his concern to establish that the earth is at rest in the centre of the universe shows that he is far from indifferent to the physical aspects of his inquiry,[372] and in XIII 2 (Iii 532.12ff), when he tells his reader not to be dismayed at the complexity of the hypotheses employed, his stand-point is again not one of indifference to the question of whether his devices represent the true system. Why, after all, should he worry over purely mathematical complexity? The source of his concern seems, in part at least, to be the implications of those complexities when translated into physical terms. But even here he does not defend his hypotheses solely on the grounds that they save the phenomena: rather he adduces physical arguments from the nature of the substance of the heavenly region (which is eternal, unchanging, homogeneous and transparent)[373] to support the possibility of the types of motion he proposes.

It is true that his astronomical model fails to deal adequately with certain problems, some at least of which he was evidently aware of:[374] and in the case of the moon, in particular, the parameters he adopted for the ratio of the epicycle and the deferent are notoriously at odds with experience in one respect, that one of their consequences should have been that the angular diameter of the moon should vary by a factor of almost two.[375] Yet whatever the imperfections and failures of his system, he clearly tried for more than a purely mathematical account. Moreover *as* a mathematical account it is fully

---

[371] Moreover he there gives absolute figures for the distances between the heavenly bodies, postulating that the greatest distance of one planet corresponds to the least distance of the next higher planet: see Goldstein 1967 and Neugebauer 1975, II pp. 917ff.

[372] *Syntaxis* I 5 and 7, Ii 16.20ff and 21.9ff.

[373] *Syntaxis* XIII 2, Iii 532.16ff, 533.3ff, 15ff.

[374] This seems clear from the embarrassment he expresses, in the passage just quoted from *Syntaxis* XIII 2 (Iii 532.12ff), concerning the complexity of the hypotheses he has to use. In other passages, e.g. *Syntaxis* III 1 and IX 5, Ii 208.13f and Iii 252.17f, he qualifies his account by remarking that his results are correct so far as present information goes.

[375] This was an objection brought by Copernicus, for example, in the *De Revolutionibus* IV 2. Similar, though less striking, difficulties arise from the parameters chosen for the epicycles of Venus and Mars.

determined: the observed positions of the heavenly bodies can be matched against the predictions, even though Ptolemy is sometimes less than energetic in submitting his theories to such tests, or at least in reporting the outcome.

Like dissection, astronomy provides clear evidence of the thoroughness, range and accuracy of the empirical investigations carried out by some Greek scientists, though – as with dissection – the early steps were hesitant and it was not until some time after the fourth century B.C. that the fullest researches were undertaken. The underlying general motives were, as we explained, complex: most investigators believed that the study of the heavenly bodies would enable them not merely to predict their movements, but also to foretell future events on earth. Moreover Ptolemy's approximation and adjustment procedures contrast with those that we should adopt or permit nowadays. Nevertheless his and Hipparchus' star catalogues presuppose careful and sustained observational work. Furthermore the epicycle–eccentric model used by Hipparchus for the sun and moon and then adapted and extended by Ptolemy provides the outstanding example, from the ancient world, of a theory that combined the mathematical rigour that Greek scientists demanded with a detailed empirical base. Even so Ptolemy at least shows more confidence in the mathematics of his theory than in its empirical support, and while some of that may be put down to his general recognition of the difficulty of accurate observation and of the doubtful reliability of his instruments, that is not the whole story. While there can be no doubt that he saw the general importance of obtaining trustworthy data, parts of the *Syntaxis* show little awareness of the need for the rigorous and repeated checking and control of results against accumulated evidence – or of the need for the meticulous recording and presentation of that evidence. As with dissection, so too in astronomy, there was no lack of observational data, but the observations were more often deployed to illustrate and support theories than to test them.

ARISTOTLE

Our studies of the biological sciences and astronomy show how, under certain conditions and within certain limits, Greek scientists were capable of undertaking detailed and fruitful empirical research. But in both cases the contrast between the quality and extent of the work carried out in classical times and in the Hellenistic period is striking. We have seen the limitations (as well as the achievements) of

Hippocratic research in the medical sciences and of what can reasonably be attributed to fourth-century astronomers, notably Eudoxus, and yet these two areas rank among the most promising instances of the empirical approach in early Greek natural science. We have, however, yet to come to terms with Aristotle's position in this regard. Various aspects of his work – his general epistemological theories, his use of dissection and the quality of the astronomical observations he reports – have been mentioned already. But we must now attempt an overview of his significance in the developments with which we are concerned.

The most important point is easily stated and uncontroversial. Although groups of Hippocratic physicians, even in certain respects the Pythagorean communities, undertook empirical inquiries in particular fields, it was Aristotle – all are agreed – who was the first to institute a comprehensive programme of research covering the natural sciences as well as many other disciplines. The school he founded, the Lyceum, owed a good deal, to be sure, to Plato's Academy,[376] but while Plato certainly stimulated work in mathematics and in the exact sciences, in political theory and moral philosophy, and in dialectic itself, the natural sciences played at most a very limited and subordinate role in the interests of the Academy. They did not figure in the programme of higher education in the *Republic*, and although Plato conducted one, somewhat idiosyncratic, foray into the field himself, in the *Timaeus*, he continued to stress, there, the inferiority of any account of the world of becoming.[377] The Lyceum, on the other hand, brought together under the leadership first of Aristotle and then of Theophrastus several of the most important figures in fourth-century natural science who worked in collaboration on a wide range of topics both in that field and in others. Here for the first time in the ancient world we can talk of a corporate research effort, planned and implemented as a whole. First there were what may be called histories of thought, surveys of the main earlier physical doctrines and of theories of sense-perception (both by Theophrastus), of medicine (by Meno) and of mathematics and astronomy (by Eudemus). Secondly there was the great collection of histories of the political constitutions of particular states.[378] Thirdly, in the domain of the natural sciences, Aristotle's own zoological treatises were complemented by Theophrastus' botanical works. In what we should call chemistry, the fourth book of the

376 On the differences in the legal status of the Academy and of the Lyceum, see Cherniss 1945, Lynch 1972.   377 E.g. *Ti.* 27 d ff, 29 bc.
378 There were 158 of these, but only one, the *Constitution of Athens*, has survived.

Meteorologica[379] was supplemented by a detailed study of minerals,
again by Theophrastus, the treatise *On Stones*. Even in statics and
dynamics, where Aristotle has no systematic discussion, but merely
introduces certain ideas on motion and weight in a variety of
contexts in the *Physics* and *De Caelo*,[380] the author of *On Mechanics*
produced a study of the lever and of three other simple machines,
the pulley, the wedge and the windlass, and the third head of the
Lyceum, Strato, appears to have been the first person to have
attempted to carry out some admittedly primitive empirical investigations on certain problems in dynamics, notably in connection with
the phenomena of acceleration.[381]

Here then was an unprecedentedly ambitious series of studies
encompassing most of the main fields of ancient φυσική and incorporating a good deal of original empirical research as well as
dialectical inquiries. Indeed apart from in the Alexandrian Museum,
the scope of natural scientific investigations undertaken in the
Lyceum was to remain unsurpassed in the whole of antiquity. Since,
in general, it is undeniable that the lack of what we may call an
institutional framework was one of the major obstacles to the
development of natural science in the ancient world, these exceptions
to that rule take on an added significance. Although, unlike the
Museum, the Lyceum was self-financing and did not enjoy royal
patronage, it provided an opportunity for collaboration in research.

Moreover it is well known that in the main the inquiries of the
Lyceum follow, and represent the full fruit of, Aristotle's own
methodological principles. Thus the systematic histories of previous
thought are a natural extension of the surveys of earlier views that
Aristotle so often presents in the extant treatises, though not, or
certainly not primarily, for historical purposes, so much as in order to
set out the accepted opinions on a subject and to identify the chief
ἀπορίαι, difficulties, that require resolution, before proceeding to his
own discussion.

As we noted before, the φαινόμενα that Aristotle generally posits

---

[379] The question of the authenticity of this work is disputed (see, for example, Hammer-
Jensen 1915, Düring 1944, Gottschalk 1961) but unlike *On Mechanics* its doctrinal
position is, on the whole, perfectly consistent with the views expressed in the indisput-
ably authentic works of the Corpus.

[380] See below, pp. 203ff.

[381] This may reasonably be inferred from the quotations from his work *On Motion* that
are preserved by Simplicius in his discussion at *In Ph.* 916.10ff, see 14ff, 21ff. Strato, who
was nicknamed ὁ φυσικός, also wrote on other aspects of physics, and on zoology,
pathology, psychology and technology, but with the exception of some arguments
concerning the existence of the void, preserved in Hero's *Pneumatics* (see Gottschalk
1965), very little of this extensive output remains.

as the starting-point of his analysis are, even in the natural sciences, far from being all the results of observation. Yet 'what appears to be the case', especially but not solely with the added specification 'according to perception', often includes what Aristotle takes to be the observed facts. The range of Aristotle's inquiries is not in doubt, nor the fact of his insistence on first obtaining the relevant data before undertaking their explanation. But the questions we must now confront are how far his actual practice in the natural sciences tallies with his methodological pronouncements, and the nature and extent of his use of observation and empirical research.

In many areas of what Aristotle terms 'physics' (the study of nature in general) the appeal to anything that could be described as the result of deliberate observation plays a minimal role. Much of the treatise that has that title consists in a highly abstract study of such concepts as time, place, infinity and the continuum. Here, naturally enough, the strengths of Aristotle's discussion lie in his scrutiny of accepted views and assumptions, in his analysis of aspects of common linguistic usage and careful distinctions between the senses of terms. The prominent example or paradigmatic case – whether a part of common experience or a feature of our way of talking about it – often acts as the starting-point, and indeed the key, to his own doctrine. Thus in his treatment of place (*Ph.* IV 1–5)[382] he first distinguishes various senses in which a thing may be said to be 'in' another, taking the strictest sense to be that in which a thing is said to be *in* a vessel (210a24). In the subsequent analysis, which starts from a series of assumptions such as that place is that which *contains* that of which it is the place, and that it is 'left behind' by what occupies it (210b34ff), the vessel example recurs. Place is indeed said to be a non-portable vessel at *Ph.* 212a15f as he works towards an eventual definition of it as the innermost static boundary of the container (212a20f).[383] Again his remarks on certain relationships between the force exerted and the movement caused are mostly based on such common experiences as that of a ship being hauled by a team of men, explicitly cited to illustrate the particular point that a single man may not be able to move the ship at all.[384]

Yet while throughout his discussion of physical problems he pays

[382] Cf. Owen 1970, pp. 252ff, who illustrates Aristotle's characteristic methods with this and other examples: cf. also Owen (1961) 1975, pp. 115ff.
[383] This has, however, the unfortunate consequence that, strictly speaking, a point cannot have a place, because a point is, and does not have, a boundary: see *Ph.* 209a7ff and 212b24f.
[384] *Ph.* 250a17ff, 253b18.

due attention to well-known data of experience as well as to features of linguistic usage, most of the data in question are clearly not the product of deliberate observation or research.[385] Even when he draws a contrast between a more abstract, logical or dialectical argument and one that he describes as more 'physical', the latter may make little use of the results of observation. Thus in *Ph.* III 5, when he refutes the suggestion that there is an infinite sensible body, he first adduces a 'logical' argument based on the definition of body,[386] and then considers the matter more from a 'physical' point of view. This argument takes the form of a multiple dilemma, but in the refutation of each of the alternatives there is only one direct reference to 'what appears', when he rejects the possibility of there being any other infinite body besides the four elements (and none of them, as he goes on to show, can be infinite).[387] Although he castigates those who deny movement on the grounds that they 'ignore' or 'do away with' perception,[388] that criticism is a purely general one. Nor, on the frequent occasions when he employs the expression 'we see' in relation to what he maintains, is it always the case that it is simply a matter of the use of the untutored eye or the mere faculty of sight. The range of things we are said to 'see' in the *Physics* alone includes, for instance, such comparatively straightforward cases as seeing that some things are sometimes in movement and sometimes at rest (254a35f) or that some things have the ability to move themselves (259b1ff), but also some much more tendentious or 'interpreted' examples (as when he says that we see that there is always some part of the animal in motion,[389] or that we see that a given weight or body

---

[385] Some of the complexities of the phenomena, or the inadequacies of some of Aristotle's generalisations about them, might have been revealed by empirical investigations. Thus Philoponus, who took some of Aristotle's admittedly rather loose statements about proportionalities (e.g. *Ph.* 216a15f) to imply that in free fall the speed varies directly with the weight, was, notoriously, to complain that this is 'completely false' and that 'this can be established by what is actually observed more powerfully than by any sort of demonstration by arguments': see *In Ph.* 683.16ff where he goes on to suggest an experiment of dropping two different weights from a considerable height, when the difference in the times will be found not to correspond to the difference in the weights.

[386] *Ph.* 204b4ff. A body, being defined as *bounded* by a surface, cannot be infinite. The argument, as Owen (1961) 1975, pp. 125f, notes, proves too much: 'starting from a definition that applies to mathematical as well as physical solids, it reaches conclusions that apply to both sciences'.

[387] *Ph.* 204b10ff, especially 35. The infinite body must be either (A) compound (but that is ruled out because the elements are finite in number and neither (i) one element nor (ii) all the elements can be infinite) or (B) simple (but that too is rejected because (i) it cannot be something other than the elements, and (ii) it cannot be one of the elements).

[388] *Ph.* 253a32ff, cf. 254a24ff: a similar charge is made again at *GC* 325a13ff.

[389] *Ph.* 253a11f. Ross explains this as a reference to growth and decay (though these are scarcely easy to observe in the process of occurring) and respiration (though not all

moves faster for two reasons, either because of a difference in the medium traversed, or – other things being equal – because of the greater weight or lightness of the body)[390] or even plainly erroneous ones: at 242 b 59ff (cf. 24ff) he says that we see that what brings about a movement is in all cases either in contact, or continuous, with what it moves.[391]

In general the empirical data marshalled in the *Physics* amount to little more than some well-known facts – or what were taken to be such – from common experience. But in the other physical treatises, especially in the *De Caelo* and the *Meteorologica*, there is a greater deployment of empirical evidence, including some evidence obtained, it would appear, from deliberate investigation. We have already mentioned some of the astronomical observations he cites (whether or not he carried them out himself).[392] Two other contexts, particularly, enable us to study the interplay of empirical and other factors in Aristotle's arguments, namely first his proofs, in the *De Caelo*, that the earth is at rest at the centre of the universe, and that it is spherical, and secondly his element theory.

The main argument for the thesis that the earth is at rest at the centre of the universe[393] depends on the doctrine of the natural movements and places of the elements. Aristotle assumes that what is true of individual parcels of earth is true also of the earth as a whole: he takes it as a fact that earth always and everywhere has a tendency to move downwards towards its natural place (the centre) and once it has reached this, it comes to rest.[394] So even if the earth as a whole were transported to the moon, he says, separate parts of it would not move towards the whole, but towards the place where the earth is now.[395] Some attempt is made to adduce specific empirical grounds for this thesis, but the attempts are sketchy. First at *Cael.* 296a34ff he develops a Modus Tollens argument based on an assumed analogy between the earth and the planets. If the earth as a whole moved,[396]

---

animals respire, see *Resp.* 470 b 9f, and again one might object that this is sometimes difficult to observe).

[390] *Ph.* 215 a 25ff, cf. 216 a 13ff (by 'other things being equal' Aristotle has in mind differences in shape). Cf. Owen 1970, p. 254, who rightly concludes that 'it was...no part of the dialectic of his argument to give these proportionalities the rigor of scientific laws or present them as the record of exact observation'.

[391] Cf. also *Ph.* 189 a 29, 203 a 24, 256 b 20ff.   [392] See above, pp. 179f.

[393] This discussion is one of the occasions when he criticises his opponents (here the Pythagoreans) for forcing the 'appearances' to fit arguments and opinions of their own, *Cael.* 293 a 25ff, see above, p. 137.

[394] The point is often repeated: see *Cael.* 295 b 19ff, 296 a 30f, b 6ff, 27ff, and cf. e.g. 270 a 3ff, 276 a 2ff.   [395] *Cael.* 310 b 3ff, cf. 294 a 17ff.

[396] He claims that the same argument applies whether the earth is thought to move (like a planet) about the centre of the universe, or about its own axis: *Cael.* 296 b 2f.

it would – like the planets – have a double motion. But if that were the case, the fixed stars would undergo 'deviations and turnings'.[397] But this does not appear to happen – the stars rise and set at the same places on the earth. The point is not developed and seemingly not thought through.[398] Secondly, at *Cael.* 297a4ff he merely states that his doctrine is supported by astronomy in that 'the appearances' of the regular changes in position of the constellations are consistent with the view that the earth is at rest at the centre, but again he does not elaborate the point. Finally it is even more remarkable that he should cite as a piece of evidence (σημεῖον) for his thesis that heavy objects fall to the earth at the same angles,[399] and not in parallel lines.[400] This was not something that he could have verified independently, but itself an inference from a point that he had not yet proved, namely that the earth is spherical.

If the empirical supports adduced for the doctrine that the earth is at rest are thin, those cited to establish its sphericity are more impressive. After various arguments relating, for example, to how heavy things would conglomerate around the centre (*Cael.* 297a8ff), he turns to new considerations at 297b23ff with the words: ἔτι δὲ καὶ διὰ τῶν φαινομένων κατὰ τὴν αἴσθησιν. First, in eclipses of the moon, the shadow of the earth is always circular.[401] Secondly, the fact that the stars that can be seen from more northerly and from more southerly positions on the earth differ shows both that the earth is spherical and that it is of no great size compared with the distance of the stars.[402] The end of the chapter gives us our first recorded estimate of the size of the earth, and although Aristotle does not tell us how the 'mathematicians' he refers to calculated this, it is perhaps a reasonable inference from the earlier mention of the differences in the positions of the stars seen from different latitudes that – like Eratosthenes some generations later – that was their method.[403]

---

[397] παρόδους καὶ τροπάς, *Cael.* 296b4. On the interpretation of these terms here, see, for example, Heath 1913, pp. 240f, Dicks 1970, pp. 196f.

[398] Aristotle appears to assume that the second postulated movement of the earth would be oblique to the first and cause differences in the observed risings and settings of the fixed stars. The possibility of a simple axial rotation accounting for the diurnal movement of the heavens is neither mentioned nor considered.

[399] That is, at right angles to the surface of the earth.

[400] *Cael.* 296b18ff, cf. 297b18f.

[401] *Cael.* 297b24ff. Aristotle says this 'always' happens. Given that lunar eclipses can occur at any point along the ecliptic, this shows that the earth is not merely an annular disk, but a sphere.

[402] *Cael.* 297b30ff. At 294a1ff he points out that it is wrong to infer that the earth is flat from the apparent straightness of the horizon (an instance where an appearance may be misleading: see φαίνεται a6, φαντασία a7).

[403] *Cael.* 298a15ff, cf. *Mete.* 365a29ff and 339b6ff. On the history of successive Greek attempts to determine the size of the earth, see for example Berger 1903, Heath 1913,

Aristotle's element theory provides an even better opportunity to compare and contrast his approach with that of his predecessors. In discussing the use of empirical methods in Presocratic natural philosophy we remarked that although most of their theories about the fundamental constituents of things attempt to accommodate what were taken to be the known 'facts', they do so only by pre-supposing a particular interpretation of the facts concerned, and little attempt is made to obtain new data, let alone to set up a situation in which the data might enable a decision to be made between two competing theories. On several occasions Aristotle criticises his predecessors' doctrines for neglecting certain of the 'appearances'. Thus Plato is taken to task for excluding one of the four elements (earth) from the natural changes that affect the other three: this is neither reasonable nor in accordance with perception.[404] Empedocles too is charged with proposing a theory that contradicts both itself and the 'appearances'.[405] Nevertheless Aristotle's fundamental objections often depend not on any appeal to what he took to be the evidence of the senses, but on abstract arguments and conceptual points. His refutation of the atomists, in particular, is based first on a distinction between actual and potential divisibility,[406] and secondly on further distinctions between coming-to-be and mere association or dissociation and between these and qualitative change and combination.[407]

As usual, his review of earlier opinions leads towards his own positive theory, which he sets out in a closely reasoned, but abstract and schematic argument in *De Generatione et Corruptione*. Coming-to-be and passing-away presuppose sensible bodies (328b32f) and these in turn presuppose sensible contrarieties: for a body must be either light or heavy, either cold or hot (329a10ff). But of the possible types of

---

pp. 147, 236, 339ff. In the *Meteorologica* Aristotle not only provides a good deal of geographical and geological data (concerning, for instance, the main rivers in the inhabited part of the world, 1 13, and changes in the relation between sea and land caused, for example, by silting, 1 14) but also outlines a theory of the main zones into which the earth's surface can be divided (*Mete.* 362a32ff).

[404] *Cael.* 306a3ff. Vlastos 1975, pp. 81ff, has recently argued that this charge against Plato is baseless: Plato's theory, like those of his predecessors, was not one that could be refuted empirically, and he would have denied that the change of earth into water or air or fire is something we see. Whatever the truth about Plato's position, it is evident, and important, that Aristotle at least treated earlier element theories as if they were empirically falsifiable.

[405] *GC* 315a3ff. Again the point is that he denied that one element can come to be from another – a view which Aristotle holds he should have been committed to, since he differentiated the elements by such qualities as white, hot, heavy and hard.

[406] See *GC* 316b19ff, 317a2ff, 12ff.

[407] *GC* 325a23ff, b29ff, 328a5ff.

contrarieties, tangible contrarieties alone will be the principles of sensible bodies (b 7ff). Already, therefore, the way he defines the problem – the search for the principles of sensible bodies – effectively rules out any quantitative or mathematical theory of the elements of physical bodies. In the argument that follows he lists the tangible contrarieties, such as rough and smooth, hard and soft, heavy and light, and then reduces these to two pairs of primary opposites, hot and cold, and wet and dry. All the other tangible contraries can, he claims, be derived from these, though these cannot be further reduced (330 a 24ff), and he then proceeds to correlate his two primary pairs with the four simple bodies.[408]

The references to what we may call empirical data in this argument are limited, and there is not a single observation that can clearly be said to involve research. Rather he builds on and appeals to certain correlations suggested by ordinary experience or by the associations of Greek terms. He is aware of the ambiguity of hot and cold and dry and wet in particular,[409] but offers his own definition of these,[410] and he does not consider alternative views of the correlations of the primary opposites and the simple bodies although we know that other opinions had been expressed.[411] He claims that the coming-to-be of the simple bodies out of one another is evident to perception:[412] yet it remains the case that in Aristotle, as in writers both before and after him, the denotations of 'earth' 'water' 'air' and even 'fire' are vague, and what is to count as a change from earth to water, for instance, is interpreted quite loosely.

Thus far there would be little reason to contrast Aristotle's

---

[408] *GC* 330 b 3ff. Fire is hot and dry, air hot and wet, water cold and wet and earth cold and dry.

[409] E.g. *GC* 330 a 12ff, and cf. *PA* 648 a 21ff, 36ff, 649 b 9ff.

[410] Hot is 'that which combines things of the same kind', cold 'that which brings together and combines homogeneous and heterogeneous things alike', wet 'that which, being readily delimited (i.e. by something else), is not determined by its own boundary', dry 'that which, not being readily determined (i.e. by something else), is determined by its own boundary', *GC* 329 b 26ff.

[411] Thus Philistion is reported in Anon. Lond. (xx 25ff) as having held a doctrine in which each of the four simple bodies is associated with a single opposite, fire being hot, air cold, water wet and earth dry. On Aristotle's theory, too, one of the two opposites associated with each simple body is primary, but on his view air is wet and water cold (see *GC* 331 a 2ff).

[412] *GC* 331 a 8f (though having said that these changes appear to occur according to *perception* he adds an argument: otherwise there would be no qualitative change, for that is change with respect to the qualities of tangible things). At *GC* 331 b 24f, too, he claims that his theory of fire coming from air and earth is in agreement with perception, arguing that flame is burning smoke and smoke consists of air and earth. Cf. also *Cael.* 302 a 21ff (fire and earth *evidently* come out of flesh, wood and suchlike), 304 b 26f (*we see* fire and water and each of the simple bodies undergo dissolution), 305 a 9f and *Mete.* 340 a 8ff.

approach to this set of problems from that of most of his predecessors even though he makes occasional direct appeals to the evidence of the senses. But if we may assume that the fourth book of the *Meteorologica* is, if not by Aristotle, at least by one of his close associates, we there have, for the first time in Greek science, an attempt to discuss a quite wide range of physical changes and phenomena. The various properties of substances are listed and discussed and a large number of compounds are classified according to the simple body that predominates in them. We are given an analysis, for example, of which substances are combustible, which incombustible, which can be melted, which solidified under the influence of either cold or heat, which are soluble in water and other liquids. Most of the theories proposed are extrapolations from prominent phenomena treated as paradigm cases. Thus when we are told that substances that are solidified by cold (and are dissoluble by fire)[413] consist predominantly of water, ice, which is cited at 388b11, is no doubt the, or at least a, paradigm. When it is suggested that those substances that are solidified by heat have a greater proportion of earth, potter's clay figures as the most prominent example.[414] Yet the theory is flexible enough to take in a considerable number of ordinary and some not so ordinary substances, and many of the properties they exhibit and the change they undergo.

The main strengths of the discussion in *Meteorologica* IV lie in the fact that it offers an account of a far greater range of phenomena than had previously been dealt with in physical theory. The behaviour of a large number of natural substances in various circumstances is discussed, and the circumstances include not only some comparatively complex technological processes[415] but also some artificially contrived situations – where the data obtained are not just familiar facts but the result of fairly deliberate investigations (whether or not it was Aristotle who carried these out). Thus among the substances that are said to freeze solid with cold are not only urine, vinegar, whey and lye, but also serum and semen.[416] Salt and soda are said to be soluble in some liquids, but not in others, where olive oil is specifically referred to (383b13ff). We are told that the blood of certain animals, and blood that has had the fibres removed, does not coagulate,[417] and different types of wines are

---

[413] See *Mete.* 383a3f.
[414] *Mete.* 388b12, cf. 383a19ff, b11ff, 23.
[415] For example iron-making at *Mete.* 383a32ff: cf. also the reference in *GA* 735b16ff to the flotation of lead ore with a mixture of water and oil.
[416] See *Mete.* 384a11ff, 389a9ff, 22f, and contrast *HA* 523a18ff, *GA* 735a35ff on semen.
[417] See *Mete.* 384a26ff, 29ff, 389a19ff, cf. *HA* 520b23ff, *PA* 650b14ff.

distinguished according to their combustibility and their readiness to freeze.[418]

Yet although a remarkable number and variety of observations are explicitly referred to and even more are presupposed by the correlations of properties that are suggested,[419] the data are, throughout, interpreted in terms of the underlying theory. In many cases the way the phenomena are *described* already incorporates the theoretical explanation, as when the curdling of milk is represented as the separating of 'the earthy part' (384a20ff). More importantly, deviant phenomena are generally dealt with by more or less *ad hoc* adjustments to the theory. Olive oil, we are told at 383b20ff, causes particular difficulty. If it contained more water, cold should solidify it, if more earth, fire should do so. In fact neither solidifies it, but both make it more dense. But the explanation given is that it contains more air – which is also why it floats on water (though this is not represented as an independent test of its containing air).[420]

It is not as if the phenomena as a whole are collected with a view to criticising the overall theory of the four primary opposites, the four simple bodies and their principal modes of interaction. The aim is, rather, to show how a rational account of the phenomena can be achieved within the given framework. We should certainly not underestimate the importance of the attempt to broaden the empirical basis of physical speculation, but the role of the data thus collected in *Meteorologica* IV was to illustrate and support the theory, not to put it to serious risk. While at the level of the distinctions between coming-to-be, qualitative change and combination such empirical research exhibited the complexities of the phenomena that any reductionist doctrine had to take into account, that research left the chief issues between quantitative and qualitative theories of matter unresolved and the problems continue to be debated, in the period after Aristotle, mainly on general grounds connected with the concepts of divisibility and the continuum.[421]

---

[418] See *Mete.* 387b9ff, 388a33ff. Cf. also the reference in *Meteorologica* II, 358b18ff, to what happens when wine is evaporated (see Lloyd 1964, pp. 64f).

[419] In chh. 6–9 especially, e.g. the discussion of the relationship between melting and softening by water, *Mete.* 385b12ff.

[420] A similar difficulty is raised about the nature of semen in *GA* 735a29ff, where, after a comparison with olive oil, Aristotle concludes that semen, too, consists of water and *pneuma* (explained as air endowed with a special form of heat, 736a1, b33ff).

[421] Both the four-element theory, and atomism, its chief rival, were sufficiently indeterminate to be incapable of conclusive corroboration or refutation by means of practical tests. In the debate between Epicureans and Stoics the empirical data cited are mostly the usual well-known phenomena. At the opposite extreme the vast array of 'facts' about natural substances that are assembled in such a writer as Pliny fails to advance the theoretical discussion materially. Nevertheless in certain areas important empirical

The range of Aristotle's investigations in zoology is such that our discussion has to be even more drastically selective than ever. Yet the need to come to some assessment of his performance in this field is all the more pressing in that it has been subject to such divergent judgements. Some of the most extravagant praise, but also some of the most damning criticisms, have been directed at his empirical researches in zoology.[422]

The massive array of information set out in the main zoological treatises[423] can hardly fail to impress at the very least as a formidable piece of organisation. But both Aristotle's sources and his principles of selection raise problems. As we have already noted in connection with his use of dissection, it is often impossible to distinguish Aristotle's personal investigations from those of his assistants, although, given the collaborative nature of the work of the Lyceum, that point is not a fundamental one. It is abundantly clear from repeated references in the text that he and his helpers consulted hunters, fishermen, horse-rearers, pig-breeders, bee-keepers, eel-breeders, doctors, veterinary surgeons, midwives and many others with specialised knowledge of animals.[424] But a second major source of information is what he has read, ranging from Homer and other poets, through Ctesias and Herodotus to many of the Hippocratic authors.[425] In general he is cautious in his evaluations of all this secondary evidence. He points out, for example, that hunters and fishermen do not observe animals from motives of research and that

work was done, even if it did not lead to major new theories. I have already noted Theophrastus' *On Stones*, which implicitly raised the problem of the status of earth as a simple body (chh. 48ff especially) and in *On Fire* Theophrastus explicitly questioned the nature of fire and drew attention to certain important respects in which it differs from the other simple bodies (notably in that it always exists in a substratum). Again on the question of the existence of the void, Strato appears to have initiated empirical investigations which were designed to provide experimental demonstration that a continuous vacuum can be created artificially (our chief evidence comes from Hero's *Pneumatica*: see 1 16.16ff especially). Some of this research remains, to be sure, fairly rudimentary, and too little use is made of quantitative measurements (even though the Greeks distinguished different kinds of waters, and in some cases solids, by their weight, that is their specific gravity). But the main shortcoming of later Greek physical speculation was not so much a lack of empirical research, nor inadequately debated theories, as a mismatch between the two, the failure to tailor the one to the other.

[422] Contrast the evaluations in, for example, Bourgey 1955 and Lewes 1864.

[423] The authenticity of some of the later books of *HA* (VII, VIII and IX) is open to doubt, but I shall treat the whole (with the exception of the patently anomalous x) as evidence for work organised and planned by Aristotle, if not carried out by him.

[424] There is a convenient analysis of Aristotle's principal sources in Manquat 1932, pp. 31ff, 49ff, 59ff. Cf. also Lones 1912, Le Blond 1939, pp. 223ff, Bourgey 1955, pp. 73ff, 83ff, Louis 1964–9, 1 pp. xxxiv ff, Preus 1975, pp. 21ff.

[425] See Manquat 1932, chh. 4 and 5, and cf. especially the analysis of the relationship between Aristotle and the Hippocratic treatises in Poschenrieder 1887.

this should be borne in mind.[426] He recognises, too, the need for experience – a trained eye will spot things that a layman will miss[427] – though even experts make mistakes.[428] He frequently expresses doubts about the reports he has received, emphasising that some stories have yet to be verified,[429] or flatly rejecting them as fictions[430] – though understandably there are tall stories that he fails to identify as such[431] – and on some occasions the different degrees of acceptance exhibited in different texts may suggest some vacillation on his part.[432] He is particularly critical of some of his literary sources, describing Ctesias as untrustworthy[433] and Herodotus as a 'mythologist'.[434] Yet he sometimes records baldly as 'what has been seen' something for which his principal or even his only evidence may be literary. Thus when at *HA* 516a 19f we read that an instance 'has been seen' of a man's skull with no suture, his (unacknowledged) source may be the famous description of such a case on the battlefield of Plataea in Herodotus (IX 83). When we are told that lions are found in Europe only in the strip of land between the rivers Achelous and Nessus[435] the authority for this may again be a passage in Herodotus (VII 126) which makes a similar suggestion, though the river Nessus appears by its alternative name Ne*s*tus.

Aristotle's use of these sources and of his own personal researches is, naturally, guided throughout by his theoretical interests and preoccupations. The very thoroughness with which he tackles the task of the description of animals reflects his declared aim to assemble the φαινόμενα, the differentiae of animals and their properties, before proceeding to state their causes.[436] As we have noted, this very programme obliges Aristotle to be comprehensive in his account, and

---

[426] See, e.g., *GA* 756a33.

[427] See, e.g., *HA* 566a6–8 (outside the breeding season, the sperm-ducts of cartilaginous fish are not obvious to the inexperienced), 573a11ff, 574b15ff.

[428] E.g. *GA* 756b3ff.

[429] E.g. *HA* 493b14ff, 580a19–22. That further research is necessary is a point repeatedly made in other contexts as well.

[430] E.g. *HA* 579b2ff, 597a32ff. *PA* 673a10–31 is a careful discussion of stories about men laughing when wounded in the midriff: he rejects as impossible the idea that a head, severed from the body, could speak (since voice depends on the windpipe) but accepts that movement of the trunk may occur after decapitation.

[431] E.g. *HA* 552b15ff on the salamander extinguishing fire.

[432] As in the notable case of his reports on the phenomenon now known as the hectocotylisation of one of the tentacles of the octopus, recorded without endorsement at *HA* 541b8ff, cf. 544a12f, and apparently accepted at 524a5ff, but rejected at *GA* 720b32ff. Such divergences may, of course, indicate not a change of mind, but inauthentic material in, or plural authorship of, the zoological works.

[433] *HA* 606a8.

[434] The context suggests that the term μυθολόγος, used of Herodotus at *GA* 756b6f, there carries pejorative undertones. 　　　[435] *HA* 579b5ff, cf. 606b14ff.

[436] See above, p. 137 and n. 64 on *PA* 639b8ff, 640a14f, *HA* 491a9ff especially.

certain features of the way he implements it stand out. His stated preference for the formal and final causes, rather than the material, dictates greater attention being paid to the functions of the organic and inorganic parts than to their material composition, although the latter is also discussed, indeed sometimes, as in the case of blood,[437] at some length. More importantly, the form of the living creature is its ψυχή, life or soul, and his psychological doctrines influence his investigations not only in insuring a detailed discussion of, for example, the presence or absence of particular senses in different species of animals in the *De Sensu*, of the different modes of locomotion in the *De Motu* and *De Incessu*, and of the fundamental problem of reproduction in the *De Generatione Animalium*, but also by providing the general framework for his description of the internal and external parts of animals in the *Historia*.

Thus at *PA* 655b29ff and *Juv.* 468a13ff he identifies the three main essential parts of animals as (1) that by which food is taken in, (2) that by which residues are discharged, and (3) what is intermediate between them – where the ἀρχή or controlling principle is located: in addition animals capable of locomotion also have organs for that purpose, and in the corresponding passage in *HA* I 2 and 3 he further adds reproductive organs where male and female are distinguished.[438] In his detailed account of the internal and external parts of the four main groups of bloodless animals (Cephalopods, Crustacea, Testacea and Insects) in *HA* IV 1–7 he evidently works quite closely to this broad and simple schema. Thus he regularly considers such questions as the position of the mouth, the presence or absence of teeth and tongue or analogous organs, the position and nature of the stomach and gut and of the vent for residue, as also the reproductive organs and differences between males and females. A series of passages shows that he actively considered whether or not certain lower groups produced residue and attempted to identify and trace the excretory vent.[439] But while the whole course of the alimentary canal is thoroughly discussed in connection with each of the bloodless groups,[440] he has little or nothing to say about the

---

[437] E.g. *PA* II 4 and cf. the subsequent chapters on fat, marrow, brain, flesh and bone. *PA* II 2 discusses the problems posed by the ambiguity of hot and cold and stresses the difficulty of determining which substances are hot and which cold.

[438] *HA* 488b29ff, especially 489a8ff, and cf. also *PA* 650a2ff.

[439] See *HA* 530a2f, 531a12ff, b8ff.

[440] *HA* 527b1ff concludes that the stomach, oesophagus and gut alone are common to bloodless and blooded groups (the passage is considered suspect by some editors, but it sums up Aristotle's position well enough). In the strictest sense in which the term σπλάγχνον is reserved for red-blooded organs, the bloodless animals have no viscera at all, but only what is analogous to them: *HA* 532b7f, *PA* 665a28ff, 678a26ff.

brain[441] or about the respiratory (or as he would say refrigeratory) system.[442] Again while the external organs of locomotion are carefully identified and classified, the internal musculature is ignored throughout.[443] It would certainly be excessive to suggest that his observations are everywhere determined by his preconceived schema: yet the influence that that schema exercised on his discussion is manifest.

In other cases too it is not hard to trace the influence of his theoretical preoccupations and preconceptions on his observational work, not only – naturally enough – on the questions he asked, but also – more seriously – on the answers he gave, that is on what he represents as the results of his research. His search for final causes is an often cited example, though it is not so much his general assumption of function and finality in biological organisms, as some of his rather crude particular suggestions, that are open to criticism: moreover he is clear that not everything in the animal serves a purpose[444] and that it is not only the final cause that needs to be considered. But many slipshod or plainly mistaken observations (or what purport to be such) relate to cases where we can detect certain underlying value judgements at work. The assumption of the superiority of right to left is one example that has been mentioned before.[445] His repeated references to the differences between man and other animals, and between males and females, are two other areas where errors and hasty generalisations are especially frequent. Man is not only marked out from the other animals by being erect – by having his parts, as Aristotle puts it, in their natural positions[446] – and by possessing the largest brain for his size, hands and a tongue adapted for speech:[447] Aristotle also claims, more doubtfully, that man's blood is the finest

---

[441] In *HA* iv 1–7 the brain is mentioned only at 524b4 and in a probably corrupt passage, 524b32. Cf. *HA* 494b27ff, and *PA* 652b23ff.

[442] In *Resp.* 475b7ff, however, he says that the Crustacea and Octopuses need little refrigeration and at 476b30ff that the Cephalopods and Crustacea effect this by admitting water, which the Crustacea expel through certain opercula, that is the gills (cf. also *HA* 524b21f).

[443] This point can be extended also to his descriptions of the blooded animals. Although his account of the external organs of locomotion in *IA* is, on the whole, quite detailed, he has almost nothing to say of the disposition and functioning of the muscles. Similarly his osteology (with the exception of his description of the limbs) is in general crude: even though he writes in praise of the hand, remarking on the importance of the opposition of thumb and fingers for prehension (see *PA* 687a7ff, b2ff, 690a30ff) he limits his account of its bones to remarks on the number of fingers and toes in the forelimbs of different species.

[444] There is an explicit statement to this effect at *PA* 677a15ff, for example.

[445] The main examples of anatomical doctrines influenced by his beliefs that right is superior to left, up to down, and front to back, are collected in Lloyd 1973.

[446] E.g. *PA* 656a10ff, *IA* 706a19f, b9f.

[447] See, e.g., *PA* 653a27ff, 687b2ff, 660a17ff.

and purest,[448] that his flesh is softest,[449] and that the male human emits more seed, and the female more menses, in proportion to their size.[450]

While his general distinction between male and female animals relates to a capacity or incapacity to concoct the blood,[451] he records largely or totally imaginary differences in the sutures of the skull,[452] in the number of teeth,[453] in the size of the brain,[454] and in the temperature,[455] of men and women. His view that in general males are better equipped with offensive and defensive weapons than females[456] is one factor that leads him to the conclusion that the worker bees are male.[457] To be sure, he sometimes notes exceptions to his general rules, as when he remarks that although males are usually bigger and stronger than females in the non-oviparous blooded animals the reverse is true in most oviparous quadrupeds, fish and insects,[458] and while he goes along with the common belief that male embryos usually move first on the right-hand side of the womb, females on the left, he remarks that this is not an exact statement since there are many exceptions.[459] But although there are certainly inaccuracies in his reported observations besides those that occur where an *a priori* assumption is at work, those where that is the case form a considerable group.[460]

We can document the influence of his over-arching theories on what he reports he has seen: but there are other occasions when the theories themselves appear to depend on, and may in some cases even be derived from, one or more observations (however accurate or inaccurate) which accordingly take on a particular significance for his argument. Undoubtedly the most striking example of this[461]

---

[448] See *HA* 521a2ff, cf. *PA* 648a9ff, *Resp.* 477a20f.
[449] *PA* 660a11, cf. *GA* 781b21f.
[450] See *HA* 521a26f, 582b28ff, 583a4ff, *GA* 728b14ff.
[451] See *GA* 728a18ff, 765b8ff.
[452] See *HA* 491b2ff, 516a18f, *PA* 653b1.   [453] See *HA* 501b19ff.
[454] See *PA* 653a28f (on which see Ogle 1882, p. 167).
[455] E.g. *GA* 765b16f, 775a5ff.
[456] E.g. *HA* 538b15ff, *PA* 661b28ff.   [457] See *GA* 759b2ff.
[458] See *HA* 538a22ff, 540b15f, *GA* 721a17ff.
[459] See *HA* 583b2ff, and 5ff, and cf. further below, p. 217, on *GA* 764a33ff. Cf. also *HA* 584a12ff on exceptions to the rule that women have easier pregnancies with male children.
[460] Another important group of mistakes relates to exotic species (for which parts of *HA* especially show some predilection) where he was, no doubt, relying more on secondary sources or hearsay. Thus many of his statements about the lion are erroneous (see Ogle 1882, p. 236). See also the mistakes mentioned by Bourgey 1955, pp. 84f (there are inaccuracies in those listed by Lewes 1864, pp. 164ff).
[461] Two others would be (1) his claim to have verified that the brain is cold to *touch*, *PA* 652a34f (not true of a recently dead warm-blooded animal) – which was no doubt a major factor in contributing to his theory that the primary function of the brain is to

is his often repeated statement that the heart is the first part of the embryo to develop. This is introduced at *Juv.* 468 b 28ff as something that is 'clear from what we have observed in those cases where it is possible to see them as they come to be'. At *PA* 666 a 18ff he says that the primacy of the heart is clear not only according to argument, but also according to perception,[462] and in reporting his investigation of the growth of hen's eggs in particular he remarks that after about three days the heart first appears as a blood spot that 'palpitates and moves as though endowed with life'.[463] The circumstantial detail of this and other accounts show that they are based on first-hand inspection, although the conclusion Aristotle arrived at is not entirely correct: as Ogle put it, 'the heart is not actually the first structure that appears in the embryo, but it is the first part to enter actively into its functions'.[464] However the consequences of Aristotle's observation were momentous. This provides the crucial empirical support for his doctrine that it is the heart – rather than say the brain – that is the principle of life, the seat not just of the nutritive soul, but also of the faculty of locomotion and of the common sensorium. As in the physical treatises, so too in his biology, Aristotle often constructs a general theory largely by extrapolation from a slight – and sometimes insufficiently secure – empirical foundation.

Destructively, however, his deployment of observations to refute opposing theories is often highly effective. This can be illustrated first with some fairly straightforward examples. (1) The idea that drink passes to the lungs is one that we know to have been widely held,[465] although it is attacked by the author of the Hippocratic treatise *On Diseases* IV.[466] At *PA* 664 b 6ff Aristotle dismisses it primarily on the simple anatomical grounds that there is no com-

---

counterbalance the heat of the heart, and (2) his reported observation that a bull that had just been castrated was able to impregnate a cow (*GA* 717 b 3ff, cf. *HA* 510 b 3f) – which presumably influenced his doctrine that the testes are mere appendages, not integral to the seminal passages (e.g. *GA* 717 a 34ff).

[462] Again at *GA* 740 a 3ff he says that not only perception but also argument shows that the heart is the first part to become distinct in actuality. Cf. also *PA* 666 a 8ff, *GA* 740 a 17f, 741 b 15f.

[463] *HA* 561 a 6ff, 11f, *PA* 665 a 33ff.

[464] Ogle 1897, p. 110 n. 24, and cf. 1882, p. 193.

[465] This we know from *Morb.* IV ch. 56, L VII 608.17ff. The view is found in Plato, *Ti.* 70 c, is attributed to Philistion by Plutarch (*Quaest. Conv.* VII 1, 698 A ff at 699 C) and after Aristotle was the subject of an attempted experimental demonstration in *Cord.* ch. 2, L IX 80.9ff (see above p. 166 n. 208).

[466] The nine proofs, ἱστόρια, that this author adduces are a very mixed bag: they include not only a reference to the epiglottis and its function (VII 608.23ff) but also arguments that if drink went to the lungs, one would not be able to breathe or speak when full, that another consequence would be that dry food would not be so easily digested, that eating garlic makes the urine smell, and other often inconclusive or question-begging considerations, see VII 606.7ff.

municating link between the lungs and the stomach (as there is between the stomach and the mouth, namely the oesophagus), and the confidence with which he rebuts the theory is clearly seen in his concluding remarks: 'but it is perhaps silly to be excessively particular in examining silly statements'.[467] (2) At *GA* 746a19ff he refutes the view that human embryos are nourished in the womb by sucking a piece of flesh: if that were true, the same would happen in other animals, but it does not, as is easy to observe by means of dissection. And while that remark is quite general,[468] he follows it with a specific reference to the membranes separating the embryo from the uterus itself.[469] Dissection again provides the evidence to refute (3) those who held that the sex of the embryo is determined by the side of the womb it is on,[470] and (4) the view that some birds copulate through their mouths.[471]

In such instances an appeal to easily verifiable points of anatomy was enough to undermine the theory. But more often no such direct refutation was possible, and Aristotle deploys a combination of empirical and dialectical arguments to attack his opponents' positions. One final example of this is his extended discussion, in *GA* I 17 and 18, of the doctrine that later came to be known as pangenesis, that is the view that the seed is drawn from the whole of the body.[472] Here most of what passed as empirical evidence was agreed on both sides, and the strengths of Aristotle's discussion lie in his acute exploration first of the coherence of his opponents' doctrine, and secondly of the inferences that could legitimately be drawn from the available data.

One of the principal arguments he mounts against pangenesis poses a dilemma:[473] the seed must be drawn either (1) from all the uniform parts (such as flesh, bone, sinew) or (2) from all the non-uniform parts (such as hand, face) or (3) from both. Against (1) he objects

---

[467] *PA* 664b18f.

[468] Some of Aristotle's general appeals to what would be shown by dissection are clearly hypothetical and were not followed up: thus at *PA* 677a5ff he dismisses the view of Anaxagoras' followers that the gall-bladder causes acute diseases with the claim that those who suffer from such diseases mostly have no gall-bladder and 'this would be clear if they were dissected'.

[469] *GA* 746a23ff says that this is true for all embryos in animals that fly, swim and walk.

[470] *GA* 764a33ff, cf. 765a16ff: yet he is prepared to allow that males often move first on the right-hand side (*HA* 583b2ff, cf. b5ff), and also that, given that the right-hand side of the body is hotter than the left, and that hotter semen is more concocted, seed from the right side is more likely to produce males (*GA* 765a34ff, but cf. b4ff).

[471] *GA* 756b16ff, especially 27ff.

[472] Our principal original sources for the pangenesis doctrine are the Hippocratic treatises, *Genit.*, *Nat. Puer.* and *Morb.* IV. See Lesky, 1951, pp. 70ff.

[473] *GA* 722a16–722b3.

that the resemblances that children bear to their parents lie rather in such features as their faces and hands, than in their flesh and bones as such. But if the resemblances in the non-uniform parts are not due to the seed being drawn from *them*, why must the resemblances in the uniform parts be explained in that way? Against (2) he points out that the non-uniform parts are composed of the uniform ones: a hand consists of flesh, bone, blood and so on. Moreover this option would suggest that the seed is not drawn from *all* the parts. He tackles (3) too by considering what must be said about the non-uniform parts. Resemblances in these must be due either to the material – but that is simply the uniform parts – or to the way in which the material is arranged or combined. But on that view, nothing can be said to be drawn from the *arrangement* to the seed, since the arrangement is not itself a material factor. Indeed a similar argument can be applied to the uniform parts themselves, since they consist of the simple bodies combined in a particular way. Yet the resemblance in the parts is due to their arrangement or combination, and has therefore to be explained in terms of what brings this about,[474] and not by the seed being drawn from the whole body.

A series of further arguments follow, for example that the seed cannot be drawn from the reproductive organs at least, because the offspring has only male or female organs, not both (*GA* 722 b 3ff), and again that the seed cannot be drawn from all the parts of both parents, for then we should have two animals (b 6ff). At *GA* 722 b 30ff he considers how the uniform and non-uniform parts are to be defined, namely in terms of certain qualities and functions respectively. Thus unless a substance has certain qualities it cannot be called 'flesh'. But it is plain that we cannot call what comes from the parent *flesh*, and we must agree that that comes from something which is not flesh.[475] But there is no reason not to agree that other substances may do the same, so again the idea that all the substances in the body are represented in the seed fails.

At the same time it is notable that he not only challenges the scope and significance of the evidence his opponents cite, but also shows some ingenuity in collecting other data that pose difficulties for them. One of the main arguments they used depended on the supposed fact that mutilated parents produce mutilated offspring, and among the evidences (μαρτύρια) they cited was that of children born with scars

---

[474] The semen has just such a function, as supplying the efficient cause, in Aristotle's own theory.

[475] He is led from this to consider Anaxagoras' theory that none of the uniform substances comes into being.

where their parents were scarred, and a case at Chalcedon of a child of a branded father born with a faint brand mark: it was claimed, as Aristotle puts it, that children resemble their parents in respect not only of congenital characteristics (τὰ σύμφυτα) but also of acquired ones (τὰ ἐπίκτητα).[476] But this he counters simply by pointing out that not all the offspring of mutilated parents are themselves mutilated, just as not all children resemble their parents.[477] Among the evidence he brings against pangenesis he cites (1) that many plants lack certain parts (they can be torn off, and yet the seed thereafter produces new plant that is identical with the old, 722 a 11ff), and (2) that plant cuttings bear seed – from which he says it is clear that even when the cutting belonged to the original plant the seed it bore did not come from the *whole* of that plant (723 b 16ff).

But the most important consideration, in his view, is (3) what he claims to have observed in insects (723 b 19ff). In most cases, during copulation the female insect inserts a part into the male, rather than the male into the female. This by itself looks quite inconclusive, but Aristotle believes that in such cases it is not semen, but simply the heat and the δύναμις (capacity) of the male that brings about generation by 'concocting' the fetation.[478] He remarks quite cautiously that not enough observations have been carried out in such cases to enable him to classify them by kinds, and his remark that the males do not have seminal passages is introduced with φαίνεται in the tentative sense, 'appears' rather than 'it is evident'.[479] Yet this erroneous observation is not only his 'strongest evidence'[480] against pangenesis, but also one of the crucial pieces of 'factual' support that he cites for his own view that the role of the male in reproduction is to supply the efficient cause, not to contribute directly to the material of the offspring.[481] In the main the arguments mounted against pangenesis are telling ones, and they draw on well known, and some not so well known, data to good effect: even so, the chief point that derives from Aristotle's personal researches in these chapters is one where, under the influence, no doubt, of his general theories, he assumed too readily that his observations yielded a conclusion that supported them.

[476] *GA* 721 b 17ff and 28ff.
[477] *GA* 724 a 3ff, cf. *HA* 585 b 35ff.
[478] See especially *GA* 729 b 21–33, and cf. other references to species that do not emit seed in copulation, 731 a 14ff, 733 b 16ff.
[479] See *GA* 721 a 12, 14ff.
[480] *GA* 723 b 19.
[481] See *GA* 729 b 8f and 21f, where in both cases there is a contrast between λόγος and ἔργα. Cf. also b 33ff with some equally doubtful evidence such as the supposed fact that a hen bird trodden twice will have eggs that resemble the second cock.

It is apparent that much of Aristotle's biology – like his physics – does not live up to his own high ideals. His drawing attention to the inadequacy of certain data, and to the need to survey all the relevant ὑπάρχοντα,[482] does not prevent him from being less than persistent in his research in some areas, nor deter him from some highly speculative theories in others. Where he remarks on other writers' inexperience of internal anatomy, for example,[483] or charges them with not taking what is familiar as the starting-point of their inquiries,[484] with generalising from a few cases or otherwise jumping to conclusions on inadequate evidence,[485] or with guessing what the result of a test would be and assuming what would happen before actually seeing it,[486] in each case similar criticisms could be levelled at him to some – if not the same – extent.

Nevertheless two simple but fundamental points remain. First, if he does not always live up to his own methodological principles, at least they are stated as the principles to follow. The end is defined in terms of giving the causes and resolving the difficulties in the common assumptions: he is no fact collector for the sake of fact collecting. But as means to his ends the appeal to the evidence of the senses is allotted its distinct role, alongside the reference to generally accepted opinions and the use of reasoned argument, and he makes it clear that in certain contexts at least it is the first of these that is to be preferred. Moreover his is the first *generalised* programme of inquiry into natural science: his doctrine of causes identifies the kind of questions to be asked, and he provides an explicit protreptic to the study of each branch of natural science as far as each is possible.[487]

Secondly the limitations of his observational work should not lead us to ignore the extraordinary scope of what he did achieve in the various departments of the inquiry concerning nature. An analysis of what he says he has observed and of how he uses this to support his theories sometimes reveals the superficiality of his empirical research. Yet as the first systematic study of animals, the zoological treatises represent a formidable achievement, not only in the individual discoveries that are recorded, but also in the patient and painstaking amassing of a vast amount of data concerning many different species and in the ingenious interplay of data and arguments in his assault on such obscure problems as those connected with reproduction.

---

[482] Apart from *PA* 639 b 8ff and other passages mentioned above (p. 137 and n. 64) see also, e.g., *GA* 735 b 7f, 748 a 14ff.   [483] See, e.g., *Resp.* 470 b 8f, 471 b 23ff.

[484] E.g. *GA* 747 b 5f, 748 a 8f, 765 b 4f.

[485] E.g. *PA* 676 b 33ff, *GA* 756 a 2ff, b 16ff, 788 b 11ff, 17ff.   [486] *GA* 765 a 25–9.

[487] *PA* I 5, 644 b 22ff: the study of the heavenly bodies and of animals each has its own attractions.

Neither the proposition that most Greek scientists ignored empirical methods, nor that they were somehow endowed with an instinctive grasp of them, stands up to scrutiny. There are important differences in the performance both between and within the various chief strands of Greek natural science – physics, astronomy, medicine and biology – and simple contrasts between 'dialectical' philosophers and 'empirical' doctors will not do. Although some of the observations in the medical literature are systematic and meticulous, we also find doctors engaging in uncontrolled and dogmatic speculation – even while they criticised other theorists on precisely that score. Conversely we have some evidence that the number-theory of the Pythagoreans stimulated some empirical research in acoustics.

We cannot attribute the infrequency of sustained empirical research in pre-Aristotelian science to a general epistemological interdiction, for although there were those who rejected or denigrated the senses, that was no simple orthodoxy. The epistemological debate was a complex one and the importance of ἱστορία was advocated by several writers. Its fruitful *practice* in any more than a purely descriptive context had, however, to be stimulated by a particular theory, as we see in the case of the Hippocratic doctrine of critical days. Moreover in both pre- and post-Aristotelian science the overall theory that played that role was sometimes one that had its mystical or fantastical aspects, as we see from Pythagorean number-theory before, and astrology after, him. Curiosity as such was no prerogative of men who single-mindedly devoted themselves to what *we* consider the lasting achievements of Greek science, and some of those lasting achievements owed much to the complex motivations of their authors. If the geometry of Hipparchus' lunar model may owe little to his astrological beliefs, the same is certainly not true of his discovery of precession, the direct outcome of observations of the stars we know to have been conducted in part with astrological considerations in mind.

Observation and research have always to be guided by theories, whether specific or general. But a particular recurrent feature of the case-studies we have examined is the way in which research served the function of corroboration. Observations are cited to illustrate and support particular doctrines, as primary or supplementary arguments for them, almost, we may say, as one of the dialectical devices available to the advocates of the thesis in question. The observations are sometimes already interpreted in the light of the theories they were meant to establish, and we find many examples where the degree of support they lend those theories was much exaggerated.

More seriously, there are occasions when, once the theory had received some empirical backing, the research was pursued no further. The theories were not put at risk by being checked against further observations carried out open-endedly and without prejudice as regards the outcome.

Furthermore a similar point applies, too, to many of the tests and experiments that are recorded. The lack of experimentation is a charge that has repeatedly been levelled at Greek science,[488] but much of the criticism has been misplaced and we must get clear where the real weaknesses lie. First of all it is worth remarking that in the most general sense the conducting of tests is certainly not confined to, nor definitive of, natural science. Evans-Pritchard's report of the operations of the poison oracle by the Zande – in which poison is given to a chicken and specific questions, expecting yes or no answers, are settled according to whether the chicken lives or dies[489] – is an example of a testing procedure for which no shortage of parallels, including Greek parallels, can be found. In such cases the rules according to which the results are interpreted are predetermined: what will count as a yes or no answer is agreed beforehand. It is notable that the Zande regularly *check* their results by putting the same question first in a positive and then in a negative form, or by following one test with a second questioning the validity of the first.[490] Yet while individual answers are thereby tested, the validity of the whole procedure is not examined or called in question.[491]

Again just as frequent in the anthropological literature and elsewhere are tests that differ from the poison oracle in that they are carried out directly on the substance or person about which or whom the question is asked, but that share the characteristic that they are used to settle issues concerning the particular case, not to arrive at explanations of classes of physical phenomena or events. Ordeals to test the guilt or innocence of individuals often take this form,[492] and would-be diagnostic tests of, for example, whether or not a woman can conceive, whether she has done so, whether she is pregnant with a boy or girl, or whether a man's semen is fertile or not, are common

---

[488] Some of the views that have been expressed on this problem are outlined and discussed in my 1964: cf. subsequently von Staden 1975.

[489] Evans-Pritchard 1937, pp. 258ff.

[490] Evans-Pritchard 1937, pp. 299ff.

[491] Cf. above, ch. 1 pp. 18f on the question of the limits of scepticism.

[492] We may compare also the type of case illustrated by the story in Herodotus 1 46ff, where Croesus first tries out the Greek oracles to see whether they can answer a question to which he knows the correct reply, in order to determine which oracle to consult on the outcome of his expedition against the Persians. The fact that this story may well be apocryphal does not alter its value as evidence for the idea of carrying out such a test.

in all kinds of medical texts, including, from the Greco-Roman world, not only writers such as Pliny, but also several of the Hippocratic authors and Aristotle.[493] Some of the particular procedures used on such occasions are better grounded than others: but they have in common not only that the rationale of the tests themselves is not examined, but also that they are directed not to investigating causes, but to resolving questions concerning individual cases.

But if the practice of testing may be seen as just one variety of universal human trial and error procedures,[494] scientific experimentation may be distinguished in the first place by its aim, to throw light on the nature and causes of physical phenomena and processes. While the major extant[495] examples of sustained and systematic experimentation in Greek science come from late antiquity,[496] the Hippocratic treatises and Aristotle especially provide instances of tests carried out in connection with physical and physio-

---

[493] See, e.g., *Aph.* v 41 and 59, L IV 546.1ff, 554.3ff, *Nat. Mul.* ch. 96, L VII 412.19ff, *Steril.* chh. 214 and 219, L VIII 414.17ff, 422.23ff, *Superf.* ch. 25, *CMG* I, 2, 2 80.28ff, and cf. Aristotle, *HA* 523a25f, 583a14ff, *GA* 747a3ff, 7ff. *Steril.* ch. 215, L VIII 416.8ff, 13ff, is a typical instance where the Hippocratic writer moves from describing certain aspects of a woman's appearance that may be used as signs that she is pregnant, to suggesting a deliberate intervention – with the administration of drugs – that purports to test for pregnancy. Cf. Soranus, *Gyn.* I 9.35, *CMG* IV 24.20ff, for a sceptical and critical response to some of the tests that had been proposed to see whether a woman can conceive.

[494] Many of the admittedly sporadic improvements made in various domains of technology in the ancient world (for example in navigation and corn-milling: see Landels 1978 for a convenient survey) may be thought to presuppose the use of such procedures. In general, however, such testing was carried out by men who left no written record of their aims and methods. The chief exception to this is the series of improvements in military technology, notably in connection with the invention and perfection of artillery of various kinds, for which we have quite extensive evidence in such writers as Philo (end of third century B.C.) and Hero (first century A.D.) (see Marsden 1971). Philo, in particular, contrasts early trial and error procedures in the construction of artillery with the more systematic experiments carried out in Alexandria by craftsmen who were 'heavily subsidised' by kings who were 'eager for fame and well disposed to the arts and crafts' (*Bel.* 50.20ff and 37ff). The circumstances in which such state support for technological research was forthcoming were, however, clearly quite exceptional.

[495] What we know of the research of Strato and of the Hellenistic biologists, Herophilus and Erasistratus, suggests that the picture might be substantially altered if more of their work had survived (see above, pp. 165ff, 167 n. 218 and 211 n. 421). One notable instance of an experimental procedure attributed to Erasistratus, for example, is that recorded in Anon. Lond. XXXIII 43ff, in which he attempted to show that there are invisible effluvia from animals by keeping a bird in a vessel without food for a period of time and then weighing the bird and its visible excreta and comparing this with the original weight: the loss in weight was taken to establish that a 'considerable emanation had taken place'. Cf. further von Staden 1975, pp. 179ff.

[496] See above, p. 167, on Galen's experiments on the digestive and nervous systems, and p. 197 on Ptolemy's in optics. Cf. also p. 144 n. 95 for evidence that the idea of varying the conditions of the test appears in some late authors even when they carried out no such tests themselves.

logical theories before the end of the fourth century.[497] It is true that the number of such examples is not great, though we must recall that in several of the topics and areas of physical inquiry in which the Greeks were interested experiments were either simply not a practical possibility (as in astronomy and meteorology) or incapable of resolving the fundamental issues (as in the debate between atomism and continuum theory). It is also true that such experiments as were conducted were often inconclusive. But more important than either of those two points (and part of the explanation of that inconclusiveness) is the fact that many tests – just like the simple observations of which they can be seen as the natural extension – were used in a corroboratory, rather than a neutrally heuristic, role. Yet that generalisation, like so many others, must be qualified: while constructively observations and tests were often partial (because carefully selected) witnesses summoned in support of theories, destructively they could be deployed, as they were by Aristotle especially, with great effect.

Our study of *On the Sacred Disease* in chapter 1 showed that the possession of extensive empirical data about the internal functioning of the body, for example, was not necessary to the rejection of magical beliefs about disease: that Hippocratic author confidently bases his own explanation of epilepsy on a very largely imaginary anatomical theory. Conversely, examples of quite sustained observations can be cited both outside, and before, Greek natural science. Yet in many areas the development of Greek science depended in part at least on the extension of its empirical base. No adequate physiology, pathology or astronomy – let alone geography, anatomy, zoology or botany – is possible without detailed information, obtainable, in many cases, only by careful and systematic observation.

We saw in chapter 2 that much of the strength of Greek science lies in its formal dialectical and demonstrative techniques, and that the definition and analysis of an axiomatic, deductive system, together with the development of the application of mathematics to the understanding of natural phenomena, occupied a considerable and productive intellectual effort. Yet despite all the limitations we have drawn attention to, the Greeks were responsible for important developments in empirical methods as well, both in the theory of

---

[497] Among the examples we have mentioned are the tests on blood described in *Carn.* (above p. 150 n. 132) and those on natural substances reported in *Mete.* IV (pp. 209f). On many occasions Hippocratic tests incorporated an element of analogy: they were carried out not on the living natural substances themselves, but on others deemed to be analogous to them, see Lloyd 1966, pp. 345ff.

research and in its practice. If one of the first contexts in which ἱστορία is successfully practised is the history (in our sense) and geography of Greece and neighbouring lands, other areas of inquiry were soon opened up: the thorough and meticulous collection of data relating to the courses of diseases in the Hippocratic *Epidemics* far surpasses anything in extant earlier Greek or non-Greek medical literature. If initially such studies were restricted to particular topics and stimulated by particular doctrines, the first generalised programme of research, including empirical research, into the various departments of the inquiry concerning nature is the work of Aristotle, and elaborate and at least to some extent coöperative projects were undertaken in his school and later in the Alexandrian Museum. The histories of dissection and of Greek observational work in astronomy illustrate how slow the full exploitation of certain techniques was in coming about. Yet the eventual successes were considerable and may rightly be ranked among the outstanding achievements of Greek science. Some of the Greeks themselves saw exact science as the only proper science: but if that idea and the concept of an axiomatic system are two of their important legacies to modern science, a third is certainly the notion of empirical research.

4

GREEK SCIENCE AND GREEK SOCIETY

THE EXPLANANDUM

We began by considering the problem posed by the criticisms advanced by certain ancient Greek writers against traditional and – as they sometimes represent them – magical beliefs and practices. Our investigation has taken us a long way round, via a discussion of the development of certain argumentative techniques and of empirical research. But we may now return to our original question, to consider now the social background to early Greek thought. How far can we go towards specifying the social and other conditions that allowed or promoted the emergence of philosophy and science? If we are to make any headway on this exceptionally difficult and much disputed problem, it is essential both to bear in mind the exact nature, and the limitations, of the intellectual developments that took place in Greece, and to press home the comparison and the contrast between our Greek data and that from other societies.

The primary task, we said, is to define the explanandum as clearly as possible. For this we have first to recapitulate, and to make more precise, some points that have come out of our earlier investigations, and secondly to come to terms more directly with the achievements of the Greeks' Near Eastern neighbours. The issue of the debts of Greek science to Egypt and Babylonia has been, since antiquity, an emotive topic; all too often it has been argued, by ancient and modern writers alike, either that the Greeks owed everything, or that they owed nothing, to Eastern wisdom, while fundamental questions relating to the processes of transmission, and to the interpretation of what was transmitted, have been ignored.[1] Aspects of the problem have been broached in previous chapters, but we must shortly attempt a rather more systematic comparison of Greek and Near

---

[1] Thus the circumstances in which a metallurgical technique may be transmitted from one culture to another are quite different from those of the transmission of a religious belief or myth, which differ in turn from those of an item of astronomical lore, which differ in turn from those of a method of cure. In particular the extent to which the transmission must be mediated *through language* differs, and so too does the extent to which the transmission is effected between individuals who already possess *specialised* knowledge.

Eastern investigations. In addition to, and in some respects more especially than, traditional pre-literate societies, the major ancient Near Eastern civilisations may be used as a control by which we may test hypotheses about the relation between Greek speculative thought and the social, political and ideological factors that obtained in Greece.

The 'emergence of philosophy and science' is a convenient short-hand term, but a vague one and one that carries the risk of distorting the focus of the problem. We have repeatedly stressed the importance of the complexity and heterogeneity of the various divergent strands of early Greek speculative thought, and to deal merely with those aspects of it that may correspond to our own notions of development is to present a hopelessly onesided picture. We may attempt to summarise the essential points under five heads.

(1) As we said at the outset, popular and traditional beliefs – including superstitions and 'magic' – were not superseded: they continued to be held not only (one presumes) by most Greeks but in particular by many highly articulate writers and they can be exemplified in prominent exponents of ἱστορίη such as Herodotus. Moreover they may be said to grow or develop in a way analogous to the development of science and philosophy, in that – partly under the influence of the models provided by science and philosophy – they become elaborated and systematised, as was the case with dream-interpretation and other forms of divination, with astrology and with 'alchemy'.

(2) What the Greeks themselves identified as progress, or at least claimed as an integral part of civilised life, could and often did include those elaborations of traditional beliefs just as much as other disciplines that we accept more readily as crafts, arts or sciences. To cite just one notable instance, the benefits that Prometheus claims to have brought mankind in Aeschylus' play include not only medicine and navigation, but also divination:[2] divination is clearly considered an important example of a successful τέχνη by the author of *On Regimen*,[3] while Plato – and many others – held it to be a divine gift.[4]

(3) By no means all those who contributed to the early development of the various branches of philosophical inquiry are remarkable

---

[2] *Pr.* 484–499. Cf., e.g., Solon Poem 1.53ff Diehl.

[3] In the writer's elaborate comparison between the τέχναι and man's nature, *Vict.* I, chh. 12–24, L vi 488.1–496.19, divination is taken as the first example, 488.2ff. Contrast *Acut.* ch. 3, L ii 240.8ff, where the disagreements among doctors are said to be a scandal for the art and one that makes laymen think medicine to be no better than divination.

[4] See, e.g., *Phdr.* 244a ff, *Ti.* 71e ff, cf. *Ti.* 24c and contrast, *R.* 364b ff.

for their sceptical, positivist attitudes. Such attitudes can – with some reservations – be attributed to Anaxagoras and Democritus, for instance. Pythagoras had the reputation of being a wonder-worker, and Empedocles laid positive claims in that direction for himself. Although the secrecy and orthodoxy of the Pythagorean sects have been exaggerated by our late secondary sources,[5] it must be acknowledged that those sects were in some respects exclusive groups and that they cultivated some esoteric doctrines and practices. If open argument can readily be illustrated in Presocratic philosophy, so too can the deliberate exploitation of ambiguity and paradox: the lesson that the practice of persuasion merges with that of ἀπατή is one that applies not just to rhetoric but also to aspects of philosophy.[6]

(4) Within mathematics, too, although we know so little about the aims and motivations of many fifth- and fourth-century mathematicians, it is clear that for some at least of the Pythagoreans their inquiries in this area were connected with, and stimulated by, a brand of number mysticism – the belief that numbers in some way hold the key not just to what we recognise as quantitative relationships but also to qualitative ones, including morality.

(5) Finally the divergent strands within early Greek medicine are particularly striking. This is a matter not merely of the contrasts between temple medicine and popular medicine on the one hand and what we find in the Hippocratic Corpus on the other, but also of those between the Hippocratic writers themselves. They differ not only in their attitudes on such questions as the proper method in medicine and its relationship to philosophy and physical speculation, but also on the fundamental issue that concerns us here, that of openness. Whereas the author of *On Ancient Medicine* insists that the doctor should explain himself clearly to the layman,[7] and several treatises exhibit a commendable frankness in acknowledging mistakes or failures in treatment,[8] the end of the work *Law* echoes the language

---

[5] This applies especially to the stories in Plutarch, Iamblichus, Pappus and elsewhere concerning the punishment of those who divulged Pythagorean mathematical learning: yet at one point where he speaks of Pythagorean secret doctrines Iamblichus purports to cite Aristotle's lost treatise on the Pythagoreans as his authority (*VP* 31 = Aristotle Fr. 192). See the balanced assessment of the problem in Burkert 1972a, pp. 178ff, 454ff.

[6] See Gorgias, *Helen*, paras. 8 and 10 (cf. above, pp. 83f).

[7] *VM* ch. 2, *CMG* I, 1 37.9ff, 17ff. One may compare Plato's demand that the 'free' doctor (at least) should converse with his freeborn patient, discuss his case with him and obtain his consent for treatment, *Lg* 720b ff, 857c ff, though we should note that Plato's distinction between free doctors treating free men and slave doctors treating slaves is contradicted by the evidence of the case-histories in the *Epidemics*, where slaves and free men and women were treated by the same physicians.

[8] This is true especially of the surgical treatises. Thus the author of *Art.* describes his own unsuccessful attempt to reduce hump-back by using an inflated bag, adding that

of the mystery religions to express the idea that medical knowledge should be revealed only to the initiated.[9] While many Hippocratic works are admirably clear and succinct, the obscurity of others, particularly of some of the aphoristic collections, appears to have been deliberately cultivated:[10] the aphorisms in question are, one presumes, shorthand formulae which are not intended to be intelligible in themselves, but only in relation to further, oral, teaching which would be available to duly apprenticed students.

THE GREEKS AND THE NEAR EAST

To these reservations concerning the heterogeneous character of early Greek speculative thought must be added others when we set the Greeks' achievements in relation to those of their ancient Near Eastern neighbours. Already in the third millennium B.C. both Egypt and Mesopotamia especially had complex and sophisticated civilisations characterised by centralised, bureaucratic government and a comparatively high level of technology. As already noted, the question of the debts of Greek science to the East has often been discussed superficially and polemically, not least by ancient writers, Greek and Roman, Christian and pagan, themselves. But thanks to the meticulous work of Egyptologists and Assyriologists, much of it of quite recent date, we are now in a much better position to define the contributions of those civilisations.

The three main fields that concern us are mathematics, astronomy and medicine, and in each case we can identify not only certain outstanding advances made by the Egyptians and Babylonians, but also certain distinctions between their work and what we find in Greek science. Thus mathematical cuneiform tablets dating from the second millennium B.C. show that the Babylonians had already attained

---

he records this deliberately because much can be learned from mistakes, ch. 47, L IV 210.9ff. *Epid.* I and III show a similar readiness to record cases where all the remedies tried were useless (e.g. *Epid.* III case 9, first series, L III 58.7, and case 5, second series, 118.8), as well as a high proportion of cases ending in death (see above, p. 154 n. 145). In the other books of the *Epidemics*, too, faulty diagnoses and treatments are frequently referred to and sometimes, at least, it is clearly the writer's own performance that is in question, e.g. *Epid.* V 27, L V 226.10 ('it escaped my notice that this lesion needed trepanning').

[9] *Lex* ch. 5, *CMG* I, I 8.15ff (see above, p. 41 n. 163). Cf. *Decent.* ch. 16, *CMG* I, I 29.13ff, 17ff, which tells the doctor to reveal nothing of the patient's condition to him for fear that this will make him worse, and ch. 18, 29.32f, which cryptically remarks that things that are 'glorious' are closely guarded. *Jusj.*, meanwhile, insists that medical knowledge should only be handed on to specified classes of individual, *CMG* I, I 4.7ff.

[10] The main examples are *Alim.*, *Hum.*, parts of *Decent.* and parts of *Praec.*, but cf. also, e.g., *Epid.* VI sec. 2, ch. 1, L V 276.3ff. Cf. especially W. H. S. Jones 1923–31, IV pp. ix ff.

considerable mastery over a range of arithmetical and algebraic techniques, for example for the solving of quadratic equations.[11] The extant remains of Egyptian geometry show a fair degree of proficiency in the manipulation of certain elementary, and of some not so elementary, procedures,[12] and the Greeks were sufficiently impressed for many of them to maintain that geometry began in Egypt.[13] Yet as we have remarked before, what is lacking from both Egyptian and Babylonian mathematics was the notion of proof. Thus although we know from a cuneiform tablet dating from around 1600 B.C. that the Babylonians were familiar with 'Pythagorean triplets' (numbers in which the squares of the first and the second equal the square of the third),[14] there is no evidence of any attempt to prove what we know as Pythagoras' theorem geometrically.

A similar point applies to astronomy, where we have commented before on the antiquity of Babylonian astronomical records and discussed the problem of the transmission of these data to Greece. The purposes for which the observations were originally undertaken were often quite different from the use the Greeks eventually made of them. The Babylonians observed what they held to be significant astronomical phenomena, such as the appearances and disappearances of particular heavenly bodies, primarily because they believed that these phenomena influenced or even determined events on earth. They constructed periodic tables from which astronomical phenomena might be predicted – and these tables must rank as, or among, the very first systematic attempts to apply mathematics to the understanding of complex phenomena:[15] but what they did not do, at least not until the Seleucid period, and then almost certainly under the influence of Greek astronomy itself, was to attempt to construct geometrical models of the movements of the heavenly bodies. Again the difference between the Greeks and the East centres on the notion of a rigorous demonstration.

In medicine, too, certain contrasts not merely of degree but also of

[11] See Neugebauer 1957, pp. 41ff.
[12] See, e.g., van der Waerden 1954, pp. 31ff.
[13] Herodotus, II 109, and Aristotle, *Metaph.* 981 b23ff, are followed by a long line of later writers, including Diodorus, Strabo, Hero, Iamblichus and Proclus (and from the classical period, cf. Isocrates XI 23 and Plato, *Phdr.* 274cd). Herodotus and Aristotle differ in their views on the stimulus to the development of geometry, Herodotus arguing that it arose from practical concerns (the measurement of the land after the Nile's flooding), Aristotle that it did so from theoretical interests. Such evidence as we have suggests that Herodotus' thesis is nearer the mark so far as Egypt is concerned.
[14] Plimpton 322, see Neugebauer and Sachs 1945, pp. 38ff.
[15] On the achievements of Babylonian astronomy, and on the differences between its aims and those of Greek astronomy, see especially Aaboe 1974, Neugebauer 1975, I pp. 347ff, 397, II pp. 589ff, 613f.

kind may be remarked, even though these are less pronounced and less clear-cut than in mathematics and astronomy. Much Egyptian and more especially Mesopotamian medicine is heavily magical in character. Yet an exception to this is the Edwin Smith papyrus, which we have cited[16] for the evidence it provides for the carrying out of clinical observations in Egyptian medicine and which is almost (though not quite)[17] free from references to charms, spells and the like. Both points are important and suggest a similarity with what we find in one strand in Greek medicine. Yet again there are differences. Egyptian medical papyri contain no deliberate arguments concerning the nature of diseases, their causes, and the constitution of the body, as a whole.[18] Moreover no Egyptian, let alone any other ancient Near Eastern, medical text contains the type of direct attack on magical practices and beliefs that we find in the Hippocratic Corpus. The Egyptian evidence shows quite clearly how even before 'magic' became an issue – as it did in Greece – the emphasis, in practice, in medicine may be very much at the empirical end of what we may call the empirical–magical spectrum.[19] But again the conclusion must be that explicit attacks on magic in medicine are, so far as our information for the ancient Mediterranean and Near East is concerned, an exclusively Greek phenomenon.

This brief excursus into some of the Near Eastern data will serve as a reminder of the misguidedness of talking about the ancient origin of science as if this were a single clearly defined intellectual event. The importance of the advances in mathematical manipulation that took place in one or other or both great ancient Near Eastern river civilisations must be given full weight. The differences between Egypt and Babylonia should not be underestimated: the way mathematics developed in the two societies was far from uniform.

[16] See above, p. 153.
[17] Case nine, which is also exceptional in containing no diagnosis and hardly any examination, ends with a reference to the charm that is to be recited to ensure that the homeopathic remedy is effective: Breasted 1930, 1 pp. 217ff.
[18] Thus although Papyrus Ebers, for instance, contains an anatomical and physiological section, setting out a schematic account of the vessels in the body (see Ebb ll 1937, pp. 114ff), which can be compared with the also often fanciful doctrines in *Morb. Sacr.* and other Greek writers (see above, pp. 21f, 157ff), there is nothing in Egyptian medicine to set beside the material in *VM* debating the relative importance of different causal factors in disease in relation to an explicit idea of what counts as a cause (see above, pp. 53f) or that in *Nat. Hom.* on the issue between monistic and pluralistic views of the constitution of the human body (pp. 92ff).
[19] Cf. G. Lewis 1975. In this important study of Gnau medicine Lewis has emphasised that many minor illnesses are *not* treated as requiring explanation in terms of spirits and he has stressed that the reporting of native accounts of illnesses in anthropological monographs is often incomplete insofar as attention is focused exclusively on major or striking conditions and minor or routine complaints are not discussed.

But both shared in certain advances in both 'pure' and 'applied' mathematics: nor is it difficult to speculate on how certain practical needs in the administration of kingdoms as large and as complex as ancient Egypt and Babylonia may have helped to stimulate the development of certain techniques in arithmetical and geometrical calculation. It is not fortuitous that such examples as determining the supplies of food necessary for a work force consisting of different grades of worker, or the numbers of workers needed to transport a given quantity of bricks, or calculating the angles of inclination of pyramids or canals of different dimensions, should figure prominently in Egyptian and Babylonian mathematical texts.[20]

Moreover in some fields we may talk if not of a continuity or a succession, at least of a direct comparability, in the data from the Near East and from Greece. The undertaking and recording of case-histories in medicine provide one instance,[21] and the sustained observation of astronomical phenomena (for whatever motive) another. Nor should we fail to notice how, in both cases, the production of *written* records transforms the situation as regards the preservation, diffusion and utilisation of the knowledge in question.[22]

So far as these three areas are concerned, some of the distinctive Greek contributions can be identified quite specifically. They were certainly not the first to develop a complex mathematics – only the first to use, and then also to give a formal analysis of, a concept of rigorous mathematical demonstration. They were not the first to carry out careful observations in astronomy and medicine, only the first – eventually – to develop an explicit notion of empirical research and to debate its role in natural science. They were not the first to diagnose and treat some medical cases without reference to postulated divine or daemonic agencies, only the first to express a category of the 'magical' and to attempt to exclude it from medicine. In each of these three instances, these differences relate to new and fundamental questions about the aims, methods and assumptions of the investigation concerned. In astronomy, mathematics and medicine, the Greeks preeminently bring into the open and discuss second-order questions concerning the nature of the inquiry itself.

[20] See, for example, Peet 1923, e.g. pp. 77ff, 97ff, Neugebauer and Sachs 1945, e.g. pp. 76ff, 91ff.

[21] That some pathological and therapeutic ideas, and many medical recipes, came to Greece from Egypt has been argued in Steuer and Saunders 1959 and in Saunders 1963 (cf. also Iversen 1939). The difficulties of establishing this are, however, formidable: independent development is often as likely a hypothesis as diffusion to explain the similarities, especially where the parallelisms are not exact.

[22] See further below, pp. 239f.

Much as the Egyptians and Babylonians contributed to the content of these studies, the investigations only acquire self-conscious methodologies for the first time with the Greeks.

This point can be elaborated and extended when we take into account other areas where the Greeks more obviously break quite new ground. Egypt and Babylonia provide little or nothing to set beside Greek work in the domain of philosophy as such, including first 'natural philosophy' (cosmology and 'physics' in the Greek sense), epistemology and ontology, and then also eventually ethics and formal logic. Now not all that is included under the rubric of Greek philosophy is open and argumentative, and in certain fields, especially psychology and cosmogony, the ideas of the philosophers owe much to religious belief and to myth, both Greek and non-Greek. Yet from early on there is a crucial difference in aim and method between one (and usually the dominant) tradition or approach in Greek philosophy on the one hand, and myth and religious belief[23] on the other, in the philosophers' concern with the question of the *grounds* for the views and theories they advanced. Story-tellers, to be sure, rival one another as story-tellers: but they do not seek to be judged according to whether they produce an account of the subjects they deal with that is better in the sense of having stronger arguments or evidence to support its claims as truth. While neither philosophers in general, nor Greek philosophers in particular, invariably provide such justification, the readiness to do so when challenged may be seen as one of the marks of the new type of inquiry initiated in Greek philosophy.

Any attempt to offer generalisations concerning Greek speculative thought as a whole must be deemed hazardous. The development of the critical approach has often been represented as its key feature, and as a preliminary statement this has the advantage both that it applies to a number of different fields of inquiry,[24] and that it can be argued to be true of Greek thought in ways that appear to distinguish it both from that of their ancient Near Eastern neighbours, and from that of traditional pre-literate societies. Yet important reservations, qualifications and additions must be made. First and most obviously, the critical approach cannot be said to charac-

---

23 If one may distinguish between religious belief and systematic theology: the latter, like philosophy, seeks to provide rational grounds for doctrines, but historically it is, in Greece at least, a later development than, and influenced by, philosophy.

24 I have mentioned those that relate to philosophy, natural science and medicine: but a similar development of critical inquiry may, of course, be illustrated by referring to ethnography, historiography and political theory.

terise the whole of Greek speculative thought, if only because it has its closed, secretive and exclusive aspects, its mystical and mythical sides. Secondly, although much Greek work is critical in the sense that the writers reject both popular assumptions and their colleagues' opinions, it is often anything but self-critical: on the contrary there is frequently a distinct incaution and dogmatism in the statement of the writer's own position, a failure not just to examine, but even to recognise, its weak points.

Nevertheless if the claim to be critical can be upheld only within certain well-defined limits, Greek philosophy, natural science and medicine are strongly characterised first by the raising of funda- mental issues – including particularly second-order questions con- cerning the nature of the inquiries themselves – secondly by the challenging of basic assumptions and by generalised, as opposed to particular, scepticism,[25] and thirdly by an argumentative, competi- tive, even combative quality, reflected not only in the rejection of rivals' views, but also in over-sanguine self-justifications. True, the Greeks argumentative flair is – as we saw in chapter 2 – not unique (though it is highly developed): but their dialectical skills are deployed over a wider range of far-reaching topics than can readily be paralleled elsewhere. It is this combination of the critical, argumentative approach and the radicalism of the questions to which it is applied that provides our best characterisation not of the whole of Greek thought, but of what is exceptional in it. Finally, the very qualifications we must thus add point to a fourth, more general, feature, the pluralism that we have emphasised under the heading of the complexity and heterogeneity of each of the main strands of inquiry.

SOME EXPLANATORY HYPOTHESES

But if we can describe some of the developments that seem important, how far can we go towards identifying the factors that may have stimulated or at least permitted them? How far can such matters as the emergence of certain kinds of inquiry in philosophy and science, and the attack on certain traditional assumptions, be 'explained'? It is as well to start with some disclaimers. If the history of ideas is always highly problematic (and it should not be forgotten that ideas themselves have no history), the study of major transformations in belief-systems is much more intractable, even when we have direct access to extensive first-hand evidence. If the reconstruction of the events, let alone of the causes, of a well-documented twentieth-

[25] See above, pp. 18f.

century political revolution is, to a greater or lesser degree, a subjective and selective matter, the discussion of what happened, and of why it happened, in a revolution in thought is liable to far less effective control; and where, as in our case, the revolution in thought is deemed to have occurred in the distant past, the situation is all too obviously a desperate one. We are reduced to speculation, and the best we can hope to do is to scrutinise possible suggestions as closely as we can. Nevertheless at a certain rather low level of explanation or elucidation we can attempt to use both other ancient civilisations and primitive societies as some sort of check for some of the hypotheses that have been or may be put forward.

We may begin with some brief negative points. First, in discussing the methods for dealing with epilepsy that were available to the author of *On the Sacred Disease* we have already given reason to doubt one much canvassed hypothesis, namely that the criticism of magical beliefs is to be directly associated with, or even stems from, an increase in technological control. Certainly the Hippocratic writer criticises his opponents on the grounds that their remedies are useless, and in this he is no doubt influenced by ideas derived from his own general medical experience concerning what an effective treatment would be like. Yet despite his own claims to the contrary, he himself had – *we* should say – no means of alleviating, let alone of curing, epilepsy.[26] Technological analogies and metaphors are undoubtedly important, both in Presocratic philosophy and in Hippocratic medicine, as a means of conveying the idea that nature in general, and particular natural phenomena or processes, are regular and orderly,[27] although Farrington's thesis that the Presocratic philosophers were themselves close to the arts and crafts seems exaggerated.[28] But there is a further overriding consideration undermining any suggestion to the effect that technological mastery is a sufficient condition for the development of critical inquiry, and this lies in the fact that throughout the crucial period from the sixth to the fourth centuries B.C. there is a general uniformity in the level of technology throughout the eastern Mediterranean and the Near East.[29] Major technological advances

[26] See above, pp. 48ff.
[27] See, e.g., Solmsen 1963, Lloyd 1966, pp. 272ff.
[28] Farrington (1944–9) 1961, pp. 40f, 80f, 104ff. His thesis, which has often been criticised, e.g. by G. Thomson 1955, pp. 171f, derives little support from the admittedly limited information we have about the lives of the Presocratic philosophers. Moreover the examples of Plato and Aristotle – both of whom draw heavily on technological analogies and metaphors to convey the idea of the purposefulness of nature – show that there is no need to be close to the arts and crafts, or to think highly of their practitioners, to make extensive use of imagery derived from these spheres.
[29] See, for example, Finley 1965, Pleket 1973.

were made in this area in the third and second millennia in metallurgy, pottery, textiles and, especially, agriculture, advances that combined to produce what Gordon Childe termed the urban revolution. But there was only limited further technological progress in the period we are chiefly concerned with, and in particular there is no important technological advance *in Greece* that can be held responsible for, or even be connected with, its distinctive intellectual developments.

A similar argument tells also against an exclusively economic interpretation. Although Aristotle associated the development of speculative thought with the leisure produced by wealth,[30] a considerable economic surplus – derived from slave-based production – can hardly be a sufficient (though it may well be a necessary) condition of the intellectual changes that took place in Greece. Again Egypt and Babylonia provide our controls, for they were, economically, incomparably more powerful than any of the Greek city-states in the period up to the end of the sixth century B.C. The invention of coinage some time in the seventh century was undoubtedly a most important factor stimulating trade and economic growth. Yet not only was this not a Greek invention (but a Lydian one),[31] but again the fact that in time coinage came into fairly common use in the eastern Mediterranean and Near East shows that we cannot appeal to it alone to explain the development of new kinds of inquiry in one particular area.

A third hypothesis will require rather more detailed consideration. It has recently been argued that knowledge of other societies and of other belief-systems is a major determining factor in the growth of an open and critical attitude towards the fundamental assumptions of one's own society.[32] We can draw on a good deal of data concerning what the Greeks knew, or imagined they knew, about other peoples. There is abundant material in the Homeric poems that illustrates both a quite extensive knowledge of, and a remarkable speculative interest in, non-Greek lands. So far as the former is concerned, it is true that the picture is patchy: a greater and firmer knowledge is displayed about the eastern sea-board of the Mediterranean and about Egypt than about the western Mediterranean or the hinter-

[30] *Metaph.* 981 b 17ff, where he is speaking of the development of the theoretical branches of knowledge, especially mathematics, which – as we have seen – he associates particularly with Egypt (see above, p. 230 n. 13).

[31] This idea, which was suggested by Herodotus (I 94) and before him (if Pollux IX 83, DK 21 B 4, is to be believed) by Xenophanes, has not been contradicted by the archaeological evidence.

[32] See Horton 1967, pp. 155ff, on what – after Popper – he calls the 'closed' and 'open' predicaments, and his subsequent discussion of 'differences connected with the presence or absence of a vision of alternatives'.

land of Asia Minor or Mesopotamia.[33] But even where, as so often in the *Odyssey*, the account of distant lands is predominantly or even purely fantastical, it is important to observe how deeply the topic of the contrasts between Greek and non-Greek societies exercised the Greek imagination from a very early period. The tales of the Laestrygonians, the Lotus-Eaters and Cyclopes are just so many vehicles for affirming – by implicit or explicit contrasts – the values of Homeric society itself.[34]

Although many of the details are obscure, there can be no doubt about the rapid expansion of Greek knowledge about other societies in the seventh and sixth centuries,[35] and indeed that expansion continued – with some fluctuations and interruptions[36] – down to the Hellenistic period, when the conquests of Alexander opened a new era in Greek–Barbarian relations. On certain points, to be sure, caution is in order. Trading relations do not necessarily imply a deep mutual understanding between, or even much mutual curiosity concerning, the societies in question,[37] and, as Momigliano has rightly emphasised,[38] the Greeks in general showed little interest and ability in learning foreign languages. Many of the stories we find in late authors about the extensive travels of philosophers or political leaders must be treated with scepticism, particularly when the writer is arguing a general thesis about the debts of Greek wisdom to the East.[39]

[33] For contrasting views on the geographical knowledge displayed in the Homeric poems, see, for example, Buchholz 1871–85, 1 Part 1 pp. 79ff and J. O. Thomson 1948, pp. 19ff.

[34] Cf. the classic studies of Greek accounts of matriarchal and gynaecocratic societies in Pembroke 1967 and Vidal-Naquet 1970.

[35] The upsurge in Greek colonisation begins in the eighth century B.C. The trading post at Naucratis, which gave the Greeks limited access to Egypt, was active in the seventh century, and the presence of Greek mercenaries in Egypt in the late seventh and early sixth centuries is attested not only by Herodotus (e.g. II 152 and 163 and cf. III 139 which mentions that some of the Greeks who accompanied Cambyses' invading force did so to trade or to see the country itself) but also by the well-known evidence of the graffiti at Abu Simbel (early sixth century). Greek penetration of Babylonia was slower, but contacts between Greeks and Persians increased once the Persians had conquered the Greek cities of the Asia Minor sea-board. Once again we can supplement the stories in Herodotus (e.g. Democedes III 131ff, Histiaeus v 23ff and Scylax's exploration of the Indus valley, IV 44) with inscriptional evidence, for example from Susa and Persepolis (late sixth/early fifth century): see Scheil 1929, Carratelli 1966, Nylander 1970, and on the whole topic, Momigliano 1975, especially pp. 74ff, 123ff.

[36] West 1971, for example, p. 239, has recently argued that there was a very sharp decrease in communications between the Greeks and the East after the 480s: 'it was as if [oriental influences] had been shut off with a tap'. But that is to be far more dogmatic than our evidence allows.

[37] In many cases where trade involved intermediaries, it was in the interests of those intermediaries to withhold information about the parties concerned.

[38] Momigliano 1975, pp. 148f.

[39] Cf. above, p. 226. This applies to many Christian polemicists such as Clement of Alexandria who, in the *Stromateis*, sets out systematically to prove that Greek philosophy

On the other hand when such reports originate in the fifth or fourth centuries B.C.,[40] even if they do not always prove that the actual visits they refer to occurred, they are at least good evidence of the fact that those who recounted the stories already assumed that the wise men of Greece were inquisitive about, and had much to learn from, non-Greek societies. With Herodotus, for one,[41] we are on firm ground. His travels, conducted, he says, for the purposes of ἱστορίη, research, into the lands he visited, were clearly extensive, even if not quite so extensive as he sometimes wishes his audience to believe.[42] By the end of the fifth century, what the Greeks knew, or what at least they thought they knew, about other societies had become a fund of argument in the debate between 'nature' and 'convention',[43] and again – as with the *Odyssey* – the fact that some of what passed for knowledge is fantastical does not diminish its importance as evidence of the Greeks' fascination with the topic of the variety of possible systems of belief.[44]

Yet an interest in, and knowledge of, other peoples were far from being confined to the Greeks.[45] First the Medes and then the Persians especially governed a wide variety of races, the differences in whose customs were remarked, no doubt, by others besides Hero-

was anticipated by the wisdom of the prophets and that Greek philosophers plagiarised their ideas from the East. But the topic is not confined to Christians. Already Diodorus (i 96–8) quotes Egyptian priests as claiming that a long list of the most notable Greek poets, law-givers, philosophers, scientists and artists visited Egypt and derived their knowledge from it (they include Orpheus, Musaeus, Melampus, Homer, Lycurgus, Solon, Plato, Pythagoras, Eudoxus, Democritus and Oenopides). Moreover the theme begins already in the classical period: for Herodotus, Egypt was the source of much Greek religion (e.g. ii 43, 49–50, 123, cf. above, p. 14 n. 26), Isocrates (xi 21f) thinks medicine and philosophy originate there, and Plato (*Ti.* 21e ff) too assumes an Egyptian origin for important Greek beliefs.

[40] E.g. Isocrates xi 28 reports that Pythagoras visited Egypt (cf. Hdt. ii 81 and 123) and cf. Plato *Ti.* 20d ff and *Criti.* 108d ff on Solon.

[41] How widely his predecessor Hecataeus had travelled is doubtful, but Herodotus (ii 143) speaks of his having gone to Thebes (cf. Pearson 1939, pp. 84, 93). Outside the historians, Aeschylus, in particular, already showed a marked interest in non-Greek lands in *Pr.* and *Suppl.* especially.

[42] On the extent of Herodotus' own travels, see, for example, Rawlinson 1880, i pp. 8ff.

[43] The origins of this debate, and the complex forms it took, have been studied by, for example, Heinimann 1945, Pohlenz 1953 and Guthrie 1969, Part i ch. 4.

[44] See especially the famous discussion of the relativity of customs in Hdt. iii 38. By the fourth century Aristotle recommends the study of γῆς περίοδοι as useful for the legislator, who may derive information from them on the νόμοι of different races (*Rh.* 1360a33ff, cf. 1365b22ff, and cf. his own series of Constitutions).

[45] We have both archaeological and literary evidence of the movements, within the Eastern Mediterranean basin, of craftsmen and artisans of all kinds, including, for example, doctors, in the sixth and early fifth centuries. On the latter, see, for instance, Hdt. iii 1 (Cyrus sends to Amasis for an eye-doctor) and iii 129ff (Darius calls in Egyptian physicians, and then his prisoner Democedes, for his dislocated foot).

dotus.[46] Of the ancient Near Eastern societies, the Phoenicians, in particular, were famous as traders and colonisers, and the Carthaginian Hanno was one of the most notable early explorers.[47] Some interest in exploration may also be attributed to the Egyptians and to the Persians. Herodotus reports that Necho, in the seventh century, dispatched a Phoenician expedition to circumnavigate Africa,[48] and also that Xerxes sent Sataspes on a similar expedition with a ship and crew he obtained from Egypt.[49] Here too, then, the control offered by Near Eastern societies suggests that we should qualify our conclusions concerning the Greeks, and that knowledge of other peoples was at most a necessary, not a sufficient, condition of the specific intellectual developments that we have identified as taking place in Greece.

A fourth and evidently more promising suggestion relates to the development of literacy. The significance of changes in the technical means by which ideas can be communicated and recorded has been the subject of seminal studies by Goody and others who have shown precisely how these means may influence and even in certain ways determine the nature of what is communicated.[50] Written records make possible the development of a distinct kind of critical evaluation of the past and what Goody calls the accumulation of scepticism in this and other regards. The use of tables and lists helps to provoke an interest in certain types of question, particularly classificatory ones. The consciousness of formal procedures in rhetoric may depend largely on the availability of texts that can be studied at leisure.

Goody's work has, as we noted, undoubtedly contributed to undermining any simplistic presentation of the so-called 'Grand Dichotomy' between 'primitive' and 'advanced' societies, and two fundamental points must be acknowledged. First the intellectual achievements of the great early Near Eastern societies evidently owe much to, and in many cases presuppose, the existence of particular means

---

46 See Hdt. vii 61ff. At i 135 Herodotus comments on the Persians' fondness for adopting foreign customs. The trilingual inscription at Susa published by Scheil 1929 (the so-called foundation charter of Darius) indicates at least a desire to impress other peoples: the languages of the inscription are Old Persian, Elamite and Babylonian.

47 A fragmentary Greek version of his 'periplous', written, it is generally thought, before 480, has survived: see C. Müller 1855–61, i pp. 1–14, Ramin 1976.

48 Hdt. iv 42: whether or not they succeeded is still disputed, but the fact that the explorers reported – to Herodotus' disbelief – that the sun passed to their right hand, i.e. northwards, shows, if true, that they penetrated south of the Tropic of Cancer.

49 Hdt. iv 43: before that, as I have already noted, Darius had sent Scylax to explore the Indus valley, Hdt. iv 44.

50 Goody and Watt 1968, Finnegan 1977 and Goody 1977 (see especially ch. 5, 'What's in a list?' and ch. 8 'The Grand Dichotomy reconsidered').

of communication, especially written records of various types. Secondly, in the complex series of changes in the modes of communication that took place in the ancient Near East a major advance that occurs not long before the period that chiefly concerns us is the invention of alphabetic systems of writing.[51] This development eventually facilitated the expansion of literacy beyond the limited classes of professional scribes to which it had generally been confined in Egypt and Babylonia. Yet here too reservations are needed. Although we can hardly doubt that the spread of critical thought depended *in part* on the availability of texts and of people to read them,[52] we have emphasised before that in the fifth and fourth centuries – and long after – the communication of ideas was still mediated principally through the spoken, rather than the written, word, even though written texts were there to be consulted. Neither the degree nor the relevance of literacy in classical Greece should be exaggerated: nor, correspondingly, should the contrasts between Greece and other ancient civilisations. Like coinage, the alphabet is not a Greek invention,[53] nor was its use by any means confined to Greece. While the study of changes in the means of communication is clearly fundamental for the understanding of the intellectual developments that took place in the ancient Near East as a whole, this can at best provide no more than a part of the solution to the problem posed by the rise of the particular kind of radical and critical investigations undertaken by the Greeks.

POLITICAL DEVELOPMENTS

To advance our inquiry further we must turn to other areas, and most obviously to social and political developments, especially those associated with the rise of the city-state. It is here that the contrast between the Greek world and the rest of the ancient Near East is generally most marked,[54] and the significance of these differences

[51] Alphabetic systems are thought to occur first in the west Semitic group of languages, and the Greek alphabet to have been developed from the north Semitic script some time about the middle of the eighth century B.C. See, for example, Jeffery 1961, pp. 21ff, Snodgrass 1971, pp. 348f, Coldstream 1977, pp. 295ff, and cf. Driver 1976.

[52] On the spread of literacy in Greece, and the changing relative importance of the written as opposed to the spoken word, see Kenyon 1951, Turner 1951, Davison 1962, Harvey 1966, Reynolds and Wilson (1968) 1974 and Pfeiffer 1968, Part I, ch. 2, but contrast Havelock 1971 and 1976 and cf. Finley 1977, especially pp. 610ff.

[53] *Pace* Havelock 1976. The distinctive features of the Greek alphabet are set out by Coldstream 1977, p. 300.

[54] This holds true as a general rule, even though one Phoenician city, Carthage, at least, developed political institutions that the Greek themselves treated as comparable with those of their own city-states: see Aristotle, *Pol.* II ch. 11 especially.

must be carefully examined. Since the main features of the growth of the city-state are well known, they may be rehearsed quite briefly.

In the late archaic and classical periods – roughly from the seventh to the fourth centuries – sovereignty lodged, in the Greek world, in a large number of autonomous and often politically dynamic, not to say unstable, units. The formation of small independent political entities was, no doubt, favoured by geographical factors in the Aegean area, and even in Homeric society the kings who fought with Agamemnon as their commander-in-chief enjoyed a fair measure of autonomy. But whereas Homeric society – as described in the *Iliad* and *Odyssey* at least – operates without any strictly formalised legal, let alone constitutional, framework,[55] the period from the seventh to the fourth century is one of unprecedented activity, throughout the Greek world, in the formulation, discussion, revision and, at times, overthrow, of legal[56] and constitutional codes.

So far as legal codes are concerned, the Greeks had, to be sure, been anticipated by many centuries by the major ancient Near Eastern states, several of whose codes are extant. The best known of these is that of Hammurabi, king of Babylon from 1792 to 1750 B.C., but some codes are much older still.[57] Their significance is two-fold: first, from the point of view of the central authority, they made the administration of the laws easier; secondly, from the point of view of members of the society concerned, justice became a somewhat less arbitrary matter, even though the interpretation of the code still rested with individuals who in turn depended ultimately on the favour of the king. Yet what these legal codes did *not* cover is as significant as what they did. They were essentially legal,[58] not constitutional, charters. While the civil and criminal laws of both Egypt and Babylonia underwent certain changes which can be documented in our sources, their political systems remained very

[55] Thus on the Shield of Achilles, the city at peace includes a scene in which a case is heard in an assembly of the people (ἀγορή) before some elders; the disputants speak before the people, who applaud both sides, and there is a prize set aside to be awarded to the elder who 'speaks the straightest δίκη' (*Il.* xviii 497–508). On the relationship between early informal, and formal, law, between prédroit and droit, see the classic papers of Gernet, notably those collected as section iii in Gernet 1968.

[56] Fragmentary Greek codes exist dating back to the seventh century: the first major code to have been preserved fairly completely (the Gortyn code) dates only from about 450, but this is thought to incorporate much earlier material. See Gernet 1955, pp. 51–9, Willetts 1967.

[57] Some of these are readily accessible in an English version in Pritchard 1969. For an analysis of Hammurabi's Code and Letters, see, for example, Gadd 1973. Cf. also, more generally, Diamond 1971.

[58] Moreover even when considered purely as practical legal codes, they display many inadequacies and omissions and their relation to existing legal documents is, to say the least, problematic.

largely static. Each state was governed by an autocratic (sometimes divine) king, supported by a strong central bureaucracy and subordinate regional authorities. But although the relationship between the king and his advisers might vary, and was certainly influenced by the personalities of the individuals concerned, the system of rule was substantially unaffected. A change of government meant, in general, merely a change of personnel, or at most minor modifications in the roles of those at the top of the chain of command, not a modification in the constitutional position. The extant Near Eastern codes do, indeed, deal with the relationships between slaves (of various kinds) and free persons. But none of them covers the *political*, as distinct from the legal, rights of the free (such questions as the right to speak or vote in assemblies) and so one cannot, strictly speaking, talk of a *constitution* as such at all.[59]

In Greece, by contrast, there is a veritable proliferation of constitutional forms, ranging from constitutional monarchy, through oligarchy, to extreme democracy, and from the seventh century many Greek states underwent several major constitutional upheavals, those at Athens being both the most striking, and the best documented, example.[60] There were, to be sure, Greek autocrats – tyrants who seized power and held it by force. But tyranny was, in most Greek states, a transient, as well as being a small-scale, phenomenon,[61] and in that it tended to undermine the power of the traditional ruling families, its long-term effect was often to favour the eventual introduction of more broadly-based constitutions.

The citizens of the emerging Greek city-states were accustomed both to participate fully in the actual government of their country and to engage in active deliberation of constitutional issues.[62]

---

[59] It was, as Finley 1973*a*, pp. 13–14, has recently put it, the Greeks who discovered not only democracy, but also politics.

[60] The openness of the political and constitutional situation in other Greek states too – at least at particular junctures in their history – is illustrated by many stories in Herodotus. Thus III 142–3 describes what happens at Samos on Polycrates' death: Maeandrius calls an assembly of the townspeople, speaks against tyranny and proclaims equal rights, only afterwards to change his mind when he is denounced by Telesarchus and he concludes that if he were to give up power, someone else would make himself tyrant. See also IV 161 (Demonax gives Cyrene a constitution), V 37–8 (Aristagoras gives the Milesians equal rights, at least λόγῳ) and VII 164 (Cadmus gives power to the Coans), while the continued factional struggles of the late fifth century are a recurrent theme in Thucydides (see especially III 82). As Vernant 1962, pp. 121ff, 1965, pp. 167ff, and Lévêque and Vidal-Naquet 1964 have rightly emphasised, the key development is often that power should be, as the Greeks put it, 'placed in the middle', ἐς μέσον or ἐς κοινόν, an idea that is expressed not just in the historians and political philosophers but also in notable passages in Greek tragedy, as, for example, in Aeschylus, *Supp.* 516ff, cf. 600ff, Euripides, *Supp.* 403ff, 426ff.

[61] Aristotle noted that most Greek tyrannies were short-lived, *Pol.* V 12, 1315b11ff.

[62] See most recently Finley 1973*a*, especially pp. 19f and 25.

Already Solon's poems testify to his concern with the question of how best to balance the rights of different groups within the state.[63] The abolition of existing debts, the Seisachtheia, and that of debt-slavery were undoubtedly two of his most important measures. But even more important, from the point of view of its consequences, was the fact that he extended the rights of the lowest class of citizens, the Thetes. In addition to taking part in meetings of the Assembly, they now had the right to participate in – and appeal to – the popular court, the Heliaea, constituted by panels of jury-men chosen by lot from the entire citizen body. Solon speaks of Justice and Good Rule in his poems:[64] but these concepts now include what we should call the political, as well as the more general legal, rights of the citizen body. As Aristotle puts it succinctly, in relation to appeal to the jury-courts or dicasteries in particular, 'Being master of the vote, the people became master of the constitution'.[65]

By the end of the fifth century, what was generally expected of a citizen of Athens can be illustrated by referring to the exceptional case of Socrates, who in Plato's *Apology* feels that he has to defend, or at least explain, his behaviour in *not* engaging in politics.[66] Yet even Socrates served on the Council, and shared executive power when it was the turn of his tribe to be Prytaneis.[67] Three features of the Athenian constitution insured a very high level of participation in the political life of the city. First, appointment to the great majority of offices was by lot;[68] secondly, most offices could not be held more than once,[69] and thirdly pay was instituted for jury

---

[63] Especially Poems 5 and 24 Diehl.   [64] See, for example, Poem 3 Diehl.

[65] *Ath.* 9.1. Aristotle also reports (*Ath.* 8.5) that Solon went so far as to pass a law to make it compulsory for every citizen to join forces with one or other faction when the city suffered stasis.   [66] *Ap.* 31c–32a.

[67] Plato (*Ap.* 32a–c) and Xenophon (*HG* I 7.15) both report that Socrates was the only Prytaneus to vote against the illegal proposal to try the generals *en bloc* after Arginusae.

[68] Writing of the fourth century, Aristotle reports that military officers were elected by show of hands, but that, with the exception of the Treasurer of Military Funds, the Commissioners of the Theoric Fund and the Superintendent of Wells, *all* the magistrates concerned with the routine administration of the city – as well as the Council of Five Hundred itself – were chosen by lot (*Ath.* 43). These included the Archons, the Treasurers of Athena, the Commissioners of Public Contracts, Public Receivers, Auditors and Assessors of Accounts, the Commissioners for the Repair of Temples, City-Commissioners, Market Commissioners, Controllers of Weights and Measures, Corn Commissioners, Port-Superintendents, the Eleven (in charge of the State prison), Introducers of Cases, the Forty (local magistrates), the Highway Commissioners and their Auditors, the Overseers of Rites and the Commissioners of Sacrifices. Even when we make due allowance for the fact that the group among whom lots were drawn was sometimes restricted, either by class (as the Treasurers of Athena to pentakosio-medimnoi, *Ath.* 47.1, cf. 8.1) or to a previously elected group (as the Archons to a panel of 40 under Solon, or to one of 500 under Cleisthenes, *Ath.* 8.1 and 22.5), the list of offices that any citizen might find himself holding is impressive.

[69] See again Aristotle on the fourth century: 'military offices may be held any number of

service, public office and membership of the Council.[70] In general, then – and this is a point we shall be returning to – we may presume that most Athenian citizens had ample opportunity to gain political experience, not only in the Assembly and serving as jurors in the various kinds of courts, but also in the Council and in one or other office or magistracy, and this is before we take into account whatever private law-suits they may have conducted.[71]

A keen interest in constitutional forms, and an insistence on the value both of freedom (ἐλευθερία) in general and of free speech (ἰσηγορία) in particular, can be documented in a wide range of fifth- and fourth-century texts. The points are so familiar that they inevitably lose some of their impact: yet they are of fundamental significance when we draw comparisons with what we know of ancient Near Eastern societies. Paradoxically, one of the best early examples of a debate on the varieties of possible constitution is an extended passage in Herodotus (III 80–3) in which he professes to report a discussion of the relative merits of democracy, oligarchy and monarchy held by Darius and the Persian leaders when they had just won power. Although the context of this exchange, in Herodotus, is Persian, the style of the discussion, and the whole idea of holding such a debate, are typically Greek – as Herodotus himself perhaps acknowledges when he introduces his account by saying that at this council 'words were spoken which to some Greeks seem incredible'.[72] Thereafter Greek theoretical analyses of constitutional forms become increasingly complex and sophisticated, as we can see from what we can reconstruct of Protagoras' political philosophy,[73] from Isocrates,[74]

times, but none of the others more than once, except membership of the Council which may be held twice' (*Ath.* 62.3).

[70] See, for example, Aristotle, *Ath.* 24 and 27.4 (it was Pericles who instituted payment for the dicasts): the situation in fourth-century Athens is set out at *Ath.* 62. The importance of payment as a factor that insured that the poor exercised their rights is repeatedly emphasised by Aristotle, who strongly disapproved of this development: see *Pol.* 1293a1ff, 1300a1ff, 1317b31ff. The converse anti-democratic devices were to make attendance at the Assembly compulsory for the rich (as in Plato, *Lg.* 764a, cf. Aristotle, *Pol.* 1266a9ff) or to fine them for non-attendance as dicasts in the courts, *Pol.* 1294a37ff, 1297a21ff, 1298b17ff.

[71] On the Athenian's reputation for litigiousness, see below, pp. 250ff.

[72] Hdt. III 80; cf. also VI 43.

[73] The so-called 'great speech' put into the mouth of Protagoras in Plato's dialogue named after him (*Prt.* 320c ff) contains the first, and one of the few, extended extant statements of the key principle underpinning the democracies, namely that *all men alike* have a share in πολιτικὴ τέχνη. In that respect, at least, (though not, no doubt, in others) this speech appears faithfully to represent Protagoras' own position. While the interpretation of the original significance of the famous dictum that 'man is the measure of all things' is highly controversial, we can hardly doubt that it had, and that Protagoras knew it had, among other things, an application in the political field.

[74] The questions of the classification of constitutions, and the differences between two kinds of equality, are broached, for example, in III 14ff, VII 20ff, and XII 130ff.

and more especially from Plato[75] and Aristotle[76] themselves. Moreover the discussions that took place on this subject were far from being all purely theoretical, for the ideas expressed could and did find practical application not only in the reform of existing constitutions, but also in framing the constitutions of new states.[77]

ἰσηγορία, like ἰσονομία, was, to be sure, a rallying-cry used by the advocates of democracy in particular.[78] Demosthenes uses the term in this way;[79] Herodotus attributes Athens' rise to power to ἰσηγορίη especially;[80] and what the democrats applauded as ἰσηγορία was castigated by their opponents as unbridled licence of tongue.[81] But oligarchic cities also deliberated on affairs of state, even though those deliberations were restricted to those with full political rights, such as those known as the ὅμοιοι, 'equals' or 'peers', for example at Sparta. Thus Thucydides reports one such discussion among the Lacedaimonians at I 79ff, where he specifies that their method of arriving at a decision was not by voting (ψήφῳ) but by acclamation (βοῇ).[82]

[75] Especially the brilliant characterisation of the main types of constitution (aristocracy, timocracy, oligarchy, democracy and tyranny) and of the corresponding types of man, in *Republic* VIII and IX. Although most Greek political analysis is set firmly in the framework of the actualities of Greek social experience, Plato, for one, is quite radical in his readiness to consider such possibilities as the community of wives or of property.

[76] Aristotle puts forward two main classifications of constitutions in *Pol.* At 1279a22ff he proposes a six-fold schema, monarchy, aristocracy and 'constitution' together with their three 'deformed' counterparts, tyranny, oligarchy and democracy; but at 1290a13ff he uses the popular classification into two main types, oligarchy and democracy, and IV 4–6 shows some sophistication in analysing the different forms of these. Following his usual practice of reviewing earlier opinions on the problems he discusses, Aristotle provides extensive information on the range of contemporary controversies in this area, not merely in his survey of ideas on the best constitution (in book II) but throughout.

[77] As in the famous example of the foundation of Thurii in 443 B.C., where, according to Heraclides Ponticus (in D.L. IX 50), Protagoras was asked to draft the laws (though cf. Diodorus Siculus XII 10f). Such cases form part of the background to Plato's discussion of the ideal state in both the *Republic* and more especially the *Laws*, which envisages the setting up of a state in Magnesia: Plato's own interests in practical politics, culminating in his disastrous experiences as adviser to Dionysius of Syracuse, are recorded in the *Seventh Letter* (which may be used as a source whether or not it is authentic).

[78] On ἰσονομία and ἰσηγορία see especially Ehrenburg 1940, Vlastos 1953 and 1964, Griffith 1966, J. D. Lewis 1971, Momigliano 1973 and Finley 1975*b*.

[79] E.g. XV 18, cf. XX 16, XXI 124, LX 28; cf. ἰσονομία e.g. at Th. III 82, IV 78, and ἰσονομεῖσθαι at Th. VI 38–9.

[80] V 78; cf. the Athenians' claims at Th. I 77, Eupolis Fr. 291, Socrates in Plato, *Grg.* 461e, and Demosthenes at IX 3.

[81] See, for example, Plato, *R.* 561de, 562b–563b. The term παρρησία was often used to express this (see Isocrates VII 20, cf. Plato, *R.* 557b), but also occurs without distinct, or any, pejorative undertones (e.g. Euripides, *Hipp.* 421–3, *Ion* 670–2): cf. Scarpat 1964, Momigliano 1973, pp. 259ff.

[82] Th. I 87. Cf. Aristotle, *Pol.* 1272a 10f who notes that in oligarchic Crete all the citizens shared in the Assembly, though this only ratified the decisions of the elders and the Cosmoi (but cf. Aeschines I 180f). The critical account of oligarchy given by Darius in Hdt III 82 incorporates the idea that each of the 'few' will try to prevail in his opinions:

Furthermore the idea that freedom in general, especially political autonomy, the right to self-government, is precisely what marks out the Greeks from most Barbarians is commonly expressed in the fifth and fourth centuries.[83] One writer who does so who is neither a historian, nor a practising politician, nor even chiefly concerned with political analysis, is the Hippocratic author of *On Airs Waters Places*, who explains the less warlike and more gentle character of Asiatic peoples (as he represents them) partly in terms of the climates of their countries,[84] but partly also in terms of their customs and institutions. It is because they are mostly ruled by despots that they lack courage and spirit.[85] The fact that in certain respects this writer, like many others, exaggerates the contrasts between Greeks and non-Greeks does not diminish the value of his testimony as evidence of the way the Greeks themselves saw those contrasts: as he views it, the chief distinguishing characteristic of Greek political life is that the Greeks are their own masters.

THE RELEVANCE OF POLITICS TO SCIENCE

It will readily be agreed both that the period from the seventh to the fifth centuries was one of a high level of political activity and involvement in Greece, and that the constitutional framework of the Greek city-states differs markedly, in certain ways, from that of the great ancient Near Eastern river civilisations: indeed some of the institutions of the democracies, such as ostracism, are unprecedented and unparalleled outside Greece. But the question of the possible relevance of these points to our own inquiry is problematical. Two suggestions, one comparatively simple, the other more complex, merit particular consideration.

cf. also the use of the adjective ἰσόνομος with ὀλιγαρχία at Th. iii 62 (on which see Vlastos 1964, pp. 13ff).

[83] See, e.g., Hdt. vii 103–4 and 135 (on the Spartans in particular) and 147 (on the Greeks in general) and cf. the many texts developing the idea that the Greeks fought for their freedom against the Persians (e.g. v 2, 49, vi 11, 109 and cf. Aeschylus, *Pers.* 241ff, 402ff, Plato, *Mx.* 239a ff especially) a theme that recurs in the different contexts of the Peloponnesian war in Thucydides (e.g. i 69, ii 8, iii 59, iv 85–6, v 9) and the confrontation with Philip in Demosthenes.

[84] He appeals especially to the 'uniformity' in the seasons: see *Aër.* ch. 16, *CMG* i, 1 70.13ff, and cf. ch. 23, 75.28ff.

[85] E.g. 'When men do not govern themselves and are not their own masters, but are ruled by despots, they do not worry so much about military exercises as about not appearing warlike' (ch. 16, *CMG* i, 1 70.21ff), and as evidence for this thesis he asserts that such Asiatics as are not so governed, are most warlike (ch. 16, 71.2ff): cf. ch. 23, 76.17ff 'Where men are ruled by kings they are necessarily most cowardly...for their souls are enslaved and they are unwilling to risk their own lives gratuitously for another's aggrandisement. On the other hand, those who govern themselves will willingly take risks because they do it for themselves.'

First, as several scholars have remarked from rather different points of view,[86] the spheres of law and justice provide important models of cosmic order. The notion that the world-whole is a cosmos, that natural phenomena are regular and subject to orderly and determinate sequences of causes and effects, is expressed partly[87] by means of images and analogies from the legal and political domain. On this view, it was the experience of regulated legal institutions that provided the necessary background against which the conception that the world as a whole is ordered could develop.

Now this suggestion seems straight away to run into a difficulty, in that – as we have seen – complex legal systems are not confined to, nor do they originate in, Greece. Certain differences in the possible attitudes towards the nature and basis of law should, however, be noted. Insofar as the ultimate sanction for the code is still the god or his representative the king – as is generally the case in the ancient Near East[88] – justice to that extent continues to depend upon a personal authority. While many Greek codes are named after their authors,[89] and we find Solon, for example, invoking Zeus in his poems,[90] there is a shift in emphasis: the notion of the abstract, impersonal character of the law, to which the lawgiver himself is subordinate, gains ground. Divine vengeance may still be mentioned; but the gods tend increasingly to become depersonalised as mere personifications of the rule of law itself. The idea that there is a higher *personal* sanction or authority for the laws is undermined as the laws themselves become the subject of open debate and depend upon public consent. The contrast between φύσις, nature, and νόμος, man-made law and convention, underlines the point, but long before that contrast had become a commonplace at the end of the fifth century, Solon shows that he knew very well that the fate of his constitution rested with the sovereign people.[91]

---

[86] See Hirzel 1907, H. Gomperz 1943, Gernet (1955) 1968, p. 19, Vlastos (1947) 1970 1953 and 1975, Vernant (1957) 1965, pp. 304f and 1962, pp. 87ff, and Vidal-Naquet 1967, pp. 58ff.

[87] Though this is far from being the only vehicle for the expression of this idea: see Solmsen 1963 and Lloyd 1966.

[88] See, for example, Pritchard 1969, pp. 159ff: Lipit-Ishtar speaks of himself as son of Enlil and invokes Utu; and pp. 164ff, where Hammurabi says 'when Marduk commissioned me to guide the people aright'.

[89] For example Draco, Solon and in the semi-mythical past Lycurgus.

[90] See Poem 1 Diehl, especially 17ff (which also illustrates how Solon connects, or rather does not clearly distinguish between, the moral/legal and the natural order of the world) and Poem 3.1. Cf. also his invocation of Earth in Poem 24, and further references to the gods in Poem 23.18 and to Zeus in Poem [28].

[91] This is illustrated by the report (Aristotle, *Ath.* 7) that he attempted to block future change by passing a law to make the laws unalterable for a hundred years. Solon's own

The thesis can, then, be upheld, though it is worth underlining two other familiar points. First, it is not just the number of legal and political images that are to be found in the scanty remains of early Greek philosophy that is striking, but also their variety.[92] The cosmos is sometimes conceived in terms of a balanced relationship, even a contract, between equal opposed forces:[93] but it is also viewed, on occasions, as a monarchy (though the king who controls the cosmos is now seen not as an arbitrary divine power, but as the personification of a quite impersonal justice);[94] and the world may even be seen in terms of a state of constant aggression or strife (though since this strife is normal, it can also be described as what is just, the divine law, which makes the world a world-order or cosmos).[95] Secondly, reflection on the idea of political and legal order is a source of models not just in cosmology, but in other areas of inquiry as well, particularly in medicine and physiology,[96] where the functioning, and the mal-functioning, of the human body were often conceived in terms of the interrelations of opposing factors, health and disease being seen as, for example, a state of equal rights, ἰσονομία, or a lack of it, between these factors.[97]

The second more complex suggestion relates to the radical examination to which both the framework of political relations, and that of beliefs about natural phenomena and the world, were submitted.[98] Just as one of the notable features of Greek political experience is the way in which, from the sixth century onwards, the questions of how society should be regulated and of the merits and demerits of different kinds of constitutions came to be a subject for open – and not merely theoretical – discussion, so too the possibility of challenging deeply held assumptions about 'nature' and of debating such issues as the origin of the world is a prominent characteristic of Greek

Poem 8 Diehl emphatically states that the Athenians should not blame the gods for their own troubles, which they are responsible for themselves (cf. also Poem 3.5ff), and the political poems as a whole may be seen as an exercise in persuasion, justifying his policies to the Athenians.   92 See Lloyd 1966, pp. 210ff.

93 See, for example, Anaximander Fr. 1, Parmenides Fr. 9, and Empedocles Fr. 17.27ff and Fr. 30.

94 See Heraclitus Fr. 53, Anaxagoras Fr. 12, Diogenes of Apollonia Frr. 5 and 8, and cf. Plato, *Ti.* 47e f, *Phlb.* 28c and *Lg.* 896de, 904a.

95 See Heraclitus Frr. 30, 80, 94 and 114.

96 Cf., e.g., Vlastos 1953, pp. 363f, Vidal-Naquet 1967, p. 58.

97 This idea is already expressed in Alcmaeon Fr. 4, and it becomes a commonplace in the Hippocratic Corpus (e.g. *VM* ch. 14, *CMG* I, 1 45.18ff); cf. also, for example, Eryximachus in Plato's *Smp.* 186d ff.

98 The theme that Greek rationality in general is the product of the city-state has been developed forcefully in works by Gernet 1917 and (1955) 1968, Vernant (1957) 1965, pp. 285ff, and 1962, Vidal-Naquet 1967, and Detienne 1967, pp. 99ff. What follows is much indebted to these studies.

speculative thought. Stated thus, there is at least a certain parallelism between the two developments. But we can perhaps go further. In some respects we appear to be dealing not just with two analogous developments, but with two aspects of the same development.

Despite the important continuities between the bronze age, the archaic period and the classical period, the city-state called for the exercise of new skills of leadership, and since these included especially skills of persuasion that were deployed in relation to a wide audience, those audiences themselves came to be keen judges of this ability. The very variety of 'wise men' active in the seventh and sixth centuries is remarkable. Apart from a Solon or a Thales (both of whom figure in the earliest list of the Seven Wise Men we have – Plato's[99]) there were many others, not just other statesmen of differing political persuasions, such as Pittacus and Periander,[100] but seers, holy men and wonder-workers, such as Epimenides of Crete,[101] Aristeas of Proconnesus, the Scythian Abaris and Hermotimus of Clazomenae.[102] Much of our information about these men and their activities is late and unreliable. But while the details are often unsure, we cannot discount the tradition of the emergence of new kinds of 'wise men' during these transitional centuries as a whole, and evidently the category of 'wise man' was a wide one and spanned both what *we* should call political, and religious and intellectual, leadership – not that the Greeks themselves drew any such hard and fast distinctions.

We may presume that those who gained a reputation for exceptional wisdom did so on a variety of grounds and appealed to different groups among their contemporaries. Some no doubt relied on a certain personal charisma: others on an implicit or explicit claim to esoteric knowledge. Yet equally clearly *in certain contexts at least*, both inside the political domain and outside it, proposals and ideas were far less likely to be accepted simply or even primarily on the say-so of some particular individual relying on his personal prestige or authority. This is easy to see in the political sphere. Far more than an Agamemnon – let alone than a Darius, an Amasis or a Croesus – Solon and Cleisthenes knew that they had to gain and

---

99 *Prt.* 343a. D.L. 141f records contrasting traditions about the membership of the Seven.
100 These two, the one a deposer of tyrants (though himself later an αἰσυμνήτης or elective monarch), the other a tyrant himself, figure in most lists of the Seven (though Periander not in Plato's): see D.L. 141ff, 74ff, 94ff.
101 Epimenides is mentioned as a θεῖος ἀνήρ by Plato, *Lg.* 642de, cf. 677de: for the tradition that he 'purified' Athens, see Aristotle, *Ath.* 1, Plutarch, *Solon* 12, D.L. 1110, cf. 112.
102 These last three have figured prominently in the debate on the question of 'shamanism' in Greece: see Meuli 1935, Gernet (1945) 1968, pp. 421ff, Dodds 1951, ch. 5, pp. 135ff, Vernant (1957) 1965, pp. 297ff, Burkert 1972a, pp. 120ff.

maintain consent for what they proposed from their fellow-citizens. It was their votes that counted and they had to be won by persuasion and argument.

But a similar point is applicable also in medicine and even in philosophy. We have noted before the competitive situation that developed in Greek medicine, not just between individuals who shared the same general approach, but between those with radically different approaches, one of the Hippocratic writers, as it might be, and one of the priests of Asclepius. But a doctor who could present plausible arguments and evidence for his theories was in a stronger position than one who was not prepared to do so – *at any rate so far as some of his potential patients*, and some of the potential audience for his lectures, *were concerned*. A similar point is true also of those who put forward physical, physiological or cosmological doctrines, whether or not they were in direct competition for pupils to teach.[103] What passed as a plausible argument, indeed what passed as 'evidence', varied considerably from one group to another: and the cases of Epimenides and Aristeas – and the whole history of the rise of the cults of Asclepius – show that there were quite other means of gaining a reputation as a wise man or as a healer. Yet the deployment of arguments and evidence of some kind came, in certain circles at least, to be what counted in other spheres of 'wisdom' besides that of statesmanship.

The degree of political involvement had important and widespread repercussions on intellectual life as a whole. The constitutional framework that guaranteed such involvement has been outlined above. In Athens, especially, where the Assembly and dicasteries were open to all citizens, and where appointment to many offices was by lot and many could not be held more than once, participation was particularly extensive. It was also particularly intense, as emerges from the frequent comments in classical literature that suggest how preoccupied the Athenians became not just with their political roles and responsibilities but also with the exercise of their legal rights. The topic is a recurrent one in Aristophanes. In the *Clouds* Strepsiades doubts that the place pointed out on a map can be Athens: 'I don't believe it: I see no dicasts sitting.'[104] In the *Wasps* we are given a

---

103 We should, however, distinguish between different modes of competition. Unlike medicine, where how to treat the sick always presented an urgent practical problem, competition between educators only began in earnest when Greek education itself had begun to expand with the sophistic movement. It is therefore hardly surprising that medicine is the field best represented in our sources for the confrontation between 'science' and 'magic': cf. above, ch. 1.

104 *Nu.* 206ff. Cf. also *Eq.* 1317 (the dicasteries are what the city delights in), *Pax* 505

graphic picture of Philocleon as a φιληλιαστής quite besotted with judging.[105] The whole play pokes fun at the dicasteries: yet in doing so Aristophanes faces a difficulty, for he knows that the dicasts are the people themselves. In the Epirrhema (*V.* 1071–90) it becomes clear that the dicasts are the autochthonous citizens of Athens, typified by the heroes of Marathon.

Prose writers make similar points. In the *Memorabilia* Xenophon refers to Criton's view that life was difficult at Athens for a man who wanted to mind his own business.[106] Socrates, who in Plato's *Apology* tells the jury that this is the first time he has appeared in a law-court, is, again, evidently an exception that illustrates, by contrast, what the usual experience was.[107] In Thucydides (1 77) the spokesman of the Athenians refers to their city's reputation for litigiousness – φιλο-δικεῖν – and they offer a defence rather than a denial. In the Funeral Speech (II 40) Pericles asserts that at Athens even those engaged in business know about politics: in Athens alone a man who takes no part in public affairs is considered not harmless, but useless, and Pericles claims that the Athenians are all sound judges of policy. In the Mytilenean debate, by a nice stroke of artistry, Cleon himself is made to chide his audience on the grounds that they are easily misled: everyone, he says, wants to be an orator and is reluctant to yield to anyone else in quickness of wit, praising sharp remarks before they are out of someone's mouth; the trouble is that they do not treat serious matters sufficiently seriously, but rather behave more *like those sitting at a performance of sophists* than like people deliberating about affairs of state.[108]

Finally there are passages that explicitly oppose domination by force and domination by reason or argument.[109] The continuation of the Athenians' speech at Thucydides 1 77 is perhaps particularly

(the Athenians do nothing but judge cases), *Av.* 1694ff and many other passages collected in De Ste Croix 1972, p. 363 nn. 8 and 10.

[105] *V.* 88ff: he hardly sleeps, and when he does his mind flutters in his dreams round the water-clock; he wakes with his fingers in the gesture of voting; while other lovers write the names of their beloved on walls, the name of his loved one is ballot-box (κημός: the top of the urn that held the votes); to insure he has enough pebbles for voting he has a whole beach-full at home.... [106] Xenophon, *Mem.* II 9.1.

[107] Plato, *Ap.* 17d 1ff. Similarly Isocrates protests his own – exceptional – lack of experience of the law-courts, while engaging in a passionate attack on the prejudices of juries and their gullibility in failing to see through the corrupt informers by whom they are surrounded (xv 15–38). At xv 295f Isocrates goes on to describe Athens, with its exceptional opportunities for the practice of rhetoric, as the teacher of orators from all over Greece.

[108] III 37ff, especially 38.7 (and cf. 38.4).

[109] Cf. above, p. 84 and n. 128 on Gorgias, *Helen* para. 12. The contrast between πειθώ ('persuasion') and βία (force) is a recurrent theme in Greek tragedy: see, for example, Detienne 1967, pp. 6off, Buxton 1977.

significant. There they contrast the Athenian empire with one based simply on military superiority. The allies, they claim, are used to dealing with Athens on equal terms (ἀπὸ τοῦ ἴσου), but when they do not get their way they are aggrieved – for men resent injustice more than violence – when they should, on the contrary, be grateful to the Athenians for their moderation. The Athenians do not conceal the fact that Athens rules from a position of superior strength,[110] but they develop an important opposition between settling disputes by violence and doing so by argument.[111] In a different context, Aristotle, too, contrasts the situation in the past, when tyrants ruled by military force, with what he says happens in his own day, where it is those who are skilled in speaking who lead the people, but their inexperience in military matters prevents them, for the most part, from attempting an armed coup.[112]

There are, of course, elements of exaggeration and of rationalisation in most of these texts. But when we have discounted these, we are still left with good evidence for some important conclusions. First and most obviously, testing arguments, weighing evidence and adjudicating between opposing points of view were, as we have said, a common part of the experience of a considerable number of Athenian citizens.[113] In the context of law and politics, when they acted as judges and voted in the assemblies, they were no mere spectators, but themselves took the effective decisions. Accountability is mentioned in Herodotus III 80 as one of the three chief marks of a democracy,[114] and the critical evaluation of testimony was a central feature not just of ordinary law-suits, but also in particular of the institutions of the δοκιμασία (which might be concerned, for example, with testing a candidate's eligibility for office)[115] and the εὔθυνα (the scrutiny of a magistrate's tenure of office, directed primarily, though not always exclusively, at its financial aspects).

That political and legal testing and scrutiny were sometimes seen as paradigmatic of testing and scrutiny of any kind is suggested first

---

[110] Cf. Cleon's remarks at Th. III 37.2.

[111] Cf. also Diodotus' remarks at Th. III 42, where he claims that a good speaker should win his case by argument not by intimidation, a point often echoed in the orators.

[112] *Pol.* 1305a10ff.

[113] Cf. Vernant, 1962, pp. 43 and 74 especially.

[114] The other two are the election of magistrates by lot, and the referring of policy decisions to the general assembly.

[115] There were many different kinds of δοκιμασία (they included one concerned with the enrolment of ephebes). From the political point of view, the most important were those of the incoming Council and of the Archons, the former undertaken by the existing Council, the latter a double δοκιμασία by the Council and the Jury-Court: see, for example, Aristotle, *Ath.* 45 and 55.

by the linguistic data. Naturally enough, Greek terminology for evidence and its examination draws heavily on words with primary meanings in the political or legal sphere. Of the words used generally for 'evidence' μαρτύριον is directly derived from the Greek word for 'witness', μάρτυς.[116] Of the terms used for testing an idea or hypothesis, ἔλεγχος and ἐλέγχειν have as their primary senses in the classical period[117] the cross-examining of witnesses and the examination – or more especially the refutation – of an opposing speaker's case.[118] Other terms with a technical legal or political application that are used more generally of testing or examining ideas are βάσανος and βασανίζειν,[119] and δοκιμασία and its cognates[120] (in both cases the extent to which these general usages were felt as 'live' metaphors is problematic), while the most common expression for 'giving an account', λόγον διδόναι, was used particularly of rendering a financial account, as in the εὔθυνα.[121]

Moreover we have good evidence that the parallelisms between political and legal debate on the one hand, and philosophical and sophistic discussions on the other, were explicitly recognised by some ancient writers. All these discussions could be referred to as ἀγῶνες, contests, and the 'agonal' or 'agonistic' features of much classical literature – the balancing of speech and counter-speech, for instance, not just in Thucydides, who presents idealisations of actual debates, but also in epic and drama – are well known.[122] But we can be more

---

[116] τεκμήριον, derived from τέκμαρ, goal, end, and so token, sign, and σημεῖον, derived from σῆμα, mark, both have extensive application in the legal context in the classical period.

[117] In Homer, however, the nouns ἔλεγχος and ἐλεγχείη are used in the sense of blame or reproach or a cause for such, e.g. *Od.* xiv 38, xxi 329.

[118] The general classical use of terms with this root can be illustrated by Hdt. ii 23 (the view that the Nile floods because it flows from Ocean is obscure and οὐκ ἔχει ἔλεγχον) and Th. i 21 (the stories of the logographers are incapable of being tested – ἀνεξέλεγκτα).

[119] βάσανος, used already in Thgn. 417 of the touchstone, was the regular term for the procedure whereby slave witnesses were tortured in Greek trials. But both the noun and the verb are also used more generally of testing hypotheses, e.g. Plato, *Ti.* 68d (the proportions of compound colours cannot be determined by testing) and Aristotle, *GA* 747a3, and 7 (testing the fertility of semen, and of women), and cf. such Hippocratic passages as *Liqu.* ch. 1, *CMG* i, 1 85.16 and *Aër.* ch. 3, *CMG* i, 1 57.11.

[120] For δοκιμάζειν, δοκιμασία and διαδοκιμάζειν used generally of testing outside the particular sphere of the institutions of the δοκιμασία, see, for example, Th. iii 38, Aristotle, *EN* 1157a22, 1162a14, Xenophon, *Oec.* 19.16. For εὔθυνα/εὐθύνειν used in a medical context, see, for example, Aristotle, *Pol.* 1282a1ff (cf. Plato, *Plt.* 299a) and in a general one, Aristotle, *de An.* 407b27ff.

[121] E.g. Lysias xxiv 26. Cf. also the use of λογιστής and ἐξεταστής (from ἐξετάζω, examine) for the auditors of public accounts, e.g. Aristotle, *Pol.* 1322b11.

[122] The importance of these elements in Greek culture as a whole has been emphasised by Burckhardt 1898–1902, and by Ehrenburg 1935, pp. 63ff, especially (and cf., e.g., Duchemin 1968 on tragedy in particular). The reservations of Huizinga, who questions how exceptional Greek culture is in this respect, should, however, be noted: see Huizinga (1944) 1970, pp. 91ff.

specific. First we saw that Gorgias juxtaposes political and legal contests with the arguments of the 'meteorologists' and the philosophers in order to illustrate different aspects of the power of persuasion.[123] Again when Cleon in Thucydides III 38 chides the Athenian assembly for behaving like an audience at a performance of sophists, the rebuke is revealing. Both types of occasion generated the same eager expectations on the part of the audience, who evidently prided themselves on their connoisseurship of the witty thrust or the telling argument. But more than that: given that it was the same body of men who constituted the Assembly and who formed the bulk of the audience at the dramatic performances in the theatre,[124] and who – with their counterparts from other states – attended the great Pan-Hellenic festivals at which public lectures were given,[125] there was a natural progression from political debate to sophistic performance.

Once again the point can be extended to medicine. We do not know precisely how those who served as public physicians – δημοσιεύοντες – were appointed,[126] but the sophist Gorgias claims, in Plato's dialogue, that the trained orator will be more successful than the doctor not only at persuading patients to submit to treatment, but also at *convincing the Assembly or any other meeting* that he should be chosen as physician.[127] Indeed Gorgias maintains that there is no subject at which the orator will not outdo the ordinary craftsman when speaking at a mass meeting. This may be – we may think – rather to exaggerate the difference between the orators and *some* doctors, at least to judge from the rhetorical skills displayed by quite a number of the medical writers.[128] But it suggests that there were overlaps not just between the doctor and the sophist in the giving of public lectures, but also between the doctor and the politician, in that if the former desired appointment as a public physician, he might well,

---

[123] *Helen*, para. 13, see above, p. 84.

[124] Cf. Demosthenes v 7–8, who remarks that the Athenian Assembly listened to Neoptolemus with as much indulgence as if they had been *attending a play* (a comparison that has added point as Neoptolemus was an actor and playwright himself, as well as being an active politician).

[125] See, for example, Plato, *Hp. Mi.* 363 cd, 364 a (cf. 368 cd), Isocrates IV 45, and cf. Dicearchus' report (in Athenaeus XIV, 620 d, cf. D.L. VIII 63) that a recitation of Empedocles' poem the *Purifications* was given at Olympia. [126] See Cohn-Haft 1956, pp. 56ff.

[127] *Grg.* 456 bc, cf. also 452 e, 459 a–c, 514 d ff. Xenophon, *Mem.* IV 2.5, also envisages a prospective public physician addressing the Athenian assembly. Cf. Socrates' remarks, at *Prt.* 319 b f, on how the Athenians, *collected in an Assembly*, call in experts in building and ship-building to advise them when they debate matters in those fields (where his point is to contrast taking 'professional' advice on technical matters with the practice of allowing anyone to speak in political discussions concerning the administration of the state). At *Plt.* 297 e ff Plato draws an elaborate picture of the consequences of allowing everyone an equal voice on such subjects as medicine and navigation, where the context is again that of an Assembly (298 c 2f). [128] See above, ch. 2, pp. 88ff.

like the latter, have to exercise his skills of persuasion before a large lay audience.

Finally there is, as it were, a qualitative point to add to the quantitative one. It is not just the frequency of the experience of debate, but also the radical nature of what might be debated, that are important. In the political sphere, as we have seen, the subjects regularly discussed by the Assembly covered every major issue of public policy, including questions of war and peace, and the laws and the constitution themselves. Where the topic of how the state should be governed could be debated openly by the citizen body as a whole, there were, we may presume, fewer inhibitions – at least in some quarters – to challenging deep-seated assumptions and beliefs about 'natural phenomena', the gods or the origin or order of things. The fate of Socrates is a reminder not to exaggerate the limits of permissible radical dissent: nor was he by any means the only intellectual leader who, for one reason or another, was prosecuted or threatened with prosecution, though we have reason to be sceptical about some of the more fanciful, late stories of trials for impiety.[129] But while free speech and free speakers came under attack in Athens as often as, or even more often, than elsewhere,[130] the scope for criticism and dissent was – normally – very wide indeed.[131]

[129] One of the first to be tried for impiety (presumably under the decree of Diopeithes) was Anaxagoras, though according to some of our sources this move was directed in part at discrediting Pericles through Anaxagoras (see Plutarch, *Pericles* 32, cf. D.L. II 12, Diodorus Siculus XII 39): thus although Anaxagoras was exiled (or left Athens before the outcome of his trial was known) he was evidently honoured by the people of Lampsacus, where he spent the last part of his life. We know from Aristophanes (*Av.* 1071ff) that Diagoras, who was regularly represented as an 'atheist', was condemned to death *in absentia*. But the reliability of some of our information about the trial of Protagoras has been doubted (particularly the report, e.g. in D.L. IX 52, that his books were burnt, and the story that he was exiled 'from the whole earth' by the Athenians, Philostratus, *VS* I 10.3). Men could be, and sometimes were, charged with 'impiety': but (*a*) the motives for such trials were often quite general, to silence or remove political rivals or undesirable characters, (*b*) the definition of 'impiety' was anything but precise, and (*c*) some famous 'atheists' were apparently never charged (see below, p. 257 n. 138). On the whole topic, see Decharme 1904, ch. 6, Lipsius 1905–15, II pp. 358ff, Drachmann 1922, Derenne 1930, especially pp. 257ff, Morrow 1960, pp. 470ff, and Dover 1975.
[130] Not much need or should be made of the statement in ps.-Xenophon, *Ath.* 2.18f that the people disapproved of attacks on itself by the comic poets, but a scholium to *Ach.* 67 refers to a decree of 440 banning comedy, though this was only a temporary measure (connected, presumably, with the political situation after the Samian revolt), since the decree was rescinded three years later. Again Aristophanes himself tells us that Cleon indicted him for defaming the state in the presence of foreigners (*Ach.* 502ff, referring to his lost play the *Babylonians*) – though that certainly did not put an end to his attacks on Cleon. It was under the Thirty that the most concerted effort at suppression of free speech seems to have occurred. At least Xenophon reports that Critias made it illegal to teach the art of speaking (*Mem.* I 2.31), a move that Xenophon interprets as being made against Socrates, though it may well have had a more general aim.
[131] As Finley 1973*a*, pp. 97ff, rightly stresses: cf. also Momigliano 1973 and Finley 1975*b*.

Once again it is useful to compare the city-state, at least when under constitutional government, with the situation that obtained outside Greece. Any member of the entourage of an autocratic ruler (whether Greek or non-Greek) was in danger as soon as he fell from favour. This applied not only to political advisers, but also to others who might be consulted on a wide range of matters, the priests and diviners, and even the doctors called in to give treatment. It is not just a Pythius or even an Artabanus who was at risk for crossing Xerxes:[132] Herodotus also reports that the Egyptian doctors who had failed to cure Darius' dislocated foot would have been impaled, but for Democedes' intercession on their behalf.[133]

Now the charge was often levelled at the democracies that the sovereign people was, collectively, just as arbitary and revengeful as any tyrant. Thus Diodotus complains, in Thucydides, that while individual speakers are accountable, the people themselves are not, a theme taken up by many other writers.[134] Again although the Athenian democracy, in particular, was generally notably lenient to its political enemies,[135] the first Mytilenean debate and the treatment of the generals after Arginusae show that it could also be vindictive. Nevertheless even those decisions were arrived at after being debated in full Assembly – indeed in both cases the Assembly met twice to consider the issue.[136] The institution of ostracism was one method at Athens by which the problem or threat posed by powerful spokesmen of minority views was dealt with. But short of situations where ostracism might be invoked, there were plenty of occasions when those who lost one particular vote might live in hopes of having the decision subsequently reversed.[137]

It is true that many political leaders were not just exiled, but done away with, in democracies and oligarchies alike: even so a remark-

---

[132] Hdt. vii 38f (Pythius) and cf. the council of war, vii 8ff. At viii 68f Herodotus refers to the risks Artemisia was assumed to run in expressing her views freely to Xerxes, though on this occasion Xerxes admired her the more for her frankness. These stories are told from a Greek perspective, to be sure: but this detracts from, rather than completely negates, the general moral they convey about the wilfulness of absolute monarchy.

[133] Hdt. iii 132: this story too has a Greek slant, but it has more or less gruesome echoes down the centuries.

[134] Th. iii 43, cf., e.g., Aristophanes, *V.* 587–8. At *Pol.* 1292 a 15ff Aristotle draws a direct comparison between one kind of democracy and tyranny. We should, however, bear in mind that most of those who emphasise such points write from a position that is either critical of, or indeed bitterly opposed to, democracy.

[135] Especially on the restoration of the democracy, where the treatment of those who had overthrown it has been hailed as the first example of a political amnesty (Acton, cited by Finley 1973*a*, p. 90).   [136] Th. iii 36ff, Xenophon, *HG* i 7.4ff and 9ff.

[137] Later Attic orators were not slow to remind their audiences that the people had made rash decisions that they had later repented: see, for example, Isocrates xv 19.

able degree of divergence of political opinion existed in many states. Equally while many intellectual leaders came under threat, others – including some with reputations for 'impiety' – were able not simply to hold their views, but also to express and teach them, both in Athens and elsewhere.[138] The trial of Anaxagoras and the execution of Socrates indicate that there were admittedly quite ill-defined limits to what a democracy would tolerate. Yet on the other side the plays of Aristophanes, several of which launch sustained attacks on the demagogue Cleon at the height of his power, show how far it was possible to go in criticising those who were in positions of great influence in the Assembly.[139] From Pericles onwards, the popular leaders themselves could not, for their part, afford to neglect public opinion, or at least to antagonise their own supporters. In this situation the critic of those in power was, comparatively speaking, much less disadvantaged than in most societies before or since. Political leaders in the democracies were frequently reminded that the days of their influence were numbered: more important still, the critic had potentially, and even sometimes in practice, equal access to the sovereign people in the Assembly, in the market-place and in the festivals, if not also in the theatre.[140]

In the admittedly speculative business of attempting to elucidate why it was that certain kinds of intellectual inquiry came to be initiated in ancient Greece, we must first take stock of certain of the economic,

---

[138] Those who were often labelled 'atheists' included Prodicus, Critias and Hippon: there is no record that the last two were prosecuted for their religious views and only untrustworthy evidence (Schol. in Plato, *R.* 600c and the Suda) that Prodicus was executed for corrupting the young (an obvious confusion with Socrates). Critias, one of the Thirty, was killed in the revolution that overthrew him, but was not, so far as we know, attacked for his views on the gods. Near contemporary sources report Prodicus visiting Athens quite freely and being in no way restricted in his teaching activities.

[139] The political courage and radicalness of Aristophanes should not be exaggerated: *Pax* was performed only very shortly before the conclusion of the peace treaty in 421 (though he also attacked the war in the less favourable climate of opinion of 425 in *Ach.* and in 411 in *Lys.*). On the other hand, *Eq.* (in 424) and *V.* (in 422) both attack Cleon soon after the apparent vindication of his policy against Sphacteria (cf. also *Ach.* 659ff), and bear out, at least to some extent, Aristophanes' own claim (*Nu.* 549f) that he 'struck Cleon in the belly at the height of his power'. Moreover it was not just in times of success or prosperity, but also at moments of crisis or considerable gloom in the fortunes of the city, that Aristophanes produced comedies satirising not only public figures, but also the gods (as Dionysius in *Ra.* in 405).

[140] The licence allowed comic poets was, no doubt, exceptional (as Isocrates, VIII 14, for one, points out when he complains that it was only they and the most reckless speakers who could criticise the democracy). Yet even critics of the democracy, such as Plato, acknowledge that there was – along with the licence – more freedom of speech, ἐξουσία τοῦ λέγειν, in Athens than anywhere else (as Socrates says when encouraging Polus to take up Gorgias' case, *Grg.* 461e).

technological and other factors we mentioned earlier as affecting not only Greece itself, but also one or more of her ancient Near Eastern neighbours, notably (1) the existence of an economic surplus and of money as a medium of exchange, (2) access to, and curiosity about, other societies, and (3) changes in the technical means of communication and the beginnings of literacy. Without the first of these, the development of the institutions of the city-state – so expensive in time and manpower – is inconceivable. The second had its positive contribution to make to the widening of mental, as well as geographical, horizons, while without the third it is hardly an exaggeration to say that the new knowledge (which must in any case have been expressed quite differently)[141] would have been stillborn.

Nevertheless the distinctive additional factors that must also be taken into account are, broadly speaking, political. Ancient Greece is marked not just by exceptional intellectual developments, but also by what is in certain respects an exceptional political situation: and the two appear to be connected. In four fundamental ways aspects of Greek political experience may be thought either to have directly influenced, or to be closely mirrored in, key features of the intellectual developments we are concerned with. First there is the possibility of radical innovation, second the openness of access to the forum of debate, third the habit of scrutiny, and fourth the expectation of justification – of giving an account – and the premium set on rational methods of doing so.

The factors that may be held responsible for the development of the particular institutions of the city-state themselves present complex problems that lie well beyond the limits of our inquiry here: apart from general economic and social considerations such as the growth of population and the agrarian crisis it provoked, the increase of wealth and the expansion of trade and industry, it has been thought that modifications in the techniques and sociology of warfare – the so-called 'hoplite reform' – have special bearing on the gradual transformation of archaic society into the city-state.[142] The group on whom the defence of the state depended came to demand full and equal political rights, although we must add (1) that intensive

---

[141] Indeed acknowledging Goody's point concerning the relationship between the means of expression and the content of what is expressed (see above, pp. 239f), we may go further and say that it would have been different knowledge.

[142] See, for example, Andrewes 1956, pp. 31ff, Vernant 1962, pp. 53ff, Snodgrass 1965, Detienne 1968, Vidal-Naquet 1968, Finley 1970. One of our most important literary documents is Aristotle, *Pol.* 1297b22ff, the implications of which are, on the whole, borne out by the material evidence for the development of the hoplite panoply and the tactics of hoplite warfare.

involvement in politics presupposed leisure and so an economic surplus guaranteed by slave production, the revenues of subject states, or both,[143] and (2) that the units in which those rights were exercised were themselves to some extent determined by geographical factors favouring the emergence of small autonomous states.[144] But from our point of view it is enough to remark that the political upheavals in this transformation created opportunities for innovation not just in the field of politics (in both practical and conceptual experimentation with constitutional forms) but in other areas as well.[145] There had, in any case, been no rigid religious orthodoxy in Homeric society, and with the advent of 'wise men' of many different kinds from the seventh century there is a high degree of pluralism in Greek religious and intellectual, as well as political leadership.

But in the competitive situation that arose those who cultivated the art of speaking and who were prepared to put their case like statesmen before an Assembly, or like advocates in a court of law, were evidently more likely to succeed in persuading certain audiences. While what seemed plausible always reflected the particular preconceptions of the individuals concerned, in some quarters at least those who could deploy evidence and arguments that appealed to common experience[146] had obvious advantages. Privilege and authority in any of their manifestations came to be open to challenge. The view that anyone was entitled to a voice and to an opinion not just on political, but also on other, matters, can be illustrated by characteristic texts from philosophy, history and medicine. Xenophanes insists that the truth is not revealed, but is found by searching.[147] Herodotus remarks that all men have the same degree of knowledge – or ignorance – about the gods,[148] and the writer of *On Ancient*

---

[143] Aristotle repeatedly points out the importance of economic factors for the question of who participated in government and how intensive that participation was. See, for example, *Pol.* 1292b25ff, 1293a1ff, 1320a17ff. At 1255b35ff he puts it that those who can put the management of their affairs in the hands of stewards, do so in order to devote themselves to either *politics or philosophy*.

[144] The idea that the size of the state should be limited is a recurrent theme in Greek political philosophy. Plato specified 5,040 households in the *Laws*, 737e ff, and Aristotle said that the citizens should not be too many for a herald to address, *Pol.* VII 4, e.g. 1326b5ff.

[145] A direct analogy between innovation in medicine and other branches of knowledge on the one hand, and in politics on the other, is drawn for example by Aristotle, *Pol.* 1268b34ff, though he insists on the differences between the two cases (1269a19ff) (cf. also Plato, *Plt.* 296bc). A number of texts indicate that the Greeks generally believed that Egyptian medicine (for example) was much more rule-bound than their own: see Aristotle, *Pol.* 1286a12ff, Diodorus 1 82, and cf. Plato, *Lg.* 656d f, on Egyptian music.

[146] As we have seen is the case, implicitly, with arguments connected with the concepts of 'nature' and 'causation', above, ch. 1, pp. 49ff.

[147] Fr. 18, see above, p. 133.   [148] Hdt. 11 3.

Medicine even insists that the doctor should express himself in such a way as to be clearly understood by laymen.[149]

Intellectual dissenters, like political ones, sometimes came to grief: but thanks in part perhaps to the close personal relationships within each city-state, and thanks more particularly to the number of such states, the Greek world tolerated a remarkable degree of divergence of opinion on many fundamental issues. While there was no strict analogue to ostracism as a means of defusing hostility in their case, philosophers or sophists who, for one reason or another, became *persona non grata* in one state could normally move to another. This is what Anaxagoras and Diagoras did, what Aristotle also was to do in the face of anti-Macedonian feeling in Athens after the death of Alexander in 323,[150] and indeed what Socrates himself had been expected by some people to do after his trial to escape execution.[151]

But if it seems possible to argue that the development of critical inquiry owed something to Greek political experience, and even that the very instability of the political situation contributed to the dynamism of that development, this thesis faces a number of obvious *prima facie* objections. Three in particular must be discussed. (1) The thesis might lead one to expect critical philosophical and scientific inquiry to be heavily concentrated in, if not confined to, the democracies, instead of being a quite widespread phenomenon where the main proponents came from, and lived in, city-states of varying political constitutions.[152] (2) It might be objected that the rise of

---

[149] *VM* ch. 2, *CMG* I, 1 37.9ff, see above, p. 95. Conversely Thucydides was in no way inhibited by being a layman from giving his own detailed description of the plague at Athens.

[150] 'To save the Athenians from sinning twice against philosophy', as it is put by Aelian, *VH* III 36, and in the lives of Aristotle (*Vita Marciana*, see Gigon 1962, p. 6.185f, the *Vita* attributed to Ammonius, see Rose 1886, p. 440.12f, and the Latin *Vita*, see Rose 1886, p. 449.18f).

[151] See Plato's *Crito*. On the whole question of the freedom of the intellectual in Greece, see Dover 1975.

[152] Important philosophers, mathematicians, sophists and doctors of the sixth and fifth centuries (to go no further) came from Miletus, Ephesus, Samos, Colophon, Elea, Croton, Clazomenae, Acragas, Athens, Cos, Cnidus, Apollonia, Abdera, Leontini, Ceos, Elis, Cyzicus and Tarentum, and we know not only that many of the sophists and doctors travelled extensively, but also that some of the philosophers (such as Xenophanes) did and that others (such as Pythagoras and Anaxagoras) did their principal work in a city other than their birthplace. But no overall pattern can be said to emerge from an analysis of our limited information concerning the political constitutions of the cities that produced, or that offered a home to, the thinkers we are chiefly interested in, and several of the cities in question had, in any case, a chequered history of stasis and revolution during those two centuries. From about the middle of the fifth century, however, Athens came to be the main centre of work of an increasing proportion of the most notable philosophers and scientists. Yet that may have had as much to do with the general power and prestige of the city as with its democratic institutions or reputation for free speech: at least it continued to attract teachers and thinkers not only through

speculative thought antedates the full development of the institutions of the city-state, for example those introduced at Athens by the reforms of Cleisthenes in 508. (3) If correct, the thesis might be thought to prove too much. If general features of Greek political and social life are invoked to help account for the emergence of critical and radical inquiry, then the continued survival of magical practices, unchallenged religious beliefs or other aspects of the irrational after, say, the fifth century, might seem hard to explain.

Each of these three points does not so much undermine the general thesis as enable us to qualify it and make it more precise. On (1) it can be argued that although certain institutions, such as ostracism, existed only in certain democratic cities, the differences between democracies and oligarchies that relate to points that concern us were, in the main, a matter of degree. While the accountability of magistrates was taken as a special mark of the democracies in Herodotus III 80, it was not confined to them.[153] The experience of deliberation on affairs of state was less widespread in the oligarchies, since those who participated in government formed a smaller percentage of the total population. But such deliberation occurred. We have mentioned Thucydides' report of one such debate among the Lacedaimonians at I 79ff, and Aristotle, in his theoretical discussion of constitutional types, sometimes distinguishes different kinds of oligarchy precisely by reference to the way the class of those who shared in deliberation was constituted and defined.[154]

Such definite information as we have – and it is very limited – suggests that the political leanings of individual philosophers, mathematicians, doctors and sophists varied. They included some, such as Empedocles, who, we are told, favoured democracy,[155] but others, such as Plato, who were its implacable opponents. But whatever their own political inclinations, they were mostly – we may assume – well aware of the character and institutions of the main types of existing constitution – just as they were also usually familiar with one another's philosophical or scientific ideas.[156] Again many,

the varying fortunes of the democracy at the end of the fifth and beginning of the fourth century, but even after it lost its political independence after Chaeronea in 338.

[153] See, for example, Aristotle, *Pol.* 1271 a 6ff, on the accountability of Spartan magistrates to the Ephors.

[154] See especially *Pol.* 1298 a 34ff, cf. 1297 a 17ff.

[155] See, for example, D.L. VIII 63–4 and 72.

[156] This is generally true not just of the doctors and sophists, who often travelled for, as it were, professional purposes, but also of the philosophers. Thus Parmenides' ideas evidently soon became familiar not just to his fellow-townsman Zeno, but also to Melissus of Samos, to Empedocles in Acragas and to Anaxagoras at Athens. Among the most important regular occasions for the exchange of ideas were the major festivals, as

indeed most, prominent philosophers and scientists, whether democrats or oligarchs, themselves belonged to a more or less distinct
elite, though, as we noted, there is an important difference between
the situation of most of the philosophers and sophists and that of the
practising physicians. The pupils of the former were no doubt drawn
mainly from the rich: the latter attempted to exercise their powers of
persuasion in relation not just to pupils or lecture audiences, but also
to a potential clientele that extended far beyond the educated
minority.[157] Yet doctors, philosophers, sophists and mathematicians
alike were all, to some degree, exposed to and influenced by the
expectations of rational discussion that were part of the common
experience of the Greek city-state. It can be represented that both
the intellectual and the political changes we have been dealing with
were *general*, even though they were not *uniform*, developments,
affecting the whole of the Greek world to a greater or lesser extent.

As to (2) it is true that the work of the first two Milesian philosophers, Thales and Anaximander, is approximately contemporary
with that of Solon and Pisistratus respectively and so antedates the
introduction of the full democracy at Athens under Cleisthenes. But
firstly we have noted that many of the features of later political
discussion are already present in, or foreshadowed by, the poems of
Solon, who is at pains to defend and justify his own policies and who
emphasises both the openness and public nature of his measures and
that the Athenians themselves are masters of their fate.[158] Secondly
and conversely we have stressed the limited nature of the achievement
of Milesian, and other sixth-century, speculative thought. Many
important areas of physical, mathematical and especially moral
inquiry only begin to be extensively debated towards the middle or
end of the fifth century. This is true, for instance, of the controversy
between 'nature' and 'convention', and our chief text that explicitly
criticises magical beliefs in medicine dates from the end of the fifth
century at the earliest. The conclusion we should draw is that the
developments we have considered are in both cases gradual ones,
which certainly took time to gather momentum and did not do so
without suffering intermittent setbacks.

(3) The most important qualification to the thesis relates to the
third difficulty or objection We have argued that the experience of
radical, critical debate in the fields of politics and law both paralleled,

we can see, for instance, from Plato's account of the visit that Parmenides and Zeno
made to Athens for the Panathenaea (*Prm.* 127 ab).

[157] See further below, pp. 263f.

[158] See especially Poems 3, 5, 8, 9, 10, 23–4 Diehl, and cf. above, pp. 243, 247 and n. 91.

and contributed to, the development of similarly radical, critical inquiry in other areas of thought. Yet the existence of certain political institutions, and of a general climate of opinion that allowed and even promoted fundamental criticism, does not mean, of course, that the entire gamut of popular beliefs would be so scrutinised or that every manifestation of the irrational – including those in philosophy and medicine themselves – would be exposed. Three reservations are relevant here. First we should remind ourselves that the power of rational arguments to uproot deepseated convictions is only a limited one. Secondly we noted that the very success of the new professionalism in the art of speaking provoked hostile reactions from such writers as Aristophanes and Plato. The citizens of Athens had ample opportunity to exercise their connoisseurship of skilful argument: but by the end of the fifth century they were also being frequently warned, by different speakers and in different contexts, not just against those who set out to make the worse appear the better cause, but also more generally against rhetoric itself.

Thirdly and more fundamentally we must recognise that there were more general restrictions to the spread of critical inquiry. We have seen that, in the political domain, public debate on affairs of state and on the best constitution was possible. Yet even in the extreme democracies that right did not extend beyond the adult male citizens. But in contrast to issues of practical politics, which were, by definition, the concern of the enfranchised alone, many of the ideas that the philosophers and the doctors attempted to combat or supplant were genuinely popular beliefs shared by men and women, free and slave, alike. The revolution in critical inquiry represented by *some* of the philosophers and Hippocratic writers is a phenomenon circumscribed by the still formidable barriers to communication that existed within Greek society, barriers created by social and political divisions as well as by illiteracy or lack of education.[159] The plausibility of the arguments in *On the Sacred Disease* is (as we have already had several occasions to stress) audience-specific: it was not that everyone in the Greek world would find them persuasive, and the set of those who fell ill was far more extensive than the set of those who participated in the political life of the city-state. The explanandum is not, in any case, the victory of rationality over magic: there was no

---

159 There were obstacles to communication in either direction, as we can see from the inhibitions of some women patients in talking to doctors about their complaints: see *Mul.* 1 ch. 62, L VIII 126.12ff, and cf. Soranus, *Gyn.* III Pref. 3, *CMG* IV 95.8f, which shows that the problems of communication between doctors and women patients persisted.

such victory: but rather how the criticism of magic got *some* purchase. The context of the attack by the Hippocratic author is a situation where one kind of healer is in competition with another. But there were many traditional beliefs which were unlikely to provide the occasion for such a confrontation, and many others – such as the belief in the superiority of the right-hand side – came to be rationalised and incorporated into natural philosophy, Hippocratic medicine, or both. Moreover while the arguments used by *On the Sacred Disease* were of a kind that seemed persuasive in some quarters, that did not mean that either the purifiers, or the exponents of temple medicine, were put out of business. *Some* of the weaknesses of *some* of their claims were exposed by writers who were highly vulnerable to a variety of other types of objection themselves: but just as we should not underestimate the exceptional nature of this achievement, so we should not exaggerate the extent to which the radical, critical approach was typical of the whole of Greek thought.

CONCLUSIONS

Some concluding remarks may help to draw together the threads of our investigations as a whole. The development of philosophy and science in ancient Greece is a unique turning-point in the history of thought. So far as the Western world goes, our science is continuous with, and may be said to originate in, that of ancient Greece. Elsewhere Western science has been imposed on (or at least imported into) other cultures from outside, even where, as in China, those cultures already possessed their own highly developed cosmologies and technologies. Even though the materials for the study of the transformations that occurred within Greek thought are limited – and they dictate an approach that differs in many respects from that of the field anthropologist – we can go some way towards defining the character, and the limits, of those changes and the circumstances in which they took place. The comparative evidence shows that what is exceptional is, first and foremost, the development of generalised scepticism and of critical inquiry directed at fundamental issues – not that that happened all at once or across the board, or that the whole of what would normally be included in Greek philosophy and science, let alone the rest of Greek thought, is radical and critical in spirit.

The growth of philosophy and science may be seen as depending partly on developments in the use (including the more self-conscious use) of 'reason' and 'experience', the elaboration of argumentative procedures and of techniques of empirical research. But in both cases the gradualness of the developments we have traced is striking. Although epistemological questions, and in particular the competing claims of reason and of sensation to be the basis of knowledge, begin to be aired with Parmenides and Heraclitus, arguments themselves do not come to be explicitly analysed and evaluated before Plato, and we have to wait until Aristotle for the first formal logic. The practice of mathematical demonstration begins in the late fifth century, but again the contributions of Plato and Aristotle in the fourth are fundamental to the development of the concept of an axiomatic system. Observation not only in physics, but also in astronomy and in biology (especially the use of dissection) was slow to develop, and the first extensive planned programme of empirical research in natural science is Aristotle's.

Moreover the *formal* continuities between philosophical/scientific and earlier thought are as remarkable as the discontinuities. If Greek philosophers and scientists eventually exploit arguments and empirical methods with great effectiveness, and on occasions both deliberately and systematically, these were developments that built on what already existed. A facility in deploying arguments of various types, and an ability to observe, even to engage in sustained observations and in trial and error procedures, are patently as old as human society itself. This part of our inquiry into the intellectual tools deployed in early Greek science, as also our discussion of what the explicit ideas of nature and causation owed to earlier *implicit* assumptions, exhibit the common ground between the work of the philosophers and scientists and earlier thought. On this score, at least, we have no cause to invoke – indeed we should rule out invoking – any talk of a different mentality, a different logic, or a totally different conceptual framework. Apart from the philosophical difficulties of such a hypothesis, it fails to account for the formal continuities we can observe.

But while philosophy and science did not involve a different mentality or a new logic, they may be represented as originating from the exceptional exposure, criticism and rejection of deep-seated beliefs. Whereas limited scepticism about traditional schemata can be paralleled in other societies readily enough, the generalised scepticism about the validity of magical procedures we find in ancient Greek authors was unprecedented. If the concepts of 'nature' and of 'causation' develop from certain implicit assumptions, those ideas had, again, to be made explicit and generalised. These conceptual moves sound simple: but they could not be made without allowing fundamental aspects of traditional beliefs to come under threat.

Philosophy and science can only begin when a set of questions is substituted for a set of vaguely assumed certainties. It is true that, the questions once posed, the answers given were sometimes not just schematic, but contained (as we have seen) elements of pure bluff. Yet while the Greeks' confidence in the rightness of their methods often outran their actual scientific performance – particularly in the matter of the collection of empirical data – those methodological ideals not only permitted, but positively promoted the further growth of the inquiry. The investigability of nature was explicitly recognised, even while the epistemological debate covered a wide spectrum of opinions on the character, aims and limits of that investigation.

The society in which these inquiries were first pursued was far from a primitive one. The level of technology and that of economic development were far in advance of those of many modern non-industrialised societies: above all literacy presents a difference not just of degree but of kind. Yet a comparison with Greece's ancient Near Eastern neighbours suggests that none of these three factors individually, nor all of them collectively, can be used to account fully for the developments we are interested in. So far as an additional distinctively Greek factor is concerned, our most promising clue (to put it no more strongly) lies in the development of a particular social and political situation in ancient Greece, especially the experience of radical political debate and confrontation in small-scale, face-to-face societies. The institutions of the city-state called for new qualities of leadership, put a premium on skill in speaking and produced a public who appreciated the exercise of that skill. Claims to particular wisdom and knowledge in other fields besides the political were similarly liable to scrutiny, and in the competition between the many and varied new claimants to such knowledge those who deployed evidence and argument were at an advantage compared with those who did not, at least – to repeat our proviso once again - so far as some audiences and contexts were concerned.

Moreover if this hypothesis helps to account for the strengths of Greek science, it also throws some light on some of its weaknesses. Although eventually Greek scientists produced lasting (if often elementary) results in areas of astronomy, mathematical geography, statics and hydrostatics, anatomy and even physiology, Greek science down to Aristotle is more notable for its achievements in second-order inquiries, in epistemology, logic, methodology and philosophy of science, in, for instance, the development of the concepts of an axiomatic system and of an exact science, and in that of the notion of empirical research. Whatever their limitations in the implementation of their ideas, the Greeks provided science with its essential framework, asserting the possibility of the inquiry and initiating the debate that continues today on its aims and methods. Yet several of the shortcomings of Greek science correspond closely to its strengths. The quest for certainty in an axiomatic system – itself in part a reaction against what was represented as the seductiveness of merely plausible arguments – was sometimes bought at the cost of a lack of empirical content. More generally, the way in which evidence and 'experiment' were often used to support, rather than to test, theories, a certain over-confidence and dogmatism, above all a certain failure in self-criticism, may all be thought to reflect the predominant tendency to view scientific debate as a contest like a political or a legal *agon*. Aristotle noted in the *De Caelo* (294 b 7ff) that 'we are all in the habit of relating an inquiry not to the subject-matter, but to our opponent in argument'. This remains true, no doubt, today, but the observation appears especially relevant to early Greek science. The sterility of much ancient scientific work is, we said, often a result of the inquiry being conducted as a dispute with each contender single-mindedly advocating his own point of view. This is easy to say with hindsight: but an examination of the Greek evidence suggests that this very paradigm of the competitive debate may have provided the essential framework for the growth of natural science.

BIBLIOGRAPHY

The bibliography aims first to provide details of all the books and articles that are cited in my text, and secondly to refer the reader to other works which, though not mentioned in my discussion, bear directly on the issues raised. The list of such works makes no claims to be comprehensive within the field of classical studies, and has necessarily been drastically selective outside that field, in such areas as anthropology and the philosophy and sociology of science.

Aaboe, A. (1955–6) 'On the Babylonian origin of some Hipparchian parameters', *Centaurus* IV, 122–5.
(1956–8) 'On Babylonian planetary theories', *Centaurus* V, 209–77.
(1964–5) 'On period relations in Babylonian astronomy', *Centaurus* X, 213–31.
(1974) 'Scientific astronomy in antiquity', in *The Place of Astronomy in the Ancient World*, edd. D. G. Kendal and others (Oxford), pp. 21–42.
Aaboe, A. and Price, D. J. de S. (1964) 'Qualitative measurement in antiquity', in *L'Aventure de la science* (Mélanges A. Koyré) (Paris), Vol. I, pp. 1–20.
Abel, K. (1958) 'Die Lehre vom Blutkreislauf im Corpus Hippocraticum', *Hermes* LXXXVI, 192–219 (reprinted in Flashar 1971, pp. 121–64).
Achinstein, P. (1963–4) 'Theoretical terms and partial interpretation', *British Journal for the Philosophy of Science* XIV, 89–105.
(1965) 'The problem of theoretical terms', *American Philosophical Quarterly* II, 193–203.
(1968) *Concepts of Science* (Johns Hopkins, Baltimore).
Ackerknecht, E. H. (1971) *Medicine and Ethnology* (Johns Hopkins, Baltimore).
Adkins, A. W. H. (1960) *Merit and Responsibility* (Oxford).
(1972) *Moral Values and Political Behaviour in Ancient Greece* (London).
Agassi, J. (1964) 'The nature of scientific problems and their roots in metaphysics', in *The Critical Approach to Science and Philosophy*, ed. M. Bunge (London), pp. 189–211.
Albright, W. F. (1972) 'Neglected factors in the Greek intellectual revolution', *Proceedings of the American Philosophical Society* CXVI, 225–42.
Alexanderson, B. (1963) *Die hippokratische Schrift Prognostikon, Überlieferung und Text* (Studia Graeca et Latina Gothoburgensia 17, Göteborg).
Allan, D. J. (1965) 'Causality, ancient and modern', *Proceedings of the Aristotelian Society, Supplements* XXXIX, 1–18.
(1970) *The Philosophy of Aristotle* (1st ed. 1952), 2nd ed. (Oxford).
Allen, R. E. (ed.) (1965) *Studies in Plato's Metaphysics* (London).
Allen, R. E. and Furley, D. J. (edd.) (1975) *Studies in Presocratic Philosophy* Vol. II (London).
Allman, G. J. (1889) *Greek Geometry from Thales to Euclid* (Dublin and London).
Anderhub, J. H. (1941) 'Genetrix Irrationalium', in *Joco-seria aus den Papieren eines reisenden Kaufmanns* (Wiesbaden), pp. 159–222.
Anderson, P. (1974) *Passages from Antiquity to Feudalism* (London).
Andrewes, A. (1956) *The Greek Tyrants* (London).

Annequin, J. (1973) *Recherches sur l'action magique et ses représentations* (Annales littéraires de l'Université de Besançon, Paris).

Anton, J. P. and Kustas, G. L. (edd.) (1971) *Essays in Ancient Greek Philosophy* (State University of New York Press).

Apostle, H. G. (1952) *Aristotle's Philosophy of Mathematics* (University of Chicago Press).

(1958) 'Methodological superiority of Aristotle over Euclid', *Philosophy of Science* xxv, 131–4.

Arnim, H. von (1905–24) *Stoicorum Veterum Fragmenta*, 4 vols. (Leipzig).

Artelt, W. (1937) *Studien zur Geschichte der Begriffe 'Heilmittel' und 'Gift'* (Leipzig).

Austin, J. L. (1962a) *How to do things with words* (Oxford).

(1962b) *Sense and sensibilia* (Oxford).

Bachelard, G. (1972) *La Formation de l'esprit scientifique* (1st ed. 1947), 8th ed. (Paris).

Balme, D. M. (1939) 'Greek science and mechanism. I Aristotle on Nature and Chance', *Classical Quarterly* xxxiii, 129–38.

(1941) 'Greek science and mechanism. II The Atomists', *Classical Quarterly* xxxv, 23–8.

(1961/1975) 'Aristotle's use of differentiae in zoology' (in *Aristote et les problèmes de méthode*, ed. S. Mansion (Louvain and Paris, 1961), 195–212) in Barnes, Schofield, Sorabji 1975, pp. 183–93.

(1962a) 'γένος and εἶδος in Aristotle's biology', *Classical Quarterly* NS xii, 81–98.

(1962b) 'Development of biology in Aristotle and Theophrastus: theory of spontaneous generation', *Phronesis* vii, 91–104.

Balss, H. (1923) 'Praeformation und Epigenese in der griechischen Philosophie', *Archivio di Storia della Scienza* iv, 319–25.

(1936) 'Die Zeugungslehre und Embryologie in der Antike', *Quellen und Studien zur Geschichte der Naturwissenschaften und der Medizin* v, 2, 193–274.

Bambrough, R. (ed.) (1965) *New Essays on Plato and Aristotle* (London).

Barnes, J. (1969/1975) 'Aristotle's theory of demonstration' (*Phronesis* xiv (1969), 123–52) in Barnes, Schofield, Sorabji 1975, pp. 65–87.

(1975) *Aristotle's Posterior Analytics* (Oxford).

Barnes, J., Schofield, M., and Sorabji, R. (edd.) (1975) *Articles on Aristotle, I Science* (London).

Barnes, S. B. (1969) 'Paradigms, scientific and social', *Man* NS iv, 94–102.

(1972) 'Sociological explanation and natural science: a Kuhnian reappraisal', *Archives Européennes de Sociologie* xiii, 373–91.

(1973) 'The comparison of belief-systems: Anomaly versus falsehood', in Horton and Finnegan 1973, pp. 182–98.

(1974) *Scientific knowledge and sociological theory* (London).

Baumann, E. D. (1925) 'Die heilige Krankheit', *Janus* xxix, 7–32.

Beare, J. I. (1906) *Greek Theories of Elementary Cognition from Alcmaeon to Aristotle* (Oxford).

Becker, O. (1931) *Die diairetische Erzeugung der platonischen Ideelzahlen* (Quellen und Studien zur Geschichte der Mathematik, Astronomie und Physik B, 1, 4, Berlin), pp. 464–501.

(1933a) *Eudoxus-Studien I. Eine voreudoxische Proportionenlehre und ihre Spuren bei Aristoteles und Euklid* (Quellen und Studien zur Geschichte der Mathematik, Astronomie und Physik B, 2, 4, Berlin), pp. 311–33.

(1933b) *Eudoxus-Studien II. Warum haben die Griechen die Existenz der vierten Proportionale angenommen?* (Quellen und Studien zur Geschichte der Mathematik, Astronomie und Physik B, 2, 4, Berlin), pp. 369–87.

(1936a) *Eudoxus-Studien III. Spuren eines Stetigkeitsaxioms in der Art des Dedekind'schen zur Zeit des Eudoxus* (Quellen und Studien zur Geschichte der Mathematik, Astronomie und Physik B, 3, 2, Berlin), pp. 236–44.

(1936*b*) *Eudoxus-Studien IV. Das Prinzip des ausgeschlossenen Dritten in der griechischen Mathematik* (Quellen und Studien zur Geschichte der Mathematik, Astronomie und Physik B, 3, 3, Berlin), pp. 370–88.

(1936*c*) *Eudoxus-Studien V. Die eudoxische Lehre von den Ideen und den Farben* (Quellen und Studien zur Geschichte der Mathematik, Astronomie und Physik B, 3, 3, Berlin), pp. 389–410.

(1936*d*) *Zur Textgestaltung des eudemischen Berichts über die Quadratur der Möndchen durch Hippokrates von Chios* (Quellen und Studien zur Geschichte der Mathematik, Astronomie und Physik B, 3, 3, Berlin), pp. 411–19.

(1936*e*) *Die Lehre vom Geraden und Ungeraden im neunten Buch der euklidischen Elemente* (Quellen und Studien zur Geschichte der Mathematik, Astronomie und Physik B, 3, 4, Berlin), pp. 533–53.

(1957) *Das mathematische Denken der Antike* (Göttingen).

Behr, C. A. (1968) *Aelius Aristides and the Sacred Tales* (Amsterdam).

Benveniste, E. (1945) 'La doctrine médicale des Indo-Européens', *Revue de l'histoire des religions* CXXX, 5–12.

Berger, H. (1903) *Geschichte der wissenschaftlichen Erdkunde der Griechen*, 2nd ed. (Leipzig).

Berka, K. (1963) 'Aristoteles und die axiomatische Methode', *Das Altertum* IX, 200–5.

Berthelot, M. (1885) *Les Origines de l'alchimie* (Paris).

Bertholet, A. (1926–7) 'Das Wesen der Magie', *Nachrichten von der Gesellschaft der Wissenschaften zu Göttingen* (Phil.-hist. Kl. 1926–7 Geschäftliche Mitteilungen), pp. 63–85.

Berti, E. (1977) *Aristotele: Dalla Dialettica alla Filosofia Prima* (Padova).

Bidez, J. and Cumont, F. (1938) *Les Mages Hellénisés*, 2 vols. (Paris).

Björnbo, A. A. (1901) 'Hat Menelaos aus Alexandria einen Fixsternkatalog verfasst?', *Bibliotheca Mathematica*, Dritte Folge II, 196–212.

Black, M. (1959) 'Linguistic relativity: the views of Benjamin Lee Whorf', *Philosophical Review* LXVIII, 228–38.

(1962) *Models and Metaphors* (Cornell University Press, Ithaca, New York).

Blass, F. (1865) *Die griechische Beredsamkeit in dem Zeitraum von Alexander bis auf Augustus* (Berlin).

(1887–93) *Die attische Beredsamkeit* (1st ed. 1868–80), 2nd ed., 3 vols. (Leipzig).

Bluck, R. S. (1961) *Plato's Meno* (Cambridge).

Blüh, O. (1949) 'Did the Greeks perform experiments?', *American Journal of Physics* XVII, 384–8.

Boas, G. (1959) 'Some assumptions of Aristotle', *Transactions of the American Philosophical Society* NS XLIX, Part 6.

(1961) *Rationalism in Greek Philosophy* (Johns Hopkins, Baltimore).

Bochner, S. (1966) *The Role of Mathematics in the Rise of Science* (Princeton University Press, Princeton, New Jersey).

Boeder, H. (1959) 'Der frühgriechische Wortgebrauch von Logos und Aletheia', *Archiv für Begriffsgeschichte* IV, 82–112.

(1968) 'Der Ursprung der "Dialektik" in der Theorie des "Seienden". Parmenides und Zenon', *Studium Generale* XXI, 184–202.

Bohannan, P. (1957) *Justice and Judgment among the Tiv* (Oxford).

Bohannan, P. (ed.) (1967) *Law and Warfare* (New York).

Boll, F. (1894) 'Studien über Claudius Ptolemäus. Ein Beitrag zur Geschichte der griechischen Philosophie und Astrologie', *Jahrbücher für classische Philologie*, Suppl. Bd XXI, 49–244.

(1899) 'Beiträge zur Ueberlieferungsgeschichte der griechischen Astrologie und Astronomie', *Sitzungsberichte der philosophisch-philologischen und der histori-*

schen Classe der k. b. Akademie der Wissenschaften zu München (1899, 1), pp. 77–140.

(1901) 'Die Sternkataloge des Hipparch und des Ptolemaios', *Bibliotheca Mathematica*, Dritte Folge II, 185–95.

Boll, F. and Bezold, C. (1917/1931) *Sternglaube und Sterndeutung. Die Geschichte und das Wesen der Astrologie* (1st ed. 1917), 4th ed. (ed. W. Gundel) (Leipzig and Berlin).

Bollack, J. (1965–9) *Empédocle*, 3 vols. in 4 (Paris).

Boncompagni, R. (1970) 'Empirismo e osservazione diretta nel ΠΕΡΙ ΔΙΑΙΤΗΣ del *Corpus Hippocraticum*', *Physis* XII, 109–32.

Bonitz, H. (1870) *Index Aristotelicus* (Berlin).

Booth, N. B. (1957a) 'Were Zeno's arguments a reply to attacks upon Parmenides?', *Phronesis* II, 1–9.

(1957b) 'Were Zeno's arguments directed against the Pythagoreans?', *Phronesis* II, 90–103.

(1978) 'Two points of interpretation in Zeno', *Journal of Hellenic Studies* XCVIII, 157–8.

Bostock, D. (1972–3) 'Aristotle, Zeno and the potential infinite', *Proceedings of the Aristotelian Society* NS LXXIII, 37–51.

Bouché-Leclercq, A. (1879–82) *Histoire de la divination dans l'antiquité*, 4 vols. (Paris).

(1899) *L'Astrologie grecque* (Paris).

Bourgey, L. (1953) *Observation et expérience chez les médecins de la collection hippocratique* (Paris).

(1955) *Observation et expérience chez Aristote* (Paris).

Bourgey, L. and Jouanna, J. (edd.) (1975) *La Collection hippocratique et son rôle dans l'histoire de la médecine* (Leiden).

Bouteiller, M. (1950) *Chamanisme et guérison magique* (Paris).

Boyancé, P. (1937) *Le Culte des Muses chez les philosophes grecs* (Bibliothèque des écoles françaises d'Athènes et de Rome 141, Paris).

Bratescu, G. (1975) 'Eléments archaïques dans la médecine hippocratique', in Bourgey and Jouanna 1975, pp. 41–9.

Breasted, J. H. (1930) *The Edwin Smith Surgical Papyrus*, 2 vols. (University of Chicago Press).

Bretschneider, C. A. (1870) *Die Geometrie und die Geometer vor Euklides* (Leipzig).

Britton, J. P. (1969) 'Ptolemy's determination of the obliquity of the ecliptic', *Centaurus* XIV, 29–41.

Bröcker, W. (1958) 'Gorgias contra Parmenides', *Hermes* LXXXVI, 425–40.

Bruin, F. and Bruin, M. (1976) 'The equator ring, equinoxes and atmospheric refraction', *Centaurus* XX, 89–111.

Brunschvicq, L. (1949) *L'Expérience humaine et la causalité physique* (1st ed. 1922), 3rd ed. (Paris).

Brunschwig, J. (1967) *Aristote, Topiques*, Vol. I (Paris).

(1973) 'Sur quelques emplois d'ΟΨΙΣ', in *Zetesis* (Festschrift de Strycker), (Antwerp), pp. 24–39.

Buchholz, E. (1871–85) *Die homerischen Realien*, 3 vols. (Leipzig).

Burckhardt, J. (1898–1902) *Griechische Kulturgeschichte*, 2nd ed., 4 vols. (Berlin and Stuttgart).

Burkert, W. (1959) 'ΣΤΟΙΧΕΙΟΝ. Eine semasiologische Studie', *Philologus* CIII, 167–97.

(1962) 'ΓΟΗΣ. Zum griechischen "Schamanismus"', *Rheinisches Museum* CV, 36–55.

(1963a) Review of Fränkel 1962, *Gnomon* XXXV, 827–8.

(1963*b*) 'Iranisches bei Anaximandros', *Rheinisches Museum* CVI, 97–134.

(1968) 'Orpheus und die Vorsokratiker. Bemerkungen zum Derveni-Papyrus und zur pythagoreischen Zahlenlehre', *Antike und Abendland* XIV, 93–114.

(1970) 'La genèse des choses et des mots', *Les Etudes Philosophiques* (1970, 4), 443–55.

(1972*a*) *Lore and Science in Ancient Pythagoreanism*, revised translation, E. L. Minar, of *Weisheit und Wissenschaft* (1962), (Harvard University Press, Cambridge, Mass.).

(1972*b*) *Homo Necans* (Berlin).

(1977) *Griechische Religion der archaischen und klassichen Epoche* (Stuttgart).

Burnet, J. (1892/1948) *Early Greek Philosophy* (1st ed. 1892), 4th ed. (London, 1948).

(1929) *Essays and Addresses* (London).

Burnyeat, M. F. (1978) 'The philosophical sense of Theaetetus' mathematics', *Isis* LXIX, 489–513.

Buxton, R. G. A. (1977) 'Peitho: its place in Greek culture and its exploration in some plays of Aeschylus and Sophocles' (unpubl. Ph.D. dissertation, Cambridge, 1977).

Byl, S. (1968) 'Note sur la place du coeur et la valorisation de la ΜΕΣΟΤΗΣ dans la biologie d'Aristote', *L'Antiquité Classique* XXXVII, 467–76.

(1977) 'Les grands traités biologiques d'Aristote et la Collection hippocratique', in R. Joly 1977, pp. 313–26.

Calogero, G. (1927/1968) *I fondamenti della logica aristotelica* (1st ed. 1927), 2nd ed. (Firenze).

(1932) *Studi sull'eleatismo* (Roma).

Cambiano, G. (1967) 'Il metodo ipotetico e le origini della sistemazione euclidea della geometria', *Rivista di Filosofia* (Torino) LVIII, 115–49.

(1977) 'Le médecin, la main et l'artisan', in R. Joly 1977, pp. 220–32.

Cantor, M. (1880–1908) *Vorlesungen über Geschichte der Mathematik*, 4 vols. (Leipzig).

Capelle, W. (1925) 'Älteste Spuren der Astrologie bei den Griechen', *Hermes* LX, 373–95.

Carnap, R. (1956) 'The methodological character of theoretical concepts', in *The Foundations of Science and the Concepts of Psychology and Psychoanalysis* (Minnesota Studies in the Philosophy of Science, 1), ed. H. Feigl and M. Scriven (Minneapolis), pp. 38–76.

Carratelli, G. P. (1966) 'Greek inscriptions of the Middle East', *East and West* XVI, 31–6.

Carteron, H. (1923) *La Notion de force dans le système d'Aristote* (Paris).

Cartledge, P. (1977) 'Hoplites and Heroes: Sparta's contribution to the technique of ancient warfare', *Journal of Hellenic Studies* XCVII, 11–27.

Cassirer, E. (1941) *Logos, Dike, Kosmos in der Entwicklung der griechischen Philosophie* (Göteborgs Högskolas Årsskrift XLVII, 6, Göteborg).

(1946) *Language and Myth* (trans. S. K. Langer of *Sprache und Mythos*, Berlin, 1925), (New York).

(1953–7) *The Philosophy of Symbolic Forms* (trans. R. Mannheim of *Philosophie der symbolischen Formen*, Berlin, 1923–9), 3 vols. (New Haven).

Chadwick, H. M. and Chadwick, N. K. (1932–40) *The Growth of Literature*, 3 vols. (Cambridge).

Chadwick, J. and Mann, W. N. (1978) 'Medicine', in *Hippocratic Writings*, ed. G. E. R. Lloyd (originally in Chadwick and Mann, *The Medical Works of Hippocrates*, Oxford, 1950), (London).

Chadwick, N. K. (1942) *Poetry and Prophecy* (Cambridge).

Cherniss, H. (1935) *Aristotle's Criticism of Presocratic Philosophy* (Baltimore).

(1944) *Aristotle's Criticism of Plato and the Academy*, Vol. I (Baltimore).

(1945) *The Riddle of the Early Academy* (Berkeley and Los Angeles).

(1951) 'The characteristics and effects of Presocratic philosophy', *Journal of the History of Ideas* XII, 319–45.

Childe, V. Gordon (1942) *What Happened in History* (London).

(1956) *Man Makes Himself* (1st ed. 1936) 3rd ed. (New York).

(1958) *The Prehistory of European Society* (London).

Clagett, M. (1957) *Greek Science in Antiquity* (London).

Clarke, E. (1963) 'Aristotelian concepts of the form and function of the brain', *Bulletin of the History of Medicine* XXXVII, 1–14.

Clarke, L. W. (1962–3) 'Greek astronomy and its debt to the Babylonians', *British Journal for the History of Science* I, 65–77.

Classen, C. J. (1959) *Sprachliche Deutung als Triebkraft platonischen und sokratischen Philosophierens* (Zetemata 22, München).

(1970) 'Anaximandros', in *Pauly–Wissowa Real-Encyclopädie der classischen Altertumswissenschaft*, Suppl. Bd XII, cols. 30–69.

Cohn-Haft, L. (1956) *The Public Physicians of Ancient Greece* (Smith College Studies in History 42, Northampton, Mass.).

Coldstream, J. N. (1977) *Geometric Greece* (London).

Cole, F. J. (1930) *Early Theories of Sexual Generation* (Oxford).

Cole, T. (1967) *Democritus and the sources of Greek Anthropology* (American Philological Association, Philological Monographs 25, Western Reserve University Press).

Cornford, F. M. (1912) *From Religion to Philosophy* (London).

(1922) 'Mysticism and science in the Pythagorean Tradition. I', *Classical Quarterly* XVI, 137–50.

(1923) 'Mysticism and science in the Pythagorean Tradition. II', *Classical Quarterly* XVII, 1–12.

(1932/1965) 'Mathematics and dialectic in the *Republic* VI–VII' (*Mind* NS XLI (1932), 37–52 and 173–90) in Allen 1965, pp. 61–95.

(1937) *Plato's Cosmology* (London).

(1938) 'Greek natural philosophy and modern science', in *Background to Modern Science*, ed. J. Needham and W. Pagel (Cambridge), pp. 3–22.

(1950) *The Unwritten Philosophy and Other Essays* (Cambridge).

(1952) *Principium Sapientiae* (Cambridge).

Crombie, I. M. (1962) *An Examination of Plato's Doctrines*, Vol. I *Plato on Man and Society* (London).

(1963) *An Examination of Plato's Doctrines*, Vol. II *Plato on Knowledge and Reality* (London).

Cumont, F. (1912) *Astrology and Religion among the Greeks and Romans* (New York and London).

Czwalina, A. (1956–8) 'Über einige Beobachtungsfehler des Ptolemäus und die Deutung ihrer Ursachen', *Centaurus* V, 283–306.

(1959) 'Ptolemaeus: Die Bahnen der Planeten Venus und Merkur', *Centaurus* VI, 1–35.

Daremberg, C. V. (1865) *La Médecine dans Homère* (Paris).

Davidson, D. and Hintikka, J. (edd.) (1969) *Words and Objections. Essays on the work of W. V. Quine* (Dordrecht).

Davison, J. A. (1962) 'Literature and literacy in ancient Greece', *Phoenix* XVI, 141–56 and 219–33.

Decharme, P. (1904) *La Critique des traditions religieuses chez les grecs* (Paris).

Deetjen, C. (1934) 'Witchcraft and medicine', *Bulletin of the Institute of the History of Medicine* II, 164–75.

De Fidio, P. (1969) 'ΑΛΗΘΕΙΑ: dal mito alla ragione', *La Parola del Passato* xxiv, 308–20.

Deichgräber, K. (1930) *Die griechische Empirikerschule: Sammlung und Darstellung der Lehre* (Berlin).

(1933a) *Die Epidemien und das Corpus Hippocraticum* (Abhandlungen der preussischen Akademie der Wissenschaften, Jahrgang 1933, 3, phil.-hist. Kl., Berlin).

(1933b) 'Die ärztliche Standesethik des hippokratischen Eides', *Quellen und Studien zur Geschichte der Naturwissenschaften und der Medizin* iii, 2, 79–99 (reprinted in Flashar 1971, pp. 94–120).

(1933c) 'ΠΡΟΦΑΣΙΣ: Eine terminologische Studie', *Quellen und Studien zur Geschichte der Naturwissenschaften und der Medizin* iii, 4, 209–25.

(1935) *Hippokrates, Über Entstehung und Aufbau des menschlichen Körpers* (περὶ σαρκῶν) (Leipzig).

(1939) 'Die Stellung des griechischen Arztes zur Natur', *Die Antike* xv, 116–38.

De Jong, H. W. M. (1959) 'Medical prognostication in Babylon', *Janus* xlviii, 252–7.

Delambre, J. B. J. (1817) *Histoire de l'astronomie ancienne*, 2 vols. (Paris).

Delatte, A. (1915) *Etudes sur la littérature Pythagoricienne* (Bibliothèque de l'école des hautes études 217, Paris).

(1922) *La vie de Pythagore de Diogène Laërce* (Mémoires de l'Académie Royale de Belgique, Classe des Lettres, Série 2, vol. 17, Bruxelles).

(1961) *Herbarius: Recherches sur le cérémonial usité chez les anciens pour la cueillette des simples et des plantes magiques* (Mémoires de l'Académie Royale de Belgique, Classe des Lettres, Série 2, vol. 54, 4, 3rd ed., Bruxelles).

Derenne, E. (1930) *Les Procès d'impiété intentés aux philosophes à Athènes au Vme et au IVme siècles avant J.-C.* (Bibliothèque de la Faculté de Philosophie et Lettres de l'Université de Liège, 45, Liège and Paris).

De Ste Croix, G. E. M. (1963) 'Commentary', in *Scientific Change*, ed. A. C. Crombie (London), pp. 79–87.

(1972) *The Origins of the Peloponnesian War* (London).

Detienne, M. (1963) *De la pensée religieuse à la pensée philosophique: La Notion de Daïmôn dans le Pythagorisme ancien* (Bibliothèque de la Faculté de Philosophie et Lettres de l'Université de Liège, 165, Paris).

(1967) *Les Maîtres de vérité dans la grèce archaïque* (Paris).

(1968) 'La Phalange: Problèmes et controverses' in *Problèmes de la guerre en Grèce ancienne*, ed. J. P. Vernant (Paris and The Hague), pp. 119–42.

Detienne, M. and Vernant, J.-P. (1978) *Cunning Intelligence in Greek Culture and Society* (trans. J. Lloyd of *Les Ruses de l'intelligence: la métis des grecs*, Paris, 1974), (Hassocks, Sussex).

Deubner, L. (1900) *De Incubatione* (Leipzig).

(1910) 'Charms and Amulets (Greek)', in *Encyclopaedia of Religion and Ethics*, ed. J. Hastings, Vol. iii (Edinburgh), pp. 433–9.

Diamond, A. S. (1935) *Primitive Law* (London).

(1971) *Primitive Law Past and Present* (London).

Di Benedetto, V. (1966) 'Tendenza e probabilità nell'antica medicina greca', *Critica Storica* v, 315–68.

Dicks, D. R. (1953–4) 'Ancient astronomical instruments', *Journal of the British Astronomical Association* lxiv, 77–85.

(1959) 'Thales', *Classical Quarterly* NS ix, 294–309.

(1960) *The Geographical Fragments of Hipparchus* (London).

(1966) 'Solstices, equinoxes and the Presocratics', *Journal of Hellenic Studies* lxxxvi, 26–40.

Bibliography 275

(1970) *Early Greek Astronomy to Aristotle* (London).

Diels, H. (1879) *Doxographi Graeci* (Berlin).

(1884) 'Gorgias und Empedokles', *Sitzungsberichte der königlich preussischen Akademie der Wissenschaften zu Berlin*, Jahrgang 1884, Berlin), pp. 343–68.

(1893*a*) 'Über die Excerpte von Menons Iatrika in dem Londoner Papyrus 137', *Hermes* xxviii, 407–34.

(1893*b*) *Anonymi Londinensis ex Aristotelis Iatricis Menoniis et aliis medicis Eclogae* (Supplementum Aristotelicum iii, 1, Berlin).

(1900) *Aristotelis qui fertur de Melisso Xenophane Gorgia* (Abhandlungen der königlichen Akademie der Wissenschaften zu Berlin, phil.-hist. Kl., Berlin, 1900).

Dieterich, A. (1888) 'Papyrus Magica Musei Lugdunensis Batavi', *Jahrbücher für classische Philologie*, Suppl. Bd xvi, 747–829.

Dihle, A. (1963) 'Kritisch-exegetische Bemerkungen zur Schrift Über die Alte Heilkunst', *Museum Helveticum* xx, 135–50.

Diller, A. (1949) 'The ancient measurements of the earth', *Isis* xl, 6–9.

Diller, H. (1932) 'ὄψις ἀδήλων τὰ φαινόμενα', *Hermes* lxvii, 14 42 (reprinted in *Kleine Schriften zur antiken Literatur* (München, 1971), pp. 119–43).

(1934) *Wanderarzt und Aitiologe* (Philologus Suppl. Bd 26, 3, Leipzig).

(1942) Review of Pohlenz 1938, *Gnomon* xviii, 65–88 (reprinted in Diller 1973, pp. 188–209).

(1952) 'Hippokratische Medizin und attische Philosophie', *Hermes* lxxx, 385–409 (reprinted in Diller 1973, pp. 46–70).

(1964) 'Ausdrucksformen des methodischen Bewusstseins in den hippokratischen Epidemien', *Archiv für Begriffsgeschichte* ix, 133 50 (reprinted in Diller 1973, pp. 106–23).

(1971) 'Der griechische Naturbegriff' in *Kleine Schriften zur antiken Literatur* (München), pp. 144–61.

(1973) *Kleine Schriften zur antiken Medizin* (Berlin).

Dodds, E. R. (1951) *The Greeks and the Irrational* (University of California Press, Berkeley and Los Angeles).

Douglas, M. (1966) *Purity and Danger* (London).

(1970) *Natural Symbols* (London).

(1975) *Implicit Meanings* (London).

Douglas, M. (ed.) (1970) *Witchcraft Confessions and Accusations* (London).

(1973) *Rules and Meanings* (London).

Dover, K. J. (1968) *Lysias and the Corpus Lysiacum* (University of California Press, Berkeley and Los Angeles).

(1974) *Greek Popular Morality in the time of Plato and Aristotle* (Oxford).

(1975) 'The freedom of the intellectual in Greek Society', *Talanta* (Proceedings of the Dutch Archaeological and Historical Society) vii, 24–54.

Drabkin, I. E. (1938) 'Notes on the laws of motion in Aristotle', *American Journal of Philology* lix, 60–84.

Drachmann, A. B. (1922) *Atheism in Pagan Antiquity* (London and Copenhagen).

(1953–4) 'The plane astrolabe and the anaphoric clock', *Centaurus* iii, 183–9.

(1967–8) 'Archimedes and the science of physics', *Centaurus* xii, 1–11.

Drayson, W. W. (1867–8) 'Remarks on the stellar longitudes assigned by Ptolemy', *Monthly Notices of the Royal Astronomical Society* xxviii, 207–10.

Drecker, J. (1927–8) 'Das Planisphaerium des Claudius Ptolemaeus', *Isis* ix, 255–78.

(1928) 'Des Johannes Philoponos Schrift über das Astrolab', *Isis* xi, 15–44.

Dreyer, J. L. E. (1906) *History of the Planetary Systems from Thales to Kepler* (Cambridge).

(1916–17) 'On the origin of Ptolemy's catalogue of stars', *Monthly Notices of the Royal Astronomical Society* LXXVII, 528–39.

(1917–18) 'On the origin of Ptolemy's catalogue of stars', *Monthly Notices of the Royal Astronomical Society* LXXVIII, 343–9.

Driver, G. R. (1976) *Semitic Writing* (1st ed. 1948), revised ed. (London).

Druart, T.-A. (1975) 'La stoicheïologie de Platon', *Revue Philosophique de Louvain* LXXIII, 243–62.

Dubs, H. H. (1927) *Hsüntze, The Moulder of Ancient Confucianism* (London).

(1928) *The Works of Hsüntze* (London).

Ducatillon, J. (1977) *Polémiques dans la Collection Hippocratique* (Lille and Paris).

Duchemin, J. (1968) *L'ΑΓΩΝ dans la tragédie grecque* (1st ed. 1945), 2nd ed. (Paris).

Duchesne-Guillemin, J. (1953) *Ormazd et Ahriman, l'aventure dualiste dans l'antiquité* (Paris).

(1956) 'Persische Weisheit in griechischem Gewande?' *Harvard Theological Review* XLIX, 115–22.

(1962) *La Religion de l'Iran ancien* (Paris).

Duckworth, W. L. H. (1962) *Galen On Anatomical Procedures, The Later Books* trans. W. L. H. Duckworth, edd. M. C. Lyons and B. Towers (Cambridge).

Düring, I. (1944) *Aristotle's Chemical Treatise, Meteorologica Book IV* (Göteborgs Högskolas Årsskrift L, Göteborg).

(1961) 'Aristotle's method in biology', in *Aristote et les problèmes de méthode*, ed. S. Mansion (Louvain and Paris), pp. 213–21.

(1966) *Aristoteles: Darstellung und Interpretation seines Denkens* (Heidelberg).

Duhem, P. (1905–6) *Les Origines de la statique*, 2 vols. (Paris).

(1908) 'ΣΩΖΕΙΝ ΤΑ ΦΑΙΝΟΜΕΝΑ', *Annales de Philosophie Chrétienne* VI, 113–39, 277–302, 352–77, 482–514, 561–92.

(1954a) *The Aim and Structure of Physical Theory* (trans. P. P. Wiener of 2nd ed. of *La Théorie physique: son objet, sa structure*, Paris, 1914) (Princeton University Press, Princeton, New Jersey).

(1954b) *Le Système du monde: Histoire des doctrines cosmologiques de Platon à Copernic*, Vols. I and II, 2nd ed. (Paris).

Dulière, W.-L. (1965) 'Les "Dictyaques" de Denys d'Egée ou les dilemmes du "sic et non" de la médecine antique. Histoire d'un procédé dialectique', *L'Antiquité Classique* XXXIV, 506–18.

Dundes, A. (1975) *Analytic Essays in Folklore* (The Hague and Paris).

Durkheim, E. (1912/1976) *The Elementary Forms of the Religious Life* (trans. J. W. Swain of *Les Formes élémentaires de la vie religieuse*, Paris, 1912), 2nd ed. (London).

Durkheim, E. and Mauss, M. (1901–2) 'De quelques formes primitives de classification', *L'Année sociologique* VI, 1–72.

East, S. P. (1958) 'De la méthode en biologie selon Aristote', *Laval Théologique et Philosophique* XIV, 213–35.

Ebbell, B. (1937) *The Papyrus Ebers* (Copenhagen and London).

Edelstein, E. J. and Edelstein, L. (1945) *Asclepius*, 2 vols. (Johns Hopkins, Baltimore).

Edelstein, L. (1931) ΠΕΡΙ ΑΕΡΩΝ *und die Sammlung der hippokratischen Schriften* (Problemata, 4, Berlin).

(1932–3/1967) 'The history of anatomy in antiquity (originally 'Die Geschichte der Sektion in der Antike', *Quellen und Studien zur Geschichte der Naturwissenschaften und der Medizin* III, 2 (1932–3), 100–56), in Edelstein 1967, pp. 247–301.

(1935) 'Hippokrates', *Pauly-Wissowa Real-Encyclopädie der classischen Altertumswissenschaft*, Suppl. Bd VI, cols. 1290–1345.

(1937/1967) 'Greek medicine in its relation to religion and magic' (*Bulletin of the Institute of the History of Medicine* v (1937), 201–46), in Edelstein 1967, pp. 205–46.

(1943/1967) *The Hippocratic Oath* (Supplements to the Bulletin of the History of Medicine, 1, Baltimore, 1943), in Edelstein 1967, pp. 3–63.

(1952a/1967) 'The relation of ancient philosophy to medicine' (*Bulletin of the History of Medicine* xxvi (1952), 299–316), in Edelstein 1967, pp. 349–66.

(1952b/1967) 'Recent trends in the interpretation of ancient science' (*Journal of the History of Ideas* xiii (1952), 573–604), in Edelstein 1967, pp. 401–39.

(1967) *Ancient Medicine*, edd. O. and C. L. Temkin (Johns Hopkins, Baltimore).

Ehrenburg, V. (1935) *Ost und West* (Schriften der philosophischen Fakultät der deutschen Universität in Prag, 15, Brünn).

(1940) 'Isonomia', *Pauly-Wissowa Real-Encyclopädie der classischen Altertumswissenschaft*, Suppl. Bd vii, cols. 293–301.

(1946) *Aspects of the Ancient World* (Oxford).

(1973) *From Solon to Socrates* (1st ed. 1968), 2nd ed. (London).

Einarson, B. (1936) 'On certain mathematical terms in Aristotle's logic', *American Journal of Philology* lvii, 33–54 and 151–72.

Eitrem, S. (1941) 'La magie comme motif littéraire chez les grecs et les romains', *Symbolae Osloenses* xxi, 39–83.

Eliade, M. (1946) 'Le problème du chamanisme', *Revue de l'histoire des religions* cxxxi, 5–52.

(1964) *Shamanism: Ancient techniques of ecstasy* (trans. W. R. Trask of *Le Chamanisme et les techniques archaïques de l'extase*, Paris, 1951) (London).

Elias, J. A. (1968) '"Socratic" vs. "Platonic" Dialectic', *Journal of the History of Philosophy* vi, 205–16.

Essertier, D. (1927) *Les Formes inférieures de l'explication* (Paris).

Eucken, R. (1872) *Die Methode der aristotelischen Forschung* (Berlin).

Evans, J. D. G. (1975) 'The codification of false refutations in Aristotle's *De Sophisticis Elenchis*', *Proceedings of the Cambridge Philological Society* NS xxi, 42–52.

(1977) *Aristotle's Concept of Dialectic* (Cambridge).

Evans, M. G. (1958–9) 'Causality and explanation in the logic of Aristotle', *Philosophy and Phenomenological Research* xix, 466–85.

Evans-Pritchard, E. E. (1937) *Witchcraft, Oracles and Magic among the Azande* (Oxford).

(1956) *Nuer Religion* (Oxford).

Farrington, B. (1939) *Science and Politics in the Ancient World* (London).

(1944–9/1961) *Greek Science* (1st ed. 2 vols. 1944–9) revised ed. (London 1961).

(1953–4) 'The rise of abstract science among the Greeks', *Centaurus* iii, 32–9.

(1957) 'The Greeks and the experimental method', *Discovery* xviii, 68–9.

Festugière, A. J. (1944–54) *La Révélation d'Hermès trismégiste*, 4 vols. (Paris).

(1948) *Hippocrate, L'Ancienne médecine* (Etudes et commentaires, 4, Paris).

Feuer, L. S. (1953) 'Sociological aspects of the relation between language and philosophy', *Philosophy of Science* xx, 85–100.

Feyerabend, P. K. (1961) *Knowledge without Foundations* (Oberlin).

(1962) 'Explanation, reduction and empiricism', in *Scientific Explanation, Space, and Time* (Minnesota Studies in the Philosophy of Science, 3), edd. H. Feigl and G. Maxwell (Minneapolis), pp. 28–97.

(1965) 'Problems of empiricism', in *Beyond the Edge of Certainty*, ed. R. G. Colodny (Englewood Cliffs, New Jersey), pp. 145–260.

(1970a) 'Problems of empiricism, Part II', in *The Nature and Function of Scientific Theories*, ed. R. G. Colodny (University of Pittsburgh), pp. 275–353.

(1970b) 'Against Method: Outline of an anarchistic theory of knowledge', in *Analyses of Theories and Methods of Physics and Psychology* (Minnesota Studies in the Philosophy of Science, 4), edd. M. Radner and S. Winokur (Minneapolis), pp. 17–130.

(1975) *Againt Method* (London).

Field, G. C. (1930) *Plato and his Contemporaries* (London).

Filliozat, J. (1943) *Magie et médecine* (Paris).

Finley, M. I (1954/1977) *The World of Odysseus* (1st ed. 1954), 2nd ed. (New York, 1977).

(1965) 'Technical innovation and economic progress in the ancient world', *Economic History Review*, 2nd ser., XVIII 29–45.

(1970) *Early Greece: the Bronze and Archaic Ages* (London).

(1973a) *Democracy Ancient and Modern* (London).

(1973b) *The Ancient Economy* (London).

(1974) 'Athenian Demagogues' (*Past and Present* XXI (1962), 3–24) in *Studies in Ancient Society*, ed. M. I Finley (London), pp. 1–25.

(1975a) 'Myth, memory and history' (*History and Theory* IV (1964–5), 281–302) in *The Use and Abuse of History* (London), pp. 11–33.

(1975b) 'The freedom of the citizen in the Greek world', *Talanta* (Proceedings of the Dutch Archaeological and Historical Society) VII, 1–23.

(1977) 'Censura nell'antichità classica', *Belfragor* XXXII, 605–22.

Finnegan, R. (1977) *Oral Poetry* (Cambridge).

Flashar, H. (ed.) (1971) *Antike Medizin* (Darmstadt).

Forke, A. (1907) *Lun-Hêng, Part I: Philosophical Essays of Wang Ch'ung* (Leipzig).

(1911) *Lun-Hêng, Part II: Miscellaneous Essays of Wang Ch'ung* (Berlin).

Fotheringham, J. K. (1915) 'The probable error of a water-clock', *Classical Review* XXIX, 236–8.

(1922–3) 'The secular acceleration of the moon's mean motion as determined from occultations and conjunctions in the *Almagest*: (a correction)', *Monthly Notices of the Royal Astronomical Society* LXXXIII, 370–3.

(1923) 'The probable error of a water-clock', *Classical Review* XXXVII, 166–7.

(1928) 'The indebtedness of Greek to Chaldaean astronomy', *The Observatory* LI, 301–15 (also in *Quellen und Studien zur Geschichte der Mathematik, Astronomie und Physik*, B, 2, 1 (1932), Berlin, 1933, pp. 28–44).

Fotheringham, J. K. and Longbottom, G. (1914–15) 'The secular acceleration of the moon's mean motion as determined from the occultations in the Almagest', *Monthly Notices of the Royal Astronomical Society* LXXV, 377–94.

Foucault, M. (1972) *The Archaeology of Knowledge* (trans. A. M. Sheridan Smith of *L'Archéologie du savoir*, Paris, 1969) (London).

Fränkel, H. (1930/1975) 'Studies in Parmenides' (originally 'Parmenidesstudien', *Nachrichten von der Gesellschaft der Wissenschaften zu Göttingen* (Phil.-hist. Kl. 1930)), in Allen and Furley 1975, pp. 1–47.

(1942/1975) 'Zeno of Elea's attacks on plurality' (*American Journal of Philology* LXIII (1942), 1–25 and 193–206), in Allen and Furley 1975, pp. 102–42.

(1960) *Wege und Formen frühgriechischen Denkens* (1st ed. 1955), 2nd ed. (München).

(1962/1975) *Early Greek Poetry and Philosophy* (trans. M. Hadas and J. Willis of 2nd ed. of *Dichtung und Philosophie des frühen Griechentums*, München, 1962) (Oxford, 1975).

Frankfort, H. (1948) *Kingship and the Gods* (Chicago).

Frankfort, H. (ed.) (1949) *Before Philosophy* (1st ed. *The Intellectual Adventure of Ancient Man*, Chicago, 1946), 2nd ed. (London).

Frazer, J. G. (1911–15) *The Golden Bough*, 12 vols., 3rd ed. (London).

Frede, M. (1974) *Die stoische Logik* (Göttingen).

Fredrich, C. (1899) *Hippokratische Untersuchungen* (Philol. Untersuch. 15, Berlin).

Freudenthal, H. (1966) 'Y avait-il une crise des fondements des mathématiques dans l'antiquité?', *Bulletin de la Société Mathématique de Belgique* XVIII, 43–55.

Friedländer, P. (1958–69) *Plato* (trans. H. Meyerhoff of 2nd ed. of *Platon*, Berlin, 1954–60), 3 vols. (London).

Frisk, H. (1935) '*Wahrheit*' *und* '*Lüge*' *in den indogermanischen Sprachen* (Göteborgs Högskolas Årsskrift XLI, 3, Göteborg).

Fritz, H. von (1934a) 'Theaitetos', *Pauly-Wissowa Real-Encyclopädie der classischen Altertumswissenschaft*, 2nd ser. 10 Halbband, v. 2, cols. 1351–72.

(1934b) 'Theodoros', *Pauly-Wissowa Real-Encyclopädie der classischen Altertumswissenschaft*, 2nd ser. 10 Halbband, v.2, cols. 1811–25.

(1943) 'νόος and νοεῖν in the Homeric poems', *Classical Philology* XXXVIII, 79–93.

(1945) 'νοῦς, νοεῖν and their derivatives in Pre-Socratic Philosophy (excluding Anaxagoras). I', *Classical Philology* XL, 223–42.

(1945/1970) 'The discovery of incommensurability by Hippasus of Metapontum' (*Annals of Mathematics*, 2nd ser., XLVI (1945), 242–64), in Furley and Allen 1970, pp. 382–412.

(1946) 'νοῦς, νοεῖν and their derivatives in Pre-Socratic Philosophy (excluding Anaxagoras). II', *Classical Philology* XLI, 12–34.

(1955/1971) 'Die ΑΡΧΑΙ in der griechischen Mathematik' (*Archiv für Begriffsgeschichte* I (1955), 13–103), in von Fritz 1971, pp. 335–429.

(1959/1971) 'Gleichheit, Kongruenz und Ähnlichkeit in der antiken Mathematik bis auf Euklid' (*Archiv für Begriffsgeschichte* IV (1959), 7–81), in von Fritz 1971, pp. 430–508.

(1960) 'Mathematiker und Akusmatiker bei den alten Pythagoreern', *Sitzungsberichte der bayerischen Akademie der Wissenschaften*, phil.-hist. Kl., 1960, 11, München.

(1964/1971) 'Die ΕΠΑΓΩΓΗ bei Aristoteles' (*Sitzungsberichte der bayerischen Akademie der Wissenschaften*, phil.-hist. Kl., 1964, 3, München, 1964), in von Fritz 1971, pp. 623–76.

(1971) *Grundprobleme der Geschichte der antiken Wissenschaft* (Berlin and New York).

Furley, D. J. (1967) *Two Studies in the Greek Atomists* (Princeton University Press, Princeton, New Jersey).

Furley, D. J. and Allen, R. E. (edd.) (1970) *Studies in Presocratic Philosophy*, Vol. 1 (London).

Gadamer, H. G., Gaiser, K., Gundert, H., Krämer, J., and Kuhn, H. (1968) *Idee und Zahl* (Heidelberg).

Gadd, C. J. (1973) 'Hammurabi and the end of his dynasty', *Cambridge Ancient History*, 3rd ed. Vol. II, 1, ch. 5 (Cambridge), pp. 176–227.

Gandt, F. de (1975) 'La *Mathésis* d'Aristote: Introduction aux *Analytiques seconds*', *Revue des sciences philosophiques et théologiques* LIX, 564–600.

(1976) 'La *Mathésis* d'Aristote: Introduction aux *Analytiques seconds*', *Revue des sciences philosophiques et théologiques* LX, 37–84.

Gatzemeier, M. (1970) *Die Naturphilosophie des Straton von Lampsakos* (Meisenheim).

Gellner, E. (1962/1970) 'Concepts and society' (from *The Transactions of the Fifth World Congress of Sociology*, 1962) in Wilson 1970, pp. 18–49 (also reprinted in *Sociological Theory and Philosophical Analysis*, edd. D. Emmet and A. MacIntyre, London, 1970, pp. 115–49).

(1973) 'The savage and the modern mind', in Horton and Finnegan 1973, pp. 162–81.

Gent, W. (1966) 'Der Begriff des Weisen', *Zeitschrift für philosophische Forschung* XX, 77–117.

Gernet, L. (1917) *Recherches sur le développement de la pensée juridique et morale en Grèce* (Paris).

(1955) *Droit et société dans la Grèce ancienne* (Paris).

(1945/1968) 'Les origines de la philosophie' (*Bulletin de l'enseignement public du Maroc* CLXXXIII (1945), 1–12) in Gernet 1968, pp. 415–30.

(1948–9/1968) 'Droit et prédroit en Grèce ancienne' (*L'Année sociologique*, 1948–9 (Paris 1951), 21–119) in Gernet 1968, pp. 175–260.

(1955/1968) 'L'anthropologie dans la religion grecque' (*Anthropologie Religieuse*, ed. C. J. Bleeker, Supplements to NUMEN, vol. 2, Leiden, 1955, 49–59) in Gernet 1968, pp. 9–19.

(1956/1968) 'Choses visibles et choses invisibles' (*Revue philosophique* CXLVI (1956), 79–86), in Gernet 1968, pp. 405–14.

(1968) *Anthropologie de la Grèce antique* (Paris).

Ghalioungui, P. (1963) *Magic and Medical Science in Ancient Egypt* (London).

Gigon, O. (1936) 'Gorgias "Über das Nichtsein"', *Hermes* LXXI, 186–213.

(1946) 'Die naturphilosophischen Voraussetzungen der antiken Biologie', *Gesnerus* III, 35–58.

(1962) *Vita Aristotelis Marciana* (Berlin).

(1968) *Der Ursprung der griechischen Philosophie* (1st ed. 1945), 2nd ed. (Basel).

(1973) 'Der Begriff der Freiheit in der Antike', *Gymnasium* LXXX, 8–56.

Gilbert, O. (1907) *Die meteorologischen Theorien des griechischen Altertums* (Leipzig).

(1910) 'Spekulation und Volksglaube in der ionischen Philosophie', *Archiv für Religionswissenschaft* XIII, 306–32.

Gillespie, C. M. (1925) 'The Aristotelian Categories', *Classical Quarterly* XIX, 75–84.

Gingerich, O. (forthcoming) 'Was Ptolemy a fraud?', *Journal of the Royal Astronomical Society*.

Gladigow, B. (1965) *Sophia und Kosmos* (Spudasmata, 1, Hildesheim).

(1967) 'Zum Makarismos des Weisen', *Hermes* XCV, 404–33.

Gluckman, M. (1965) *Politics, Law and Ritual in Tribal Society* (Oxford).

(1967) *The Judicial Process among the Barotse of Northern Rhodesia* (1st ed. 1955), 2nd ed. (Manchester University Press).

(1972) *The Ideas in Barotse Jurisprudence* (1st ed. 1965) 2nd ed. (Manchester University Press).

Götze, A. (1923) 'Persische Weisheit in griechischem Gewande: Ein Beitrag zur Geschichte der Mikrokosmos-Idee', *Zeitschrift für Indologie und Iranistik* II, 60–98 and 167–77.

Gohlke, P. (1924) 'Die Entstehungsgeschichte der naturwissenschaftlichen Schriften des Aristoteles', *Hermes* LIX, 274–306.

(1936) *Die Entstehung der aristotelischen Logik* (Berlin).

Goldschmidt, V. (1947*a*) *Le Paradigme dans la dialectique platonicienne* (Paris).

(1947*b*) *Les Dialogues de Platon: structure et méthode dialectique* (Paris).

(1970) *Questions Platoniciennes* (Paris).

Goldstein, B. R. (1967) 'The Arabic version of Ptolemy's *Planetary Hypotheses*', *Transactions of the American Philosophical Society* LVII, 4.

Goldstein, B. R. and Swerdlow, N. (1970) 'Planetary distances and sizes in an Anonymous Arabic treatise preserved in Bodleian MS Marsh 621', *Centaurus* XV, 135–70.

Goltz, D. (1974) *Studien zur altorientalischen und griechischen Heilkunde, Therapie, Arzneibereitung, Rezeptstruktur* (Sudhoffs Archiv Beiheft 16, Wiesbaden).

Gomme, A. W. (1945–70) *A Historical Commentary on Thucydides*, 4 vols. (Vol. IV with A. Andrewes and K. J. Dover) (Oxford).

Gomperz, H. (1912) *Sophistik und Rhetorik* (Leipzig).

(1943) 'Problems and methods of early Greek science', *Journal of the History of Ideas* IV, 161–76.

Gomperz, T. (1910) *Die Apologie der Heilkunst*, 2nd ed. (Leipzig).

Goody, J. (1977) *The Domestication of the Savage Mind* (Cambridge).
Goody, J. and Watt, I. P. (1968) 'The consequences of literacy', in *Literacy in Traditional Societies*, ed. J. Goody (Cambridge), pp. 27–68.
Gottschalk, H. B. (1961) 'The authorship of *Meteorologica*, Book IV', *Classical Quarterly* NS XI, 67–79.
  (1965) *Strato of Lampsacus: Some texts* (Proceedings of the Leeds Philosophical and Literary Society, Literary and Historical Section XI (1964–6), Part VI, 1965).
Gourevitch, D. (1969) 'Déontologie médicale: quelques problèmes, I', *Mélanges d'Archéologie et d'Histoire* LXXXI, 519–36.
  (1970) 'Déontologie médicale: quelques problèmes, II', *Mélanges d'Archéologie et d'Histoire* LXXXII, 737–52.
Granet, M. (1934) *La Pensée chinoise* (Paris).
Grensemann, H. (1968) *Die hippokratische Schrift 'Über die heilige Krankheit'* (Ars Medica, Abt. II Bd 1, Berlin).
Griffith, G. T. (1966) 'Isegoria in the assembly at Athens', in *Ancient Society and Institutions* (Studies presented to V. Ehrenburg) (Oxford), pp. 115–38.
Grimaldi, W. M. A. (1972) *Studies in the Philosophy of Aristotle's Rhetoric* (Hermes Einzelschriften 25, Wiesbaden).
Gueroult, M. (1963) 'Logique, argumentation, et histoire de la philosophie chez Aristote', *Logique et Analyse* VI, 431–49.
Güterbock, H. G. (1962) 'Hittite medicine', *Bulletin of the History of Medicine* XXXVI, 109–13.
Gullini, G. (1972) 'Tradizione e originalità nell'architettura Achemenide a Pasargade', *Parola del Passato* XXVII, 13–39.
Gundel, H. G. (1968) *Weltbild und Astrologie in den griechischen Zauberpapyri* (München).
Gundel, W. (1922) *Sterne und Sternbilder im Glauben des Altertums und der Neuzeit* (Leipzig).
  (1936) *Neue astrologische Texte des Hermes Trismegistos* (Abhandlungen der bayerischen Akademie der Wissenschaften, phil.-hist. Abt., NF 12, 1935, München).
Gundel, W. and Gundel, H. G. (1966) *Astrologumena: Die astrologische Literatur in der Antike und ihre Geschichte* (Sudhoffs Archiv Beiheft 6, Wiesbaden).
Gundert, H. (1971) *Dialog und Dialektik: Zur Struktur des platonischen Dialogs* (Amsterdam).
  (1973) '"Perspektivische Täuschung" bei Platon und die Prinzipienlehre', in *Zetesis* (Festschrift de Strycker) (Antwerp), pp. 80–97.
Guthrie, W. K. C. (1950) *The Greeks and their Gods* (London).
  (1962) *A History of Greek Philosophy, Vol. I, The Earlier Presocratics and the Pythagoreans* (Cambridge).
  (1965) *A History of Greek Philosophy, Vol. II, The Presocratic Tradition from Parmenides to Democritus* (Cambridge).
  (1969) *A History of Greek Philosophy, Vol. III, The Fifth-Century Enlightenment* (Cambridge).
Hahm, D. E. (1972) 'Chrysippus' solution to the Democritean dilemma of the cone', *Isis* LXIII, 205–20.
Halliday, W. R. (1913) *Greek Divination* (London).
  (1936) 'Some notes on the treatment of disease in antiquity', in *Greek Poetry and Life* (Essays presented to G. Murray) (Oxford), pp. 277–94.
Hamblin, C. L. (1970) *Fallacies* (London).
Hamelin, O. (1907) *Essai sur les éléments principaux de la représentation* (Paris).
  (1931) *Le Système d'Aristote* (1st ed. 1920), 2nd ed. (Paris).

Hamilton, M. (1906) *Incubation or the cure of disease in pagan temples and Christian churches* (St Andrews and London).

Hamlyn, D. W. (1961) *Sensation and Perception* (London).

(1976) 'Aristotelian Epagoge', *Phronesis* XXI, 167–84.

Hammer-Jensen, I. (1915) 'Das sogenannte IV. Buch der *Meteorologie* des Aristoteles', *Hermes* L, 113–36.

Hanson, N. R. (1958) *Patterns of Discovery* (Cambridge).

Hare, R. M. (1965) 'Plato and the Mathematicians', in Bambrough 1965, pp. 21–38.

Harig, G. (1977) 'Bemerkungen zum Verhältnis der griechischen zur altorientalischen Medizin', in R. Joly 1977, pp. 77–94.

Harris, C. R. S. (1973) *The Heart and the Vascular System in Ancient Greek Medicine from Alcmaeon to Galen* (Oxford).

Harrison, A. R. W. (1968–71) *The Laws of Athens*, 2 vols. (Oxford).

Harrison, J. E. (1908) *Prolegomena to the Study of Greek Religion* (1st ed. 1903), 2nd ed. (Cambridge).

Hartner, W. (1968) *Oriens–Occidens* (Hildesheim).

Harvey, F. D. (1966) 'Literacy in the Athenian Democracy', *Revue des études grecques* LXXIX, 585–635.

Hase, H. (1839) 'Joannis Alexandrini, cognomine Philoponi, de usu astrolabii ejusque constructione libellus', *Rheinisches Museum* VI, 127–71.

Hasse, H. and Scholz, H. (1928) *Die Grundlagenkrisis der griechischen Mathematik* (Berlin) (also in *Kant-Studien* XXXIII (1928), 4–34).

Havelock, E. A. (1963) *Preface to Plato* (Oxford).

(1966) 'Pre-Literacy and the Pre-Socratics', *Bulletin of the Institute of Classical Studies* XIII, 44–67.

(1971) *Prologue to Greek Literacy* (University of Cincinnati).

(1976) *Origins of Western Literacy* (Ontario Institute for Studies in Education 14, Toronto).

Heath, T. E. (1913) *Aristarchus of Samos* (Oxford).

(1921) *A History of Greek Mathematics*, 2 vols. (Oxford).

(1926) *The Thirteen Books of Euclid's Elements*, 3 vols. (1st ed. 1908), 2nd ed. (Cambridge).

(1949) *Mathematics in Aristotle* (Oxford).

Heiberg, I. L. (1925) *Geschichte der Mathematik und Naturwissenschaften im Altertum* (München).

Heidel, W. A. (1909–10) 'περὶ φύσεως: A study of the conception of nature among the Pre-Socratics', *Proceedings of the American Academy of Arts and Sciences* XLV, 77–133.

(1933) *The Heroic Age of Science* (Baltimore).

(1937) *The Frame of the Ancient Greek Maps* (New York).

(1940) 'The Pythagoreans and Greek mathematics', *American Journal of Philology* LXI, 1–33.

(1941) *Hippocratic Medicine: its spirit and method* (New York).

Heim, R. (1893) 'Incantamenta Magica Graeca Latina', *Jahrbücher für classische Philologie*, Suppl. Bd XIX, 463–576.

Heinimann, F. (1945) *Nomos und Physis* (Schweizerische Beiträge zur Altertumswissenschaft 1, Basel).

(1961) 'Eine vorplatonische Theorie der τέχνη', *Museum Helveticum* XVIII, 105–30.

(1975) 'Mass–Gewicht–Zahl', *Museum Helveticum* XXXII, 183–96.

Heitsch, E. (1962) 'Die nicht-philosophische ΑΛΗΘΕΙΑ', *Hermes* XC, 24–33.

(1963) 'Wahrheit als Erinnerung', *Hermes* XCI, 36–52.

(1970) *Gegenwart und Evidenz bei Parmenides* (Akademie der Wissenschaften und der Literatur, Mainz, Abhandlungen der geistes- und sozialwissenschaftlichen Klasse, Jahrgang 1970, 4, Wiesbaden).

(1974) 'Evidenz und Wahrscheinlichkeitsaussagen bei Parmenides', *Hermes* CII, 411–19.

Heller, S. (1956–8) 'Eine Beitrag zur Deutung der Theodoros-Stelle in Platons Dialog "Theaetet"', *Centaurus* V, 1–58.

(1958) *Die Entdeckung der stetigen Teilung durch die Pythagoreer* (Abhandlungen der deutschen Akademie der Wissenschaften zu Berlin, Klasse für Mathematik, Physik und Technik, Jahrgang 1958, 6, Berlin).

(1967) 'Theaetets Bedeutung als Mathematiker', *Sudhoffs Archiv* LI, 55–78.

Hempel, C. G. (1958) 'The theoretician's dilemma: a study in the logic of theory construction', in *Concepts, Theories, and the Mind–Body Problem* (Minnesota Studies in the Philosophy of Science, 2), ed. H. Feigl, M. Scriven and G. Maxwell (Minneapolis), pp. 37–98.

(1973) 'The meaning of theoretical terms: a critique of the standard empiricist construal', in *Logic, Methodology and Philosophy of Science*, IV, cd. P. Suppes and others (Amsterdam and New York), pp. 367–78.

Hempel, C. G. and Oppenheim, P. (1948) 'Studies in the logic of explanation', *Philosophy of Science* XV, 135–75 and 350–2.

Henrichs, A. (1975) 'Two doxographical notes: Democritus and Prodicus on religion', *Harvard Studies in Classical Philology* LXXIX, 93–123.

Herter, H. (1963) 'Die kulturhistorische Theorie der hippokratischen Schrift von der alten Medizin', *Maia* XV, 464–83.

Herzog, R. (1931) *Die Wunderheilungen von Epidauros* (Philologus Suppl. Bd XXII, 3, Leipzig).

Hess, W. (1970) 'Erfahrung und Intuition bei Aristoteles', *Phronesis* XV, 48–82.

Hesse, M. (1961) *Forces and Fields: The Concept of Action at a Distance in the history of physics* (London).

(1963) *Models and Analogies in Science* (London and New York).

(1974) *The Structure of Scientific Inference* (London).

Hintikka, J. (1972) 'On the ingredients of an Aristotelian science', *Nous* VI, 55–69.

(1973) *Time and Necessity* (Oxford).

(1974) *Knowledge and the Known* (Dordrecht and Boston).

Hintikka, J. and Remes, U. (1974) *The Method of Analysis* (Boston Studies in the Philosophy of Science 25, Dordrecht and Boston).

Hirzel, R. (1903) Ἄγραφος Νόμος (Abhandlungen der phil.-hist. Classe der königlich sächsischen Gesellschaft der Wissenschaften XX, 1, 1900, Leipzig, 1903).

(1907) *Themis, Dike und Verwandtes* (Leipzig).

Hocutt, M. (1974) 'Aristotle's Four Becauses', *Philosophy* XLIX, 385–99.

Hoebel, E. Adamson (1964) *The Law of Primitive Man* (Harvard University Press, Cambridge, Mass.).

Hölscher, U. (1953/1970) 'Anaximander and the beginnings of Greek philosophy' (originally 'Anaximander und die Anfänge der Philosophie', *Hermes* LXXXI (1953), 257–77 and 385–418), in Furley and Allen 1970, pp. 281–322.

(1968) *Anfängliches Fragen* (Göttingen).

Hoffmann, E. (1925) *Die Sprache und die archaische Logik* (Tübingen).

Hofmann, J. E. (1956–8) 'Ergänzende Bemerkungen zum "geometrischen" Irrationalitätsbeweis der alten Griechen', *Centaurus* V, 59–72.

Hoijer, H. (1954) 'The Sapir–Whorf hypothesis', in *Language and Culture*, ed. H. Hoijer (University of Chicago Press), pp. 92–105.

Holwerda, D. (1955) ΦΥΣΙΣ (Groningen).

Hopfner, T. (1925) *Orient und griechische Philosophie* (Leipzig).
  (1928) 'Mageia', *Pauly-Wissowa Real-Encyclopädie der classischen Altertumswissenschaft*, 27 Halbband, XIV, 1, cols. 301–93.
  (1937) 'Traumdeutung', *Pauly-Wissowa Real-Encyclopädie der classischen Altertumswissenschaft*, 2nd ser., 12 Halbband, VI, 2, cols. 2233–45.
Horne, R. A. (1966) 'Aristotelian Chemistry', *Chymia* XI, 21–7.
Horton, R. (1967) 'African traditional thought and western science', *Africa* XXXVII, 50–71 and 155–87 (abbreviated version reprinted in Wilson 1970, pp. 131–71).
Horton, R. and Finnegan, R. (edd.) (1973) *Modes of Thought* (London).
Hubert, H. (1904) 'Magia', in *Dictionnaire des antiquités grecques et romaines*, ed. C. Daremberg, E. Saglio, E. Pottier, Vol. III (Paris), pp. 1494–1521.
Hudson-Williams, H. L. (1950) 'Conventional forms of debate and the Melian dialogue', *American Journal of Philology* LXXI, 156–69.
Huizinga, J. (1944/1970) *Homo Ludens* (trans. R. F. C. Hull of 1944 German edition (original Dutch 1938), 2nd ed. London, 1970).
Hume, R. E. (1931) *The Thirteen Principal Upanishads* (1st ed. 1921), 2nd ed. (Oxford).
Hussey, E. (1972) *The Presocratics* (London).
Ilberg, J. (1931) *Rufus von Ephesos. Ein griechischer Arzt in trajanischer Zeit* (Abhandlungen der phil.-hist. Klasse der sächsischen Akademie der Wissenschaften XLI, 1, 1930, Leipzig, 1931).
Itard, J. (1961) *Les Livres arithmétiques d'Euclide* (Paris).
Iversen, E. (1939) *Papyrus Carlsberg No. VIII* (Det kgl. Danske Videnskabernes Selskab. Historisk-filologiske Meddelelser XXVI, 5, Copenhagen).
Jacobsen, T. (1949) 'Mesopotamia', in Frankfort 1949, pp. 137–234.
Jacoby, F. (1923–58) *Die Fragmente der griechischen Historiker* (Berlin 1923–30, Leiden 1940–58).
Jaeger, W. (1939–45) *Paideia: the Ideals of Greek Culture*, trans. G. Highet, 3 vols. (Oxford).
  (1947) *The Theology of the Early Greek Philosophers* (Gifford Lectures 1936), trans. E. S. Robinson (Oxford).
  (1948) *Aristotle: Fundamentals of the History of his Development*, trans. R. Robinson (1st ed. 1934), 2nd ed. (Oxford).
Janssens, E. (1949–50) 'Platon et les sciences d'observation'. *Revue de l'Université de Bruxelles* II, 249–68.
Jarvie, I. C. (1976) 'On the limits of symbolic interpretation in anthropology', *Current Anthropology* XVII, 687–91.
Jarvie, I. C. and Agassi, J. (1967/1970) 'The problem of the rationality of magic' (*British Journal of Sociology* XVIII (1967), 55–74), in Wilson 1970, pp. 172–93.
Jeanmaire, H. (1939) *Couroi et Courètes* (Lille).
Jeffery, L. H. (1961) *The Local Scripts of Archaic Greece* (Oxford).
Joachim, H. H. (1904) 'Aristotle's conception of chemical combination', *Journal of Philology* XXIX, 72–86.
  (1922) *Aristotle, On Coming-to-be and Passing-away* (Oxford).
Johann, H.-T. (1973) 'Hippias von Elis und der Physis-Nomos-Gedanke', *Phronesis* XVIII, 15–25.
Joly, H. (1974) *Le Renversement platonicien: Logos, Episteme, Polis* (Paris).
Joly, R. (1966) *Le Niveau de la science hippocratique* (Paris).
  (1968) 'La biologie d'Aristote', *Revue philosophiqie* CLVIII, 219–53.
Joly, R. (ed.) (1977) *Corpus Hippocraticum* (Editions Universitaires de Mons, Série Sciences Humaines IV, Université de Mons).
Jones, J. W. (1956) *Law and legal theory of the Greeks* (Oxford).

Jones, W. H. S. (1923-31) *Hippocrates*, Loeb. ed. 4 vols. (Vol. III with E. T. Withington) (London and Cambridge, Mass.).

(1946) *Philosophy and Medicine in Ancient Greece* (Suppl. to the Bulletin of the History of Medicine 8, Baltimore).

(1947) *The Medical Writings of Anonymus Londinensis* (Cambridge).

Jope, J. (1972) 'Subordinate demonstrative science in the Sixth Book of Aristotle's *Physics*', *Classical Quarterly* NS XXII, 279-92.

Jouanna, J. (1961) 'Présence d'Empédocle dans la Collection Hippocratique', *Bulletin de l'Association Guillaume Budé* (1961), 452-63.

Jürss, F. (1967) 'Über die Grundlagen der Astrologie', *Helikon* VII, 63-80.

Junge, G. (1958) 'Von Hippasus bis Philolaus. Das Irrationale und die geometrischen Grundbegriffe', *Classica et Mediaevalia* XIX, 41-72.

Kahn, C. H. (1960) *Anaximander and the Origins of Greek Cosmology* (New York).

(1966a) 'Sensation and consciousness in Aristotle's psychology', *Archiv für Geschichte der Philosophie* XLVIII, 43-81.

(1966b) 'The Greek verb "to be" and the concept of being', *Foundations of Language* II, 245-65.

(1970) 'On early Greek astronomy', *Journal of Hellenic Studies* XC, 99-116.

(1971) 'Religion and natural philosophy in Empedocles' doctrine of the soul' (*Archiv für Geschichte der Philosophie* XLII (1960), 3-35), in Anton and Kustas 1971, pp. 3-38.

(1973) *The Verb 'Be' in Ancient Greek* (Foundations of Language Suppl. 16, Dordrecht).

Kapp, E. (1931/1975) 'Syllogistic' (originally 'Syllogistik', in *Pauly-Wissowa Real-Encyclopädie der classischen Altertumswissenschaft*, 2nd ser., 7 Halbband, IV 1 (1931), cols. 1046-67), in Barnes, Schofield, Sorabji 1975, pp. 35-49.

(1942) *Greek Foundations of Traditional Logic* (New York).

Kattsoff, L. O. (1947-8) 'Ptolemy and scientific method', *Isis* XXXVIII, 18-22.

Keith, A. B. (1925) *The Religion and Philosophy of the Veda and Upanishads*, 2 vols. (Harvard University Press, Cambridge, Mass.).

Keller, O. (1909-13) *Die antike Tierwelt*, 2 vols. (Leipzig).

Kennedy, G. (1963) *The Art of Persuasion in Greece* (London).

Kenyon, F. G. (1951) *Books and Readers in Ancient Greece and Rome* (1st ed. 1932), 2nd ed. (Oxford).

Kerferd, G. (1955) 'Gorgias on Nature or that which is not', *Phronesis* I, 3-25.

(1956-7) 'The moral and political doctrines of Antiphon the sophist. A reconsideration', *Proceedings of the Cambridge Philological Society* NS IV, 26-32.

Kerschensteiner, J. (1945) *Platon und der Orient* (Stuttgart).

(1962) *Kosmos: quellenkritische Untersuchungen zu den Vorsokratikern* (Zetemata 30, München).

Kessels, A. H. M. (1969) 'Ancient systems of dream-classification', *Mnemosyne* 4th ser. XXII, 389-424.

Kirk, G. S. (1960/1970) 'Popper on science and the Presocratics' (*Mind* NS LXIX (1960), 318-39), in Furley and Allen 1970, pp. 154-77.

(1961) 'Sense and common-sense in the development of Greek philosophy', *Journal of Hellenic Studies* LXXXI, 105-17.

Kirk, G. S. and Raven, J. E. (1957) *The Presocratic Philosophers* (Cambridge).

Klein, J. (1968) *Greek Mathematical Thought and the Origin of Algebra* (trans. E. Brann of *Die griechische Logistik und die Entstehung der Algebra*, Quellen und Studien zur Geschichte der Mathematik, Astronomie und Physik B 3, 1 (1934), pp. 18-105 and B 3, 2 (1936), pp. 122-235) (M.I.T. Press, Cambridge, Mass.).

Kleingünther, A. (1933) ΠΡѠΤΟΣ ΕΥΡΕΤΗΣ *Untersuchungen zur Geschichte einer Fragestellung* (Philologus, Suppl. Bd XXVI, 1, Leipzig).

Klowski, J. (1966) 'Der historische Ursprung des Kausalprinzips', *Archiv für Geschichte der Philosophie* XLVIII, 225–66.

(1970) 'Zum Entstehen der logischen Argumentation', *Rheinisches Museum* CXIII, 111–41.

Kneale, W. and Kneale, M. (1962) *The Development of Logic* (Oxford).

Knorr, W. R. (1975) *The Evolution of the Euclidean Elements* (Dordrecht and Boston).

Knutzen, G. H. (1964) *Technologie in den hippokratischen Schriften* περὶ διαίτης ὀξέων, περὶ ἀγμῶν, περὶ ἄρθρων ἐμβολῆς (Akademie der Wissenschaften und der Literatur, Mainz, Abhandlungen der geistes- und sozialwissenschaftlichen Klasse, Jahrgang 1963, 14, Wiesbaden, 1964).

König, E. (1970) 'Aristoteles' erste Philosophie als universale Wissenschaft von der ΑΡΧΑΙ', *Archiv für Geschichte der Philosophie* LII, 225–46.

Körner, O. (1929) *Die ärztlichen Kenntnisse in Ilias und Odyssee* (München).

(1930) *Die homerische Tierwelt* (1st ed. 1880), 2nd ed. (München).

Koller, H. (1959–60) 'Das Modell der griechischen Logik', *Glotta* XXXVIII, 61–74.

Kollesch, J. (1974) 'Die Medizin und ihre sozialen Aufgaben zur Zeit der Poliskrise', in *Hellenische Poleis*, Vol. IV, ed. E. C. Welskopf (Berlin), pp. 1850–71.

Kosman, L. A. (1973) 'Understanding, explanation, and insight in the *Posterior Analytics*', in *Exegesis and Argument*, ed. E. N. Lee, A. P. D. Mourelatos and R. M. Rorty (Assen), pp. 374–92.

Kranz, W. (1938a) 'Gleichnis und Vergleich in der frühgriechischen Philosophie', *Hermes* LXXIII, 99–122.

(1938b) 'Kosmos und Mensch in der Vorstellung frühen Griechentums', *Nachrichten von der Gesellschaft der Wissenschaften zu Göttingen* (phil.-hist. Kl. NF II, 1938), pp. 121–61.

(1938–9) 'Kosmos als philosophischer Begriff frühgriechischer Zeit', *Philologus* XCIII, 430–48.

Krischer, T. (1965) 'ΕΤΥΜΟΣ und ΑΛΗΘΗΣ', *Philologus* CIX, 161–74.

Kroll, W. (1897) *Antiker Aberglaube* (Hamburg).

(1940) *Zur Geschichte der aristotelischen Zoologie* (Akademie der Wissenschaften in Wien, phil.-hist. Kl., Sitzungsberichte, 218, 2, Wien).

Kucharski, P. (1949) *Les Chemins du savoir dans les derniers dialogues de Platon* (Paris).

(1965) 'Sur l'évolution des méthodes du savoir dans la philosophie de Platon', *Revue philosophique* CLV, 427–40.

Kudlien, F. (1964) 'Herophilos und der Beginn der medizinischen Skepsis', *Gesnerus* XXI, 1–13 (reprinted in Flashar 1971, pp. 280–95).

(1966) Review of Alexanderson 1963, *Göttingische Gelehrte Anzeigen* CCXVIII, 36–42.

(1967) *Der Beginn des medizinischen Denkens bei den Griechen* (Zürich and Stuttgart).

(1968) 'Early Greek primitive medicine', *Clio Medica* III, 305–36.

(1974) 'Dialektik und Medizin in der Antike', *Medizinhistorisches Journal* IX, 187–200.

Kühn, J.-H. (1956) *System- und Methodenprobleme im Corpus Hippocraticum* (Hermes Einzelschriften 11, Wiesbaden).

Kugler, F. X. (1900) *Die babylonische Mondrechnung* (Freiburg).

(1907–35) *Sternkunde und Sterndienst in Babel*, 4 vols. with supplements (ed. J. Schaumberger) (Münster).

Kuhn, T. S. (1957) *The Copernican Revolution* (Harvard University Press, Cambridge, Mass.).

(1962/1970) *The Structure of Scientific Revolutions* (1st ed. 1962), 2nd ed. (University of Chicago, 1970).

(1970a) 'Postscript-1969', in *The Structure of Scientific Revolutions*, 2nd ed. (University of Chicago), pp. 174–210.

(1970*b*) 'Reflections on my critics', in Lakatos and Musgrave 1970, pp. 231–78.

(1974) 'Second thoughts on paradigms', in *The Structure of Scientific Theories*, ed. F. Suppe (University of Illinois, Urbana, Chicago), pp. 459–82.

Kullmann, W. (1974) *Wissenschaft und Methode: Interpretationen zur aristotelischen Theorie der Naturwissenschaft* (Berlin and New York).

Kurz, D. (1970) AKPIBEIA: *Das Ideal der Exaktheit bei den Griechen bis Aristoteles* (Göppingen).

Kutsch, F. (1913) *Attische Heilgötter und Heilheroen* (Religionsgeschichtliche Versuche und Vorarbeiten XII, 3 (1912–13), Giessen, 1913).

Lämmli, F. (1962) *Vom Chaos zum Kosmos* (Schweizerische Beiträge zur Altertumswissenschaft 10, Basel).

Laín Entralgo, P. (1970) *The Therapy of the Word in Classical Antiquity* (trans. L. J. Rather and J. M. Sharp of *La curación por la palabra en la Antigüedad clásica*, Madrid, 1958) (New Haven and London).

(1975) 'Quaestiones hippocraticae disputatae tres', in Bourgey and Jouanna 1975, pp. 305–19.

Lakatos, I. (1970) 'Falsification and the methodology of scientific research programmes', in Lakatos and Musgrave 1970, pp. 91–195 (reprinted in Lakatos 1978*a*, pp. 8–101).

(1976) *Proofs and Refutations* (revised version of *British Journal for the Philosophy of Science* XIV (1963–4), 1–25, 120–39, 221–45, 296–342) edd. J. Worrall and E. G. Zapar (Cambridge).

(1978*a*) *The Methodology of Scientific Research Programmes, Philosophical Papers*, Vol. I, edd. J. Worrall and G. Currie (Cambridge).

(1978*b*) *Mathematics, Science and Epistemology, Philosophical Papers*, Vol. II, edd. J. Worrall and G. Currie (Cambridge).

Lakatos, I. and Musgrave, A. (edd.) (1970) *Criticism and the Growth of Knowledge* (Cambridge).

Lanata, G. (1967) *Medicina Magica e Religione Popolare in Grecia* (Roma).

Landels, J. G. (1978) *Engineering in the Ancient World* (London).

Lanza, D. and Vegetti, M. (1975) 'L'ideologia della città', *Quaderni di Storia* II, 1–37.

Lasserre, F. (1964) *The Birth of Mathematics in the Age of Plato* (trans. H. Mortimer) (London).

(1966) *Die Fragmente des Eudoxos von Knidos* (Texte und Kommentare 4, Berlin).

Le Blond, J. M. (1938) *Eulogos et l'argument de convenance chez Aristote* (Paris).

(1939) *Logique et méthode chez Aristote* (Paris).

(1945) *Aristote, Philosophe de la vie* (Paris).

Lee, H. D. P. (1935) 'Geometrical method and Aristotle's account of first principles', *Classical Quarterly* XXIX, 113–24.

(1936) *Zeno of Elea* (Cambridge).

(1948) 'Place-names and the date of Aristotle's biological works', *Classical Quarterly* XLII, 61–7.

(1962) *Aristotle, Meteorologica*, Loeb. ed. (1st ed. 1952) 2nd ed. (London and Cambridge, Mass.).

Lejeune, A. (1956) *L'Optique de Claude Ptolémée* (Louvain).

(1957) *Recherches sur la catoptrique grecque* (Mémoires de l'Académie Royale de Belgique, Classe des Lettres, Série 2, vol. 52, 2, 1954, Bruxelles, 1957).

Lesher, J. H. (1973) 'The meaning of NOYΣ in the *Posterior Analytics*', *Phronesis* XVIII, 44–68.

Lesky, E. (1951) *Die Zeugungs- und Vererbungslehren der Antike und ihr Nachwirken* (Akademie der Wissenschaften und der Literatur, Mainz, Abhandlungen der geistes- und sozialwissenschaftlichen Klasse, Jahrgang 1950, 19, Wiesbaden, 1951).

Leszl, W. (1972-3) 'Knowledge of the universal and knowledge of the particular in Aristotle', *Review of Metaphysics* xxvi, 278-313.

Lévêque, P. and Vidal-Naquet, P. (1964) *Clisthène l'Athénien* (Annales Littéraires de l'Université de Besançon) (Paris).

Lévi-Strauss, C. (1963) *Structural Anthropology* (trans. C. Jacobson and B. G. Schoepf of *Anthropologie structurale*, Paris, 1958) (New York and London).

  (1966) *The Savage Mind* (trans. of *La Pensée Sauvage*, Paris, 1962) (London).

  (1973) *Anthropologie structurale Deux* (Paris).

Lévy-Bruhl, L. (1923) *Primitive Mentality* (trans. L. A. Clare of *La Mentalité primitive*, Paris, 1922) (London).

  (1926) *How Natives Think* (trans. L. A. Clare of *Les Fonctions mentales dans les sociétés inférieures*, Paris, 1910) (London).

  (1936) *Primitives and the Supernatural* (trans. L. A. Clare of *Le Surnaturel et la nature dans la mentalité primitive*, Paris, 1931) (London).

  (1947) 'Les *Carnets* de Lucien Lévy-Bruhl', *Revue philosophique* cxxxvii, 257-81.

Lewes, G. H. (1864) *Aristotle: A chapter from the history of science* (London).

Lewis, G. (1975) *Knowledge of Illness in a Sepik Society* (London).

Lewis, J. D. (1971) 'Isegoria at Athens. When did it begin?', *Historia* xx, 129-40.

Liao, W. K. (1939) *The Complete Works of Han Fei Tzŭ*, Vol. i (London).

  (1959) *The Complete Works of Han Fei Tzŭ*, Vol. ii (London).

Lichtenthaeler, C. (1948) *La Médecine Hippocratique: I Méthode expérimentale et méthode hippocratique* (Lausanne).

  (1957) 'De l'origine sociale de certains concepts scientifiques et philosophiques grecs', in *La Médecine Hippocratique* ii-v (Neuchâtel), pp. 91-114.

Lienhardt, G. (1961) *Divinity and Experience: The Religion of the Dinka* (Oxford).

Linforth, I. M. (1941) *The Arts of Orpheus* (University of California, Berkeley and Los Angeles).

Lippmann, E. O. von (1910) 'Chemisches und Alchemisches aus Aristoteles', *Archiv für die Geschichte der Naturwissenschaften und der Technik* ii (1909-10), 233-300.

  (1919) *Entstehung und Ausbreitung der Alchemie* (Berlin).

Lipsius, J. H. (1905-15) *Das attische Recht und Rechtsverfahren*, 3 vols. (Leipzig).

Littré, E. (1839-61) *Oeuvres complètes d'Hippocrate*, 10 vols. (Paris).

Lloyd, G. E. R. (1963) 'Who is attacked in *On Ancient Medicine*?', *Phronesis* viii, 108-26.

  (1964) 'Experiment in early Greek philosophy and medicine', *Proceedings of the Cambridge Philological Society* NS x, 50-72.

  (1964/1970) 'Hot and cold, dry and wet in early Greek thought' (*Journal of Hellenic Studies* lxxxiv (1964), 92-106) in Furley and Allen 1970, pp. 255-80.

  (1966) *Polarity and Analogy* (Cambridge).

  (1967) 'Popper versus Kirk: a controversy in the interpretation of Greek science', *British Journal for the Philosophy of Science* xviii, 21-38.

  (1968) 'Plato as a Natural Scientist', *Journal of Hellenic Studies* lxxxviii, 78-92.

  (1973) 'Right and left in Greek philosophy' (*Journal of Hellenic Studies* lxxxii (1962), 56-66) in *Right and Left*, ed. R. Needham (Chicago), pp. 167-86.

  (1975a) 'Alcmaeon and the early history of dissection', *Sudhoffs Archiv* lix, 113-47.

  (1975b) 'The Hippocratic Question', *Classical Quarterly* NS xxv, 171-92.

  (1975c) 'Aspects of the interrelations of medicine, magic and philosophy in ancient Greece', *Apeiron* ix, 1-16.

  (1978a) 'The empirical basis of the physiology of the *Parva Naturalia*', in *Aristotle on Mind and the Senses*, edd. G. E. R. Lloyd and G. E. L. Owen (Cambridge), pp. 215-39.

Bibliography 289

(1978b) 'Saving the appearances', *Classical Quarterly* NS XXVIII, 202–22.
Lones, T. E. (1912) *Aristotle's Researches in Natural Science* (London).
Long, A. A. (1966) 'Thinking and sense-perception in Empedocles: Mysticism or Materialism?', *Classical Quarterly* NS XVI, 256–76.
Longrigg, J. (1963) 'Philosophy and medicine, some early interactions', *Harvard Studies in Classical Philology* LXVII, 147–75.
   (1975) 'Elementary Physics in the Lyceum and Stoa', *Isis* LXVI, 211–29.
Lonie, I. M. (1973) 'The paradoxical text "On the Heart"', *Medical History* XVII, 1–15 and 136–53.
   (1978) 'Cos versus Cnidus and the historians', *History of Science* XVI, 42–75 and 77–92.
Lorenzen, P. (1960) *Die Entstehung der exakten Wissenschaften* (Berlin, Göttingen, Heidelberg).
   (1975) 'L'établissement constructif des fondements des sciences exactes', *Bulletin de l'Association Guillaume Budé* (1975), 467–77.
Louis, P. (1945) *Les Métaphores de Platon* (Paris).
   (1955a) 'Remarques sur la classification des animaux chez Aristote', in *Autour d'Aristote* (receuil d'études... offert à M. A. Mansion) (Louvain), pp. 297–304.
   (1955b) 'Le mot ΙΣΤΟΡΙΑ chez Aristote', *Revue de Philologie* XXIX, 39–44.
   (1956) *Aristote: Les Parties des animaux* (Paris).
   (1961) *Aristote: De la Génération des animaux* (Paris).
   (1964–9) *Aristote: Histoire des animaux*, 3 vols. (Paris).
   (1967) 'Les animaux fabuleux chez Aristote', *Revue des études grecques* LXXX, 242–6.
   (1973) *Aristote: Marche des animaux, Mouvement des Animaux* (Paris).
   (1975) 'Monstres et monstruosités dans la biologie d'Aristote' in *Le Monde Grec, Hommages à Claire Préaux*, edd. J. Bingen, G. Cambier, G. Nachtergael (Bruxelles), pp. 277–84.
Lovejoy, A. O. (1909) 'The meaning of Φύσις in the Greek physiologers', *Philosophical Review* XVIII, 369–83.
Lukasiewicz, J. (1957) *Aristotle's Syllogistic* (1st ed. 1951), 2nd ed. (Oxford).
Lukes, S. (1967/1970) 'Some problems about rationality' (*Archives Européennes de Sociologie* VIII (1967), 247–64) in Wilson 1970, pp. 194–213.
   (1973) 'On the social determination of truth', in Horton and Finnegan 1973, pp. 230–48.
Luria, S. (1932–3) *Die Infinitesimaltheorie der antiken Atomisten* (Quellen und Studien zur Geschichte der Mathematik, Astronomie und Physik B, 2, 2 (1932), Berlin, 1933), pp. 106–85.
   (1963) *Anfänge griechischen Denkens* (Berlin).
Luther, W. (1935) *"Wahrheit" und "Lüge" im ältesten Griechentum* (Göttingen).
   (1958) 'Der frühgriechische Wahrheitsgedanke im Lichte der Sprache', *Gymnasium* LXV, 75–107.
Lycos, K. (1964) 'Aristotle and Plato on "Appearing"', *Mind* NS LXXIII, 496–514.
Lynch, J. P. (1972) *Aristotle's School* (University of California Press, Berkeley and Los Angeles).
Maass, E. (1898) *Commentariorum in Aratum Reliquiae* (Berlin).
MacDermot, V. (1971) *The Cult of the Seer in the Ancient Middle East* (London).
Mach, E. (1893) *The Science of Mechanics* (trans. T. J. McCormack of 2nd German edition 1888) (Chicago).
MacIntyre, A. (1967/1970) 'The idea of a social science' (*Proceedings of the Aristotelian Society Suppl.* XLI (1967)) in Wilson 1970, pp. 112–30.
McKeon, R. (1947) 'Aristotle's conception of the development and the nature of scientific method', *Journal of the History of Ideas* VIII, 3–44.

MacKinney, L. (1964) 'The concept of isonomia in Greek medicine', in *Isonomia*, edd. J. Mau and E. G. Schmidt (Berlin), pp. 79–88.

McKirahan, R. D. (1978) 'Aristotle's Subordinate Sciences', *British Journal for the History of Science* XI, 197–220.

Macran, H. S. (1902) *The Harmonics of Aristoxenus* (Oxford).

Mahoney, M. S. (1968–9) 'Another look at Greek geometrical analysis', *Archive for History of Exact Sciences* V, 318–48.

Manetti, D. (1973) 'Valore semantico e risonanze culturali della parola ΦΥΣΙΣ', *La Parola del Passato* XXVIII, 426–44.

Manitius, K. (1963) *Ptolemäus Handbuch der Astronomie*, ed. O. Neugebauer, 2 vols. (Leipzig).

Mannheim, K. (1936) *Ideology and Utopia* (trans. L. Wirth and E. Shils) (London).

Manquat, M. (1932) *Aristote naturaliste* (Paris).

Mansfeld, J. (1964) *Die Offenbarung des Parmenides und die menschliche Welt* (Assen).
  (1971) *The Pseudo-Hippocratic Tract* ΠΕΡΙ ἙΒΔΟΜΑΔΩΝ *ch. 1–11 and Greek Philosophy* (Assen).
  (1975) 'Alcmaeon: "Physikos" or Physician?', in *Kephalaion* (Studies... offered to C. J. de Vogel), edd. J. Mansfeld and L. M. de Rijk (Assen), pp. 26–38.

Mansion, A. (1946) *Introduction à la physique aristotélicienne* (1st ed. 1913), 2nd ed. (Louvain and Paris).
  (1956) 'L'objet de la science philosophique suprême d'après Aristote, Métaphysique E, 1', in *Mélanges de Philosophie grecque offerts à Mgr Diès* (Paris), pp. 151–68.

Mansion, S. (1946) *Le Jugement d'existence chez Aristote* (Louvain and Paris).
  (1969) 'L'objet des mathématiques et l'objet de la dialectique selon Platon', *Revue philosophique de Louvain* LXVII, 365–88.

Marcovich, M. (1967) *Heraclitus editio maior* (Merida, Venezuela).

Marignac, A. de (1951) *Imagination et dialectique* (Paris).

Marrou, H. I. (1956) *A History of Education in Antiquity* (trans. G. Lamb) (London).

Marsden, E. W. (1969) *Greek and Roman Artillery: Historical Development* (Oxford).
  (1971) *Greek and Roman Artillery: Technical Treatises* (Oxford).

Marshall, L. (1957) '"N!ow"', *Africa* XXVII, 232–40.

Marten, R. (1965) *Der Logos der Dialektik* (Berlin).
  (1968) 'Die Methodologie der platonischen Dialektik', *Studium Generale* XXI, 218–49.

Masson-Oursel, P. (1916) 'La Sophistique: étude de philosophie comparée', *Revue de Métaphysique et de Morale* XXIII, 343–62.
  (1917a) 'Etudes de logique comparée, I: évolution de la logique indienne', *Revue philosophique* LXXXIII, 453–69.
  (1917b) 'Etudes de logique comparée, II: évolution de la logique chinoise', *Revue philosophique* LXXXIV, 59–76.
  (1918) 'Etudes de logique comparée, III: confrontations et analyse comparative', *Revue philosophique* LXXXV, 148–66.

Masterman, M. (1970) 'The nature of a paradigm', in Lakatos and Musgrave 1970, pp. 59–89.

Mates, B. (1961) *Stoic Logic* (University of California, Berkeley and Los Angeles).

Matson, W. I. (1952–3) 'The naturalism of Anaximander', *Review of Metaphysics* VI, 387–95.
  (1954–5) Review of Cornford 1952, *Review of Metaphysics* VIII, 443–54.

Mau, J. (1954) *Zum Problem des Infinitesimalen bei den antiken Atomisten* (Berlin).

Maula, E. (1974) *Studies in Eudoxus' Homocentric Spheres* (Helsinki).
  (1975–6) 'The spider in the sphere: Eudoxus' Arachne', *Philosophia* V–VI, 225–57.
  (1977) 'Man's orientation in time and place: the discovery of the relation

between temporal and geographical measurement', in *Abstracts of Scientific Section Papers of XVth International Congress of the History of Science* (Edinburgh, 1977), p. 7.

Mauss, M. (1950/1972) *A General Theory of Magic* (trans. R. Brain, from *Sociologie et Anthropologie*, Paris, 1950, originally 'Esquisse d'une théorie générale de la magie' (with H. Hubert) in *L'Année Sociologique* VII (1902–3) 1904, 1–146) (London, 1972).

Mead, H. L. (1975) 'The methodology of Ptolemaic astronomy: an aristotelian view', *Laval Théologique et Philosophique* XXXI, 55–74.

Méautis, G. (1922) *Recherches sur le Pythagorisme* (Neuchâtel).

Meritt, B. D. (1928) *The Athenian Calendar in the Fifth Century* (Harvard University Press, Cambridge, Mass.).

(1961) *The Athenian Year* (University of California, Berkeley and Los Angeles).

Merlan, P. (1953) 'Ambiguity in Heraclitus', *Proceedings of the 11th International Congress of Philosophy* (Bruxelles, 1953), vol. XII (Louvain–Amsterdam), pp. 56–60.

(1963) *Monopsychism, mysticism, metaconsciousness* (Archives internationales d'histoire des idées 2, The Hague).

(1968) *From Platonism to NeoPlatonism* (1st ed. 1953), 3rd ed. (The Hague).

Meuli, K. (1935) 'Scythica', *Hermes* LXX, 121–76.

Meyer, J. B. (1855) *Aristoteles Thierkunde: ein Beitrag zur Geschichte der Zoologie, Physiologie und alten Philosophie* (Berlin).

Meyerson, I. (1954) 'Thèmes nouveaux de psychologie objective: l'histoire, la construction, la structure', *Journal de Psychologie normale et pathologique* (47–51 Years) (1954), 3–19.

Meyer-Steineg, T. (1912) *Chirurgische Instrumente des Altertums. Ein Beitrag zur antiken Akiurgie* (Jenaer medizin-historische Beiträge 1, Jena).

Michel, P.-H. (1950) *De Pythagore à Euclide* (Paris).

Michler, M. (1962) 'Das Problem des westgriechischen Heilkunde', *Sudhoffs Archiv für Geschichte der Medizin und der Naturwissenschaften* XLVI, 137–52.

Mignucci, M. (1965) *La teoria aristotelica della scienza* (Firenze).

(1975) *L'argomentazione dimostrativa in Aristotele* (Padova).

Milhaud, G. (1900) *Les Philosophes-Géomètres de la Grèce* (Paris).

(1903) 'Aristote et les mathématiques', *Archiv für Geschichte der Philosophie* NF IX, 367–92.

(1906) *Etudes sur la pensée scientifique chez les Grecs et chez les modernes* (Paris).

Miller, H. W. (1952) '*Dynamis* and *Physis* in *On Ancient Medicine*', *Transactions and Proceedings of the American Philological Association* LXXXIII, 184–97.

(1953) 'The concept of the divine in *De Morbo Sacro*', *Transactions and Proceedings of the American Philological Association* LXXXIV, 1–15.

(1966) '*Dynamis* and the seeds', *Transactions and Proceedings of the American Philological Association* XCVII, 281–90.

Miller, J. Innes (1969) *The Spice Trade of the Roman Empire 29 B.C. to A.D. 641* (Oxford).

Minar, E. L. (1942) *Early Pythagorean Politics in Practice and Theory* (Baltimore).

Mittelstrass, J. (1962) *Die Rettung der Phänomene* (Berlin).

Moesgaard, K. P. (1976) 'The bright stars of the zodiac: a catalogue for historical use', *Centaurus* XX, 129–58.

Momigliano, A. (1930) 'Sul pensiero di Antifonte il Sofista', *Rivista di filologia e di istruzione classica*, NS, VIII, 129–40.

(1973) 'Freedom of speech in antiquity', in *Dictionary of the History of Ideas*, ed. P. P. Wiener, Vol. 2 (New York), pp. 252–62.

(1975) *Alien Wisdom: the Limits of Hellenization* (Cambridge).

Mondolfo, R. (1936) *Problemi del pensiero antico* (Bologna).
(1956) *L'Infinito nel pensiero dell'antichità classica* (1st ed. *L'Infinito nel pensiero dei Greci*, 1934), 2nd ed. (Firenze).
Moraux, P. (1968) 'La joute dialectique d'après le huitième livre des *Topiques*', in ed. Owen 1968, pp. 277–311.
Moreau, J. (1959) 'L'éloge de la biologie chez Aristote', *Revue des études anciennes* LXI, 57–64.
(1968) 'Aristote et la dialectique platonicienne', in ed. Owen 1968, pp. 80–90.
Morrison, J. S. (1941) 'The place of Protagoras in Athenian public life (460–415 B.C.)', *Classical Quarterly* XXXV, 1–16.
(1961) 'Antiphon', *Proceedings of the Cambridge Philological Society* NS VII, 49–58.
(1963) 'The *Truth* of Antiphon', *Phronesis* VIII, 35–49.
Morrow, G. R. (1960) *Plato's Cretan City* (Princeton University Press, Princeton, New Jersey).
(1970) 'Plato and the mathematicians: an interpretation of Socrates' dream in the *Theaetetus* (201 e–206 c)', *Philosophical Review* LXXIX, 309–33.
Moulinier, L. (1952) *Le Pur et l'impur dans la pensée des Grecs d'Homère à Aristote* (Etudes et commentaires 12, Paris).
Mourelatos, A. P. D. (1967) 'Aristotle's "Powers" and modern empiricism', *Ratio* IX, 97–104.
(1970) *The Route of Parmenides* (Yale University Press, New Haven and London).
Müller, C. (1855–61) *Geographi Graeci Minores*, 2 vols. (Paris).
Müller, C. W. (1965a) *Gleiches zu Gleichem. Ein Prinzip frühgriechischen Denkens* (Wiesbaden).
(1965b) 'Die Heilung "durch das Gleiche" in den hippokratischen Schriften *De morbo sacro* und *De locis in homine*', *Sudhoffs Archiv für Geschichte der Medizin und der Naturwissenschaften* XLIX, 225–49.
(1967) 'Protagoras über die Götter', *Hermes* XCV, 140–59.
Müller, F. Max (1879) *The Upanishads, Part I* (*Sacred Books of the East*, Vol. 1) (Oxford).
(1884) *The Upanishads, Part II* (*Sacred Books of the East*, Vol. XV) (Oxford).
Mueller, I. (1969) 'Euclid's *Elements* and the Axiomatic Method', *British Journal for the Philosophy of Science* XX, 289–309.
(1974) 'Greek mathematics and Greek logic', in *Ancient logic and its Modern Interpretations*, ed. J. Corcoran (Dordrecht and Boston), pp. 35–70.
Mugler, C. (1948) *Platon et la recherche mathématique de son époque* (Strasbourg–Zurich).
(1973) 'Sur quelques points de contact entre la magie et les sciences appliquées des anciens', *Revue de Philologie* XLVII, 31–7.
Mure, G. R. G. (1975) 'Cause and because in Aristotle', *Philosophy* L, 356–7.
Needham, J. (1954–) *Science and civilisation in China* (in progress) (Cambridge, 1954–) (Vol. I, *Introductory Orientations*, 1954, Vol. II, *History of Scientific Thought*, 1956, Vol. III, *Mathematics and the Sciences of the Heavens and the Earth*, 1959, Vol. V, 2, *Chemistry and Chemical Technology*, 1974).
(1959) *A History of Embryology* (1st ed. 1934), 2nd ed. (Cambridge).
Needham, J., Ling, W., and Price, D. J. de S. (1960) *Heavenly Clockwork* (Cambridge).
Needham, R. (1973) *Right and Left, Essays on Dual Symbolic Classification* (Chicago).
Nelson, A. (1909) *Die hippokratische Schrift* περὶ φυσῶν, *Text und Studien* (Uppsala).
Nestle, W. (1903) 'Kritias', *Neue Jahrbücher für das klassische Altertum, Geschichte und deutsche Literatur* XI, 81–107 and 178–99 (reprinted in Nestle 1948, pp. 253–320).
(1922) 'Die Schrift des Gorgias "über die Natur oder über das Nichtseiende"', *Hermes* LVII, 551–62 (reprinted in Nestle 1948, pp. 240–52).

(1938) 'Hippocratica', *Hermes* LXXIII, 1–38 (reprinted in Nestle 1948, pp. 517–66).

(1942) *Vom Mythos zum Logos*, 2nd ed. (Stuttgart).

(1948) *Griechische Studien* (Stuttgart).

Neugebauer, O. (1928) 'Zur Geschichte des pythagoräischen Lehrsatzes', *Nachrichten von der Gesellschaft der Wissenschaften zu Göttingen*, math.-phys. Kl. (Berlin), pp. 45–8.

(1942) 'Egyptian planetary texts', *Transactions of the American Philosophical Society* NS XXXII, 2, 209–50.

(1945) 'The history of ancient astronomy: problems and methods', *Journal of Near Eastern Studies* IV, 1–38.

(1947) 'The water-clock in Babylonian astronomy', *Isis* XXXVII, 37–43.

(1949) 'The early history of the astrolabe', *Isis* XL, 240–56.

(1953) 'On the "hippopede" of Eudoxus', *Scripta Mathematica* XIX, 225–9.

(1955) *Astronomical Cuneiform Texts*, 3 vols. (Princeton).

(1957) *The Exact Sciences in Antiquity* (1st ed. 1952), 2nd ed. (Providence, R.I.).

(1959) 'The equivalence of eccentric and epicyclic motion according to Apollonius', *Scripta Mathematica* XXIV, 5–21.

(1972) 'On some aspects of early Greek astronomy', *Proceedings of the American Philosophical Society* CXVI, 243–51.

(1975) *A History of Ancient Mathematical Astronomy*, 3 vols. (Berlin and New York).

Neugebauer, O. and van Hoesen, H. B. (1959) *Greek Horoscopes* (American Philosophical Society, Memoirs 48, Philadelphia).

Neugebauer, O. and Sachs, A. (1945) *Mathematical Cuneiform Texts* (American Oriental Series 29, New Haven).

Newton, R. R. (1973) 'The authenticity of Ptolemy's parallax data, Part I', *Quarterly Journal of the Royal Astronomical Society* XIV, 367–88.

(1974a) 'The authenticity of Ptolemy's parallax data, Part II', *Quarterly Journal of the Royal Astronomical Society* XV, 7–27.

(1974b) 'The authenticity of Ptolemy's eclipse and star data', *Quarterly Journal of the Royal Astronomical Society* XV, 107–21.

(1977) *The Crime of Claudius Ptolemy* (Johns Hopkins, Baltimore).

Nilsson, M. P. (1907) *Die Kausalsätze im griechischen bis Aristoteles* (Würzburg).

(1940) *Greek Popular Religion* (New York).

(1955–61) *Geschichte der griechischen Religion*, 2nd ed., 2 vols. (München).

Nittis, S. (1940) 'The authorship and probable date of the Hippocratic Oath', *Bulletin of the History of Medicine* VIII, 1012–21.

Nock, A. D. (1972) *Essays on Religion and the Ancient World*, 2 vols. (Oxford).

Nock, A. D. and Festugière, A. J. (1945–54) *Corpus Hermeticum*, 4 vols. (Paris).

Nörenberg, H. W. (1968) *Das Göttliche und die Natur in der Schrift über die heilige Krankheit* (Bonn).

Nylander, C. (1970) *Ionians in Pasargadae* (Acta Universitatis Upsaliensis–Boreas–1, Uppsala).

O'Brien, D. (1969) *Empedocles' Cosmic Cycle* (Cambridge).

(1977) 'Heavy and light in Democritus and Aristotle: two conceptions of change and identity', *Journal of Hellenic Studies* XCVII, 64–74.

Ogle, W. (1882) *Aristotle on the Parts of Animals* (London).

(1897) *Aristotle on Youth and Old Age, Life and Death and Respiration* (London).

Ohlert, K. (1912) *Rätsel und Rätselspiele der alten Griechen* (1st ed. 1886), 2nd ed. (Berlin).

Onians, R. B. (1951) *The Origins of European Thought* (Cambridge).

Oppenheim, A. Leo (1962) 'Mesopotamian medicine', *Bulletin of the History of Medicine* XXXVI, 97–108.

(1964) *Ancient Mesopotamia. Portrait of a Dead Civilisation* (Chicago).

Oppenheimer, J. M. (1971) 'Aristotle as a biologist', *Scientia* CVI, 649–58.

Osler, W. (1947) *The Principles and Practice of Medicine*, 16th ed. (ed. H. A. Christian) (New York and London).

Ostwald, M. (1969) *Nomos and the Beginnings of the Athenian Democracy* (Oxford).

Owen, G. E. L. (1953/1965) 'The place of the *Timaeus* in Plato's dialogues' (*Classical Quarterly* NS III (1953), 79–95) in Allen 1965, pp. 313–38.

(1957/1965) 'A proof in the περὶ ἰδεῶν' (*Journal of Hellenic Studies* LXXVII (1957), 103–11) in Allen 1965, pp. 295–312.

(1957–8/1975) 'Zeno and the Mathematicians' (*Proceedings of the Aristotelian Society* NS LVIII (1957–8), 199–222) in Allen and Furley 1975, pp. 143–65.

(1960) 'Logic and metaphysics in some earlier works of Aristotle', in *Aristotle and Plato in the Mid-Fourth Century*, edd. I. Düring and G. E. L. Owen (Göteborg), pp. 163–90.

(1960/1975) 'Eleatic questions' (*Classical Quarterly* NS X (1960), 84–102) in Allen and Furley 1975, pp. 48–81.

(1961/1975) 'Tithenai ta phainomena' (*Aristote et les problèmes de méthode*, ed. S. Mansion (Louvain and Paris, 1961), pp. 83–103) in Barnes, Schofield, Sorabji 1975, pp. 113–26 (also in *Aristotle*, ed. J. M. E. Moravcsik, London, 1968, pp. 167–90).

(1965) 'Aristotle on the snares of ontology', in Bambrough 1965, pp. 69–95.

(1965/1975) 'The Platonism of Aristotle' (*Proceedings of the British Academy* LI (1965), 125–50) in Barnes, Schofield, Sorabji 1975, pp. 14–34.

(1968) 'Dialectic and eristic in the treatment of the Forms', in ed. Owen 1968, pp. 103–25.

(1970) 'Aristotle: Method, Physics, and Cosmology', in *Dictionary of Scientific Biography*, ed. C. C. Gillispie (New York), Vol. I, pp. 250–8.

(1976) 'Aristotle on time', in *Motion and Time, Space and Matter*, edd. P. K. Machamer and R. G. Turnbull (Ohio State University Press), pp. 3–27.

Owen, G. E. L. (ed.) (1968) *Aristotle on Dialectic* (Oxford).

Page, D. L. (1955) *Sappho and Alcaeus* (Oxford).

(1959) Review of Oxyrhynchus Papyri XXIV, *Classical Review* NS IX, 15–23.

Palter, R. (1970–1) 'An approach to the history of early astronomy', *Studies in History and Philosophy of Science* I, 93–133.

Pannekoek, A. (1955) 'Ptolemy's precession', in *Vistas in Astronomy*, ed. A. Beer, Vol. I (London and New York), pp. 60–6.

Parker, R. A. (1959) *A Vienna Demotic Papyrus on Eclipse- and Lunar-Omina* (Brown Egyptological Studies 2, Providence, R.I.).

Parker, R. C. T. (1977) 'Miasma: Pollution and purification in early Greek Religion' (unpubl. D.Phil. diss. Oxford).

Patzig, G. (1968) *Aristotle's Theory of the Syllogism* (trans. J. Barnes of *Die aristotelische Syllogistik*, 2nd ed. 1963) (Dordrecht).

Pearson, L. (1939) *Early Ionian Historians* (Oxford).

Peck, A. L. (1937) *Aristotle, Parts of Animals*, Loeb ed. (London and Cambridge, Mass.).

(1943) *Aristotle, Generation of Animals*, Loeb ed. (London and Cambridge, Mass.).

Pedersen, O. (1974) *A Survey of the Almagest* (Odense University Press).

Pedersen, O. and Pihl, M. (1974) *Early Physics and Astronomy* (London and New York).

Peet, T. E. (1923) *The Rhind Mathematical Papyrus* (London).

Pembroke, S. (1967) 'Women in charge: the function of alternatives in early Greek tradition and the ancient idea of matriarchy', *Journal of the Warburg and Courtauld Institutes* XXX, 1–35.

Penn, J. M. (1972) *Linguistic Relativity versus Innate Ideas* (Paris and The Hague).

Penwill, J. (1974) 'Alkman's cosmogony', *Apeiron* VIII, 13–39.

Perelman, C. (1970) *Le Champ de l'argumentation* (Bruxelles).

Perelman, C. and Olbrechts-Tyteca, L. (1969) *The New Rhetoric. A treatise on argumentation* (trans. J. Wilkinson and P. Weaver of *La nouvelle rhétorique*, Paris, 1958) (University of Notre Dame Press, Notre Dame and London).

Peters, C. H. F. (1877) 'Ueber die Fehler des Ptolemäischen Sternverzeichnisses', *Vierteljahrsschriften der astronomischen Gesellschaft*, 12 Jahrgang (1877), 296–9.

Peters, C. H. F. and Knobel, E. B. (1915) *Ptolemy's Catalogue of Stars: A Revision of the Almagest* (Carnegie Institution of Washington).

Petersen, V. M. (1969) 'The three lunar models of Ptolemy', *Centaurus* XIV, 142–71.

Petersen, V. M. and Schmidt, O. (1967–8) 'The determination of the longitude of the apogee of the orbit of the sun according to Hipparchus and Ptolemy', *Centaurus* XII, 73–96.

Pfeiffer, R. (1968) *History of Classical Scholarship from the beginnings to the end of the Hellenistic age* (Oxford).

(1976) *History of Classical Scholarship from 1300 to 1850* (Oxford).

Pfister, F. (1935) 'Katharsis', *Pauly-Wissowa Real-Encyclopädie der classischen Altertumswissenschaft*, Suppl. Bd. VI, cols. 146–62.

Philip, J. A. (1966) *Pythagoras and early Pythagoreanism* (Phoenix Suppl. 7, Toronto).

Plamböck, G. (1964) *Dynamis im Corpus Hippocraticum* (Akademie der Wissenscaften und der Literatur, Mainz, Abhandlungen der geistes- und sozialwissenschaftlichen Klasse, Jahrgang 1964, 2, Wiesbaden).

Platt, A. (1912) *Aristotle, De Generatione Animalium*, Oxford trans. in *The Works of Aristotle translated into English*, ed. W. D. Ross, Vol. V (Oxford).

(1921) 'Aristotle on the heart', in *Studies in the History and Method of Science*, ed. C. Singer, Vol. II (Oxford), pp. 521–32.

Pleket, H. W. (1973) 'Technology in the Greco-Roman world: a general report', *Talanta* (Proceedings of the Dutch Archaeological and Historical Society) V, 6–47.

Plochmann, G. K. (1953) 'Nature and the living thing in Aristotle's biology', *Journal of the History of Ideas* XIV, 167–90.

Pohle, W. (1971) 'The mathematical foundations of Plato's atomic physics', *Isis* LXII, 36–46.

Pohlenz, M. (1937) 'Hippokratesstudien', *Nachrichten von der Gesellschaft der Wissenschaften zu Göttingen*, phil.-hist. Kl., NF II, 4 (Göttingen, 1937), pp. 67–101 (reprinted in *Kleine Schriften*, Vol. II, ed. H. Dörrie (Hildesheim, 1965), pp. 175–209).

(1938) *Hippokrates und die Begründung der wissenschaftlichen Medizin* (Berlin).

(1939) 'Hippokrates', *Die Antike* XV, 1–18.

(1953) 'Nomos und Physis', *Hermes* LXXXI, 418–38 (reprinted in *Kleine Schriften*, Vol. II, ed. H. Dörrie (Hildesheim, 1965), pp. 341-60).

Pokora, T. (1962) 'The necessity of a more thorough study of philosopher Wang Ch'ung and of his predecessors', *Archiv Orientalni* XXX, 231–57.

(1975) *Hsin-Lun (New Treatise) and Other Writings by Huan T'an (43 B.C.– 28 A.D.)* (Michigan Papers in Chinese Studies 20, Ann Arbor).

Popper, K. R. (1958–9/1963) 'Back to the Presocratics' (*Proceedings of the Aristotelian Society* NS LIX (1958–9), 1–24), in *Conjectures and Refutations* (London, 1963), pp. 136–65 (also reprinted in Furley and Allen 1970, pp. 130–53).

(1959) *The Logic of Scientific Discovery* (trans. J. and L. Freed of *Logik der Forschung*, 1935) (London).

(1962) *The Open Society and its Enemies*, 2 vols. (1st ed. 1945), 4th ed. (London).

(1963) *Conjectures and Refutations* (London).

(1970) 'Normal science and its dangers', in Lakatos and Musgrave 1970, pp. 51–8.

(1972) *Objective Knowledge: an evolutionary approach* (Oxford).

Poschenrieder, F. (1887) *Die naturwissenschaftlichen Schriften des Aristoteles in ihrem Verhältnis zu den Büchern der hippokratischen Sammlung* (Bamberg).

Prauss, G. (1966) *Platon und der logische Eleatismus* (Berlin).

Préaux, C. (1973) *La Lune dans la pensée grecque* (Académie Royale de Belgique, Mémoires de la Classe des Lettres, 2nd ser. 61, 4, Bruxelles).

Precope, J. (1954) *Medicine, Magic and Mythology* (London).

Preisendanz, K. (1973–4) *Papyri Graecae Magicae*, 2nd ed., ed. A. Henrichs (Stuttgart).

Preus, A. (1975) *Science and Philosophy in Aristotle's Biological Works* (Hildesheim and New York).

Price, D. J. de S. (1957) 'Precision instruments: to 1500' in *A History of Technology*, Vol. III, edd. C. Singer and others (Oxford), pp. 582–619.

(1964–5) 'The Babylonian "Pythagorean Triangle" Tablet' *Centaurus* X, 1–13.

(1974) 'Gears from the Greeks. The Antikythera Mechanism – a calendar computer from *ca*. 80 B.C.', *Transactions of the American Philosophical Society* NS LXIV, 7.

Pritchard, J. B. (1969) *Ancient Near Eastern Texts* (1st ed. 1955) 3rd ed. (Princeton).

Pritchett, W. K. (1957) 'Calendars of Athens Again', *Bulletin de correspondance Hellénique* LXXXI, 269–301.

(1964) 'Thucydides V 20', *Historia* XIII, 21–36.

(1970) *The Choiseul Marble* (University of California Publications: Classical Studies 5, Berkeley and Los Angeles).

Pritchett, W. K. and Neugebauer, O. (1947) *The Calendars of Athens* (Harvard University Press, Cambridge, Mass.).

Pritchett, W. K. and van der Waerden, B. L. (1961) 'Thucydidean time-reckoning and Euctemon's seasonal calendar', *Bulletin de correspondance Hellénique* LXXXV, 17–52.

Putnam, H. (1962/1975) 'What theories are not' (in *Logic, Methodology and Philosophy of Science*, edd. E. Nagel and others (Stanford University Press, 1962), pp. 240–51), in *Mathematics, Matter and Method, Philosophical Papers*, Vol. I (Cambridge, 1975), pp. 215–27.

Quimby, R. W. (1974) 'The Growth of Plato's perception of Rhetoric', *Philosophy and Rhetoric* VII, 71–9.

Quine, W. Van O. (1953) *From a Logical Point of View* (Harvard University Press, Cambridge, Mass.).

(1960) *Word and Object* (M.I.T. Press, Cambridge, Mass.).

Radermacher, L. (1951) *Artium Scriptores* (Österreichische Akademie der Wissenschaften, phil.-hist. Kl., Sitzungsberichte 227, 3, Abhandlung, Wien).

Radin, M. (1927) 'Freedom of speech in ancient Athens', *American Journal of Philology* XLVIII, 215–30.

Ramin, J. (1976) *Le Périple d'Hannon* (BAR Supplementary Series 3, Oxford).

Ramnoux, C. (1970) *Etudes Présocratiques* (Paris).

Randall, J. H. (1960) *Aristotle* (New York).

Ranulf, S. (1924) *Der eleatische Satz vom Widerspruch* (Copenhagen).

Raphael, S. (1974) 'Rhetoric, dialectic and syllogistic argument: Aristotle's position in *Rhetoric* I–II', *Phronesis* XIX, 153–67.

Raven, J. E. (1948) *Pythagoreans and Eleatics* (Cambridge).

Rawlings, H. R. (1975) *A semantic study of PROPHASIS to 400 B.C.* (Hermes Einzelschriften 33, Wiesbaden).

Rawlinson, G. (1880) *The History of Herodotus*, 4th ed., 4 vols. (London).

Reale, G. (1970) *Melisso, Testimonianze e frammenti* (Firenze).

Regenbogen, O. (1931) *Eine Forschungsmethode antiker Naturwissenschaft* (Quellen und Studien zur Geschichte der Mathematik, Astronomie und Physik, B, 1, 2 (1930) Berlin, 1931), pp. 131–82 (reprinted in *Kleine Schriften* (München, 1961), pp. 141–94).

Régis, L.-M. (1935) *L'Opinion selon Aristote* (Paris and Ottawa).

Rehm, A. (1899) 'Zu Hipparch und Eratosthenes', *Hermes* xxxiv, 251–79.

  (1938) 'Zur Rolle der Technik in der griechisch-römischen Antike', *Archiv für Kulturgeschichte* xxviii, 135–62.

  (1941) *Parapegmastudien* (Abhandlungen der bayerischen Akademie der Wissenschaften, phil.-hist. Abteilung, NF 19, München).

Rehm, A. and Vogel, K. (1933) *Exakte Wissenschaften*, 4th ed. (Leipzig and Berlin).

Reiche, H. A. T. (1960) *Empedocles' Mixture, Eudoxan Astronomy and Aristotle's Connato Pneuma* (Amsterdam).

Reidemeister, K. (1949) *Das exakte Denken der Griechen* (Hamburg).

Reinhardt, K. (1916) *Parmenides und die Geschichte der griechischen Philosophie* (Bonn).

  (1926) *Kosmos und Sympathie* (München).

Rey, A. (1930–48) *La Science dans l'antiquité*, 5 vols. (Paris).

Reymond, A. (1927) *History of the Sciences in Greco-Roman Antiquity* (trans. R. G. de Bray of 1st ed. of *Histoire des sciences exactes et naturelles dans l'antiquité gréco-romaine*, Paris, 1924) (London).

Reynolds, L. D. and Wilson, N. G. (1968/1974) *Scribes and Scholars* (1st ed. 1968), 2nd ed. (Oxford, 1974).

Rhodes, P. J. (1972) *The Athenian Boule* (Oxford).

Richard, G. (1935) 'L'impureté contagieuse et la magie dans la tragédie grecque', *Revue des études anciennes* xxxvii, 301–21.

Richardson, N. J. (1975) 'Homeric professors in the age of the Sophists', *Proceedings of the Cambridge Philological Society* NS xxi, 65–81.

Richter, G. M. A. (1946) 'Greeks in Persia', *American Journal of Archaeology* L, 15–30.

Riondato, G. (1954) 'ἱστορία ed ἐμπειρία nel pensiero aristotelico', *Giornale di Metafisica* ix, 303–35.

Robin, L. (1928) *Greek Thought* (trans. M. R. Dobie) (London).

Robinson, J. M. (1971) 'Anaximander and the problem of the earth's immobility', in Anton and Kustas 1971, pp. 111–18.

  (1973) 'On Gorgias', in *Exegesis and Argument*, edd. E. N. Lee, A. P. D. Mourelatos, R. M. Rorty (Assen), 49–60.

Robinson, R. (1936/1969) 'Analysis in Greek Geometry' (*Mind* NS xlv (1936), 464–73) in *Essays in Greek Philosophy* (Oxford, 1969), pp. 1–15.

  (1942/1969) 'Plato's consciousness of fallacy' (*Mind* NS li (1942), 97–114) in *Essays in Greek Philosophy* (Oxford, 1969), pp. 16–38.

  (1953) *Plato's Earlier Dialectic* (1st ed. 1941), 2nd ed. (Oxford).

Roebuck, C. (1959) *Ionian Trade and Colonization* (New York).

Rohde, E. (1925) *Psyche* (trans. W. B. Hillis) (London).

Rome, A. (1937) 'Les observations d'équinoxes et de solstices dans le chapitre 1 du livre 3 du Commentaire sur l'Almageste par Théon d'Alexandrie, I', *Annales de la société scientifique de Bruxelles*, Sér. 1, Sciences mathématiques et physiques LVII, 213–36.

  (1938) 'Les observations d'équinoxes et de solstices dans le chapitre 1 du livre 3 du Commentaire sur l'Almageste par Théon d'Alexandrie, II', *Annales de la société scientifique de Bruxelles*, Sér. 1, Sciences mathématiques et physiques LVIII, 6–26.

Romilly, J. de (1956) *Histoire et raison chez Thucydide* (Paris).

(1971) *La Loi dans la pensée grecque des origines à Aristote* (Paris).

(1975) *Magic and Rhetoric in Ancient Greece* (Harvard University Press, Cambridge, Mass.).

Roscher, W. H. (1913) *Die hippokratische Schrift von der Siebenzahl in ihrer vierfachen Überlieferung* (Studien zur Geschichte und Kultur des Altertums 6, Paderborn).

Rose, V. (1886) *Aristotelis qui ferebantur librorum fragmenta* (Leipzig).

Rosenmeyer, T. G. (1960/1971) 'Plato's hypothesis and the upward path' (*American Journal of Philology* LXXXI (1960), 393–407) in Anton and Kustas 1971, pp. 354–66.

Ross, W. D. (1924/1953) *Aristotle, Metaphysics*, 2 vols. (1st ed. 1924) revised ed. (Oxford).

(1936) *Aristotle, Physics* (Oxford).

(1953) *Plato's Theory of Ideas* (1st ed. 1951) 2nd ed. (Oxford).

Rostagni, A. (1924) *Il verbo di Pitagora* (Torino).

Rothschuh, K. E. (1962) 'Idee und Methode in ihrer Bedeutung für die geschichtliche Entwicklung der Physiologie', *Sudhoffs Archiv für Geschichte der Medizin und der Naturwissenschaften* XLVI, 97–119.

Ruben, W. (1929) 'Über die Debatten in den alten Upaniṣad's', *Zeitschrift der deutschen morgenländischen Gesellschaft* NF VIII, 238–55.

(1954) *Geschichte der indischen Philosophie* (Berlin).

(1971) *Die Entwicklung der Philosophie im alten Indien* (Berlin).

(1973) 'Der Charakter der Weltanschauung im alten China, in Indien und in Griechenland', *Klio* LV, 5–41.

Rudio, F. (1907) *Die Bericht des Simplicius über die Quadraturen des Antiphon und des Hippokrates* (Leipzig).

Rüsche, F. (1930) *Blut, Leben und Seele* (Paderborn).

Ryle, G. (1965) 'Dialectic in the Academy', in Bambrough 1965, pp. 39–68.

(1966) *Plato's Progress* (Cambridge).

Sachs, A. (1948) 'A classification of the Babylonian astronomical tablets of the Seleucid period', *Journal of Cuneiform Studies* II, 271–90.

(1952) 'Babylonian horoscopes', *Journal of Cuneiform Studies* VI, 49–75.

(1974) 'Babylonian observational astronomy', in *The Place of Astronomy in the Ancient World*, edd. D. G. Kendal and others (Oxford), pp. 43–50.

Sachs, E. (1917) *Die fünf platonischen Körper* (Philol. Untersuch. 24, Berlin).

Salmon, J. (1977) 'Political Hoplites?', *Journal of Hellenic Studies* XCVII, 84–101.

Salmon, W. C. (ed.) (1970) *Zeno's Paradoxes* (Indianapolis and New York).

Salomon, M. (1911) 'Die Begriff das Naturrechts bei den Sophisten', *Zeitschrift der Savigny-Stiftung für Rechtsgeschichte* (Romanistiche Abteilung) XXXII, 129–67.

Sambursky, S. (1956) *The Physical World of the Greeks* (trans. M. Dagut) (London).

(1958) 'Conceptual developments in Greek atomism', *Archives Internationales d'Histoire des Sciences* XI, 251–61.

(1959) *Physics of the Stoics* (London).

(1961) 'Atomism versus continuum theory in ancient Greece', *Scientia* XCVI, 376–81.

(1962) *The Physical World of Late Antiquity* (London).

(1963) 'Conceptual developments and modes of explanation in later Greek scientific thought', in *Scientific Change*, ed. A. C. Crombie (London), pp. 61–78.

(1966) 'Phänomen und Theorie. Das physikalische Denken der Antike im Licht der modernen Physik', *Eranos-Jahrbuch* XXXV, 303–48.

Santillana, G. de (1953) *Galileo Galilei: Dialogue on the Great World Systems in the Salusbury translation* (Chicago).

Sapir, E. (1949) *Selected Writings of Edward Sapir in Language, Culture, and Personality* (University of California Press, Berkeley and Los Angeles).

Sarton, G. (1954) *Ancient Science and Modern Civilization* (London).

Saunders, J. B. de C. M. (1963) *The Transitions from ancient Egyptian to Greek Medicine* (University of Kansas Press, Lawrence).

Sayre, K. M. (1969) *Plato's Analytic Method* (Chicago).

Scarpat, G. (1964) *Parrhesia: storia del termine e delle sue traduzioni in latino* (Brescia).

Schaerer, R. (1938) *La Question platonicienne* (Mémoires de l'Université de Neuchâtel 10, Neuchâtel).

(1958) *L'Homme antique et la structure du monde interieur d'Homère à Socrate* (Paris).

Scheffler, I. (1956–7a) 'Prospects of a modest empiricism', *Review of Metaphysics* x, 385–400 and 602–25.

(1956–7b) 'Explanation, prediction, and abstraction', *British Journal for the Philosophy of Science* vii, 293–309.

(1967) *Science and Subjectivity* (Indianapolis and New York).

Scheil, V. (1929) *Inscriptions des Achéménides à Suse* (Mémoires de la Mission Archéologique de Perse 21, Paris).

Schiaparelli, G. V. (1873) *I precursori di Copernico nell'antichità* (Memorie del reale istituto Lombardo di scienze e lettere, Classe di scienze matematiche e naturali, Vol. xii (Ser. 3, Vol. iii) (Milano), pp. 381–432) (reprinted in *Scritti sulla storia della astronomia antica*, Part i, Vol. i, Bologna, 1925, pp. 361–458).

(1877) *Le sfere omocentriche di Eudosso, di Callippo e di Aristotele* (Memorie del reale istituto Lombardo di scienze e lettere, Classe di scienze matematiche e naturali, Vol. xiii (Ser. 3, Vol. iv) (Milano), pp. 117–79) (reprinted in *Scritti sulla storia della astronomia antica*, Part i, Vol. ii, Bologna, 1926, pp. 1–112).

Schjellerup, H. C. F. C. (1881) 'Recherches sur l'Astronomie des Anciens, I', *Urania* ii, 25–39.

Scholz, H. (1928) 'Warum haben die Griechen die Irrationalzahlen nicht aufgebaut?', *Kant-Studien* xxxiii, 35–72.

(1930/1975) 'The ancient axiomatic theory' (originally 'Die Axiomatik der Alten', *Blätter für deutsche Philosophie* iv (1930), 259–78) in Barnes, Schofield, Sorabji 1975, pp. 50–64.

Schramm, M. (1962) *Die Bedeutung der Bewegungslehre des Aristoteles für seine beiden Lösungen der zenonischen Paradoxie* (Philosophische Abhandlungen 19, Frankfurt).

Schuhl, P. M. (1949) *Essai sur la formation de la pensée grecque* (1st ed. 1934), 2nd ed. (Paris).

Schultz, W. (1914) 'Rätsel', *Pauly-Wissowa Real-Encyclopädie der classischen Altertumswissenschaft*, 2nd ser., 1 Halbband, 1, 1, cols. 62–125.

Schumacher, J. (1963) *Antike Medizin*, 2nd ed. (Berlin).

Scolnicov, S. (1975) 'Hypothetical method and rationality in Plato', *Kant-Studien* lxvi, 157–62.

Scot, R. (1584/1964) *The Discoverie of Witchcraft* (1st ed. London 1584), introduced by H. R. Williamson (Arundel, 1964).

Seeck, G. A. (1964) *Über die Elemente in der Kosmologie des Aristoteles* (Zetemata 34, München).

(1965) '*Nachträge*' im achten Buch der Physik des Aristoteles (Akademie der Wissenschaften und der Literatur, Mainz, Abhandlungen der geistes- und sozialwissenschaftlichen Kl., Jahrgang 1965, 3, Wiesbaden).

Segal, C. P. (1962) 'Gorgias and the psychology of the Logos', *Harvard Studies in Classical Philology* lxvi, 99–155.

Seidl, H. (1971) *Der Begriff des Intellekts* (νοῦς) *bei Aristoteles* (Meisenheim).

Senn, G. (1929) 'Über Herkunft und Stil der Beschreibungen von Experimenten im Corpus Hippocraticum', *Sudhoffs Archiv für Geschichte der Medizin* XXII, 217–89.

(1933) *Die Entwicklung der biologischen Forschungsmethode in der Antike und ihre grundsätzliche Förderung durch Theophrast von Eresos* (Veröffentlichungen der schweizerischen Gesellschaft für Geschichte der Medizin und der Naturwissenschaften 8, Aarau and Leipzig).

Sesonke, A. (1968) 'To make the weaker argument defeat the stronger', *Journal of the History of Philosophy* VI, 217–31.

Shapere, D. (1964) 'The structure of scientific revolutions', *Philosophical Review* LXXIII, 383–94.

(1966) 'Meaning and scientific change', in *Mind and Cosmos*, ed. R. G. Colodny (University of Pittsburgh Press), pp. 41–85.

Shaw, J. R. (1972) 'Models for cardiac structure and function in Aristotle', *Journal of the History of Biology* V, 355–88.

Sheppard, H. J. (1970) 'Alchemy: origin or origins?', *Ambix* XVII, 69–84.

Shorey, P. (1927) 'Platonism and the History of Science', *Proceedings of the American Philosophical Society* LXVI, 159–82.

Sichirollo, L. (1966) Διαλέγεσθαι-*Dialektik* (Hildesheim).

Sicking, C. M. J. (1964) 'Gorgias und die Philosophen', *Mnemosyne*, 4th ser. XVII, 225–47.

Sigerist, H. E. (1951–61) *A History of Medicine*, 2 vols. (Oxford).

Singer, C. (1928) *From Magic to Science* (London).

Skorupski, J. (1976) *Symbol and Theory* (Cambridge).

Snell, B. (1924) *Die Ausdrücke für den Begriff des Wissens in der vorplatonischen Philosophie* (Philol. Untersuch. 29, Berlin).

(1953) *The Discovery of the Mind* (trans. T. G. Rosenmeyer of *Die Entdeckung des Geistes*, 2nd ed. Hamburg, 1948) (Oxford).

Snodgrass, A. (1965) 'The hoplite reform and history', *Journal of Hellenic Studies* LXXXV, 110–22.

(1971) *The Dark Age of Greece* (Edinburgh).

Solmsen, F. (1929) *Die Entwicklung der aristotelischen Logik und Rhetorik* (Neue Philol. Untersuch. 4, Berlin).

(1931) *Platos Einfluss auf die Bildung der mathematischen Methode* (Quellen und Studien zur Geschichte der Mathematik, Astronomie und Physik B, I, 1 (1929), Berlin; 1931, pp. 93–107).

(1940) 'Plato and the unity of science', *Philosophical Review* XLIX, 566–71.

(1955) 'Antecedents of Aristotle's psychology and scale of beings', *American Journal of Philology* LXXVI, 148–64.

(1957) 'The vital heat, the inborn pneuma and the aether', *Journal of Hellenic Studies* LXXVII, 119–23.

(1960) *Aristotle's System of the Physical World* (Cornell University Press, Ithaca, New York).

(1961) 'Greek Philosophy and the Discovery of the Nerves', *Museum Helveticum* XVIII, 150–67 and 169–97.

(1963) 'Nature as craftsman in Greek thought', *Journal of the History of Ideas* XXIV, 473–96.

(1968) 'Dialectic without the Forms', in ed. Owen 1968, pp. 49–68.

(1975) *Intellectual Experiments of the Greek Enlightenment* (Princeton University Press, Princeton, New Jersey).

Sorabji, R. (1969) 'Aristotle and Oxford philosophy', *American Philosophical Quarterly* VI, 129–35.

(1972) 'Aristotle, Mathematics, and Colour', *Classical Quarterly* NS xxii, 293–308.

Souilhé, J. (1919) *Etude sur le terme* δύναμις *dans les dialogues de Platon* (Paris).

Souques, A. (1935) 'Connaissances neurologiques d'Hérophile et d'Erasistrate', *Revue neurologique* LXIII, 145–76.

Spengel, L. (1828) ΣΥΝΑΓΩΓΗ ΤΕΧΝΩΝ *sive artium scriptores* (Stuttgart).

Sprague, R. K. (1962) *Plato's Use of Fallacy* (London).

Sprute, J. (1962) *Der Begriff der DOXA in der platonischen Philosophie* (Hypomnemata 2, Göttingen).

Staden, H. von (1975) 'Experiment and experience in Hellenistic medicine', *Bulletin of the Institute of Classical Studies* xxii, 178–99.

Stahlman, W. D. (1953) 'An astronomical note on the two systems', in G. de Santillana 1953, pp. 475–96.

Stannard, J. (1965) 'The Presocratic origin of explanatory method', *Philosophical Quarterly* xv, 193–206.

Starr, C. (1968) 'Ideas of truth in early Greece', *La Parola del Passato* xxiii, 348–59.

Stein, W. (1931) *Der Begriff des Schwerpunktes bei Archimedes* (Quellen und Studien zur Geschichte der Mathematik, Astronomie und Physik B, 1, 2 (1930), Berlin, 1931, pp. 221–44).

Steinmetz, P. (1964) *Die Physik des Theophrastos von Eresos* (Bad Homburg, Berlin, Zürich).

Stemplinger, E. (1919) *Sympathieglaube und Sympathiekuren in Altertum und Neuzeit* (München).

(1922) *Antiker Aberglaube in modernen Ausstrahlungen* (Das Erbe der Alten 7, Leipzig).

(1925) *Antike und moderne Volksmedizin* (Das Erbe der Alten 10, Leipzig).

Stenzel, J. (1921) 'Über den Einfluss der griechischen Sprache auf die philosophische Begriffsbildung', *Neue Jahrbücher für das klassische Altertum, Geschichte und deutsche Literatur* XLVII, 152–64.

(1931) *Zur Theorie des Logos bei Aristoteles* (Quellen und Studien zur Geschichte der Mathematik, Astronomie und Physik B, 1, 1 (1929), Berlin, 1931, pp. 34–66).

(1940) *Plato's Method of Dialectic* (trans. D. J. Allan of *Die Entwicklung der platonischen Dialektik*, 2nd ed., Leipzig, 1931) (Oxford).

Steuer, R. O. and Saunders, J. B. de C. M. (1959) *Ancient Egyptian and Cnidian Medicine* (University of California Press, Berkeley and Los Angeles).

Sticker, G. (1933) 'Hiera Nousos', *Quellen und Studien zur Geschichte der Naturwissenschaften und der Medizin* iii, 4, 347–58.

Stiebitz, F. (1930) 'Über die Kausalerklärung der Vererbung bei Aristoteles', *Sudhoffs Archiv für Geschichte der Medizin* xxiii, 332–45.

Stokes, M. C. (1962) 'Hesiodic and Milesian cosmogonies, I', *Phronesis* vii, 1–37.

(1963) 'Hesiodic and Milesian cosmogonies, II', *Phronesis* viii, 1–34.

(1971) *One and Many in Presocratic Philosophy* (Washington).

Stratton, G. M. (1917) *Theophrastus and the Greek Physiological Psychology before Aristotle* (London).

Suppe, F. (1974) 'The search for philosophic understanding of scientific theories', in *The Structure of Scientific Theories*, ed. F. Suppe (University of Illinois Press, Urbana, Chicago), pp. 2–241.

Swerdlow, N. (1969) 'Hipparchus on the distance of the sun', *Centaurus* xiv, 287–305.

Szabó, Á. (1951–2) 'Beiträge zur Geschichte der griechischen Dialektik', *Acta Antiqua Academiae Scientiarum Hungaricae* i, 377–406.

(1954a) 'Zur Geschichte der Dialektik des Denkens', *Acta Antiqua Academiae Scientiarum Hungaricae* II, 17–57.

(1954b) 'Zum Verständnis der Eleaten', *Acta Antiqua Academiae Scientiarum Hungaricae* II, 243–86.

(1955) 'Eleatica', *Acta Antiqua Academiae Scientiarum Hungaricae* III, 67–102.

(1958) 'δείκνυμι, als mathematischer Terminus für "beweisen"', *Maia* x, 106–31.

(1960–2) 'Anfänge des euklidischen Axiomensystems', *Archive for History of Exact Sciences* I, 37–106.

(1964–6) 'The transformation of mathematics into deductive science and the beginnings of its foundation on definitions and axioms', *Scripta Mathematica* XXVII, 27–48a and 113–39.

(1966) 'Theaitetos und das Problem der Irrationalität in der griechischen Mathematikgeschichte', *Acta Antiqua Academiae Scientiarum Hungaricae* XIV, 303–58.

(1969) *Anfänge der griechischen Mathematik* (Wien and München).

(1970) 'Ein Lob auf die altpythagoreische Geometrie', *Hermes* XCVIII, 405–21.

Tambiah, S. J. (1968) 'The magical power of words', *Man* NS III, 175–208.

(1973) 'Form and meaning of magical acts: a point of view', in Horton and Finnegan 1973, pp. 199–229.

Tambornino, J. (1909) *De antiquorum daemonismo* (Religionsgeschichtliche Versuche und Vorarbeiten VII, 3 (1908–9), Giessen, 1909).

Tannery, P. (1887) *La Géométrie grecque* (Paris).

(1893) *Recherches sur l'histoire de l'astronomie ancienne* (Paris).

(1912–43) *Mémoires scientifiques*, 16 vols. (Paris).

(1930) *Pour l'histoire de la science Hellène*, 2nd ed. (Paris).

Tarán, L. (1965) *Parmenides: A Text with translation, commentary, and critical essays* (Princeton University Press, Princeton, New Jersey).

Tasch, P. (1947–8) 'Quantitative measurements and the Greek atomists', *Isis* XXXVIII, 185–9.

Taylor, A. E. (1928) *A Commentary on Plato's Timaeus* (Oxford).

Taylor, C. C. W. (1967) 'Plato and the Mathematicians: an examination of Professor Hare's views', *Philosophical Quarterly* XVII, 193–203.

Temkin, O. (1933a) 'Views on epilepsy in the Hippocratic period', *Bulletin of the Institute of the History of Medicine* I, 41–4.

(1933b) 'The doctrine of epilepsy in the Hippocratic writings', *Bulletin of the Institute of the History of Medicine* I, 277–322.

(1971) *The Falling Sickness* (1st ed. 1945), 2nd ed. (Baltimore and London).

Theiler, W. (1924) *Zur Geschichte der teleologischen Naturbetrachtung bis auf Aristoteles* (Zürich).

(1967) 'Historie und Weisheit', in *Festgabe H. von Greyerz* (edd. E. Walder and others) (Bern), pp. 69–81 (reprinted in Theiler 1970, pp. 447–59).

(1970) *Untersuchungen zur antiken Literatur* (Berlin).

Thivel, A. (1975) 'Le "divin" dans la médecine hippocratique', in Bourgey and Jouanna 1975, pp. 57–76.

Thomas, K. (1971) *Religion and the Decline of Magic* (London).

Thompson, A. R. (1952) 'Homer as a surgical anatomist', *Proceedings of the Royal Society of Medicine* XLV (Section of the History of Medicine), 765–7.

Thompson, D'A. W. (1910) *Aristotle, Historia Animalium*, in *The Works of Aristotle translated into English* (ed. W. D. Ross), Vol. IV (Oxford).

(1913) *On Aristotle as a Biologist* (Oxford).

(1940) 'Aristotle the Naturalist', in *Science and the Classics* (London), pp. 37–78.

(1942) *On Growth and Form* (1st ed. 1917), 2nd ed. (Cambridge).

Thompson, R. Campbell (1923–4) 'Assyrian medical texts', *Proceedings of the Royal Society of Medicine* XVII, Section of the History of Medicine, 1–34.

(1925–6) 'Assyrian Medical texts', *Proceedings of the Royal Society of Medicine* XIX, Section of the History of Medicine, 29–78.

Thomson, G. (1946) *Aeschylus and Athens* (1st ed. 1941), 2nd ed. (London).

(1954) *Studies in Ancient Greek Society, I The Prehistoric Aegean* (1st ed. 1949), 2nd ed. (London).

(1955) *Studies in Ancient Greek Society, II The First Philosophers* (London).

Thomson, J. O. (1948) *History of Ancient Geography* (Cambridge).

Thoren, V. E. (1971) 'Anaxagoras, Eudoxus, and the regression of the lunar nodes', *Journal for the History of Astronomy* II, 23–8.

Thorndike, L. (1923–58) *A History of Magic and Experimental Science*, 8 vols. (New York).

Thrämer, E. (1913) 'Health and Gods of Healing (Greek)', in *Encyclopaedia of Religion and Ethics*, ed. J. Hastings, Vol. VI (Edinburgh), pp. 540–53.

Tigner, S. (1974) 'Empedocles' twirled ladle and the vortex-supported earth', *Isis* LXV, 433–47.

Toeplitz, O. (1931) *Das Verhältnis von Mathematik und Ideenlehre bei Plato* (Quellen und Studien zur Geschichte der Mathematik, Astronomie und Physik B, I, I (1929), Berlin, 1931, pp. 3–33).

Toomer, G. J. (1967–8) 'The size of the lunar epicycle according to Hipparchus', *Centaurus* XII, 145–50.

(1973–4) 'The chord table of Hipparchus and the early history of Greek trigonometry', *Centaurus* XVIII, 6–28.

(1974) 'Meton', in *Dictionary of Scientific Biography*, ed. C. C. Gillispie, Vol. IX (New York), pp. 337–40.

(1974–5) 'Hipparchus on the distances of the sun and moon', *Archive for History of Exact Sciences* XIV, 126–42.

(1975) 'Ptolemy', in *Dictionary of Scientific Biography*, ed. C. C. Gillispie, Vol. XI (New York), pp. 186–206.

Tracy, T. (1969) *Physiological Theory and the Doctrine of the Mean in Plato and Aristotle* (Studies in Philosophy 17, The Hague, and Paris).

Turner, E. G. (1951) 'Athenian Books in the Fifth and Fourth Centuries B.C.' (Inaugural lecture, University College London).

Unguru, S. (1975–6) 'On the need to rewrite the history of Greek mathematics', *Archive for History of Exact Sciences* XV, 67–114.

Usener, M. (1896) *Götternamen* (Bonn).

Vegetti, M. (1973) 'Nascita dello scienzato', *Belfragor* XXVIII, 641–63.

Verdenius, W. J. (1962) 'Science grecque et science moderne', *Revue Philosophique* CLII, 319–36.

(1966) 'Der Logosbegriff bei Heraklit und Parmenides', *Phronesis* XI, 81–98.

(1967) 'Der Logosbegriff bei Heraklit und Parmenides', *Phronesis* XII, 99–117.

Vernant, J. P. (1957/1965) 'La formation de la pensée positive dans la Grèce archaïque' (*Annales* XII (1957), 183–206), in Vernant 1965, pp. 285–314.

(1962) *Les Origines de la pensée grecque* (Paris).

(1963/1965) 'Géométrie et astronomie sphérique dans la première cosmologie grecque' (*La Pensée* CIX (1963), 82–92), in Vernant 1965, pp. 145–58.

(1965) *Mythe et pensée chez les grecs*, 2nd ed. (Paris).

(1970) 'Thétis et le poème cosmogonique d'Alcman', in *Hommages à Marie Delcourt* (Collection Latomus 114, Bruxelles), pp. 38–69.

Viano, C. A. (1958) 'La dialettica in Aristotele', *Rivista di Filosofia* (Torino) XLIX, 154–78.

(1965) 'Retorica, magia e natura in Platone', *Rivista di Filosofia* (Torino) LVI, 411–53.

Vicaire, P. (1970) 'Platon et la divination', *Revue des études grecques* LXXXIII, 333–50.

Vidal-Naquet, P. (1967) 'La raison grecque et la cité', *Raison Présente* II, 51–61.
(1968) 'La tradition de l'hoplite athénien', in *Problèmes de la guerre en Grèce ancienne*, ed. J. P. Vernant (Paris and The Hague), pp. 161–81.
(1970) 'Esclavage et gynécocratie dans la tradition, le mythe, l'utopie', in *Recherches sur les structures sociales dans l'antiquité classique*, edd. C. Nicolet and C. Leroy (Paris), pp. 63–80.
Vlastos, G. (1947/1970) 'Equality and justice in early Greek cosmologies' (*Classical Philology* XLII (1947), 156–78), in Furley and Allen 1970, pp. 56–91.
(1952/1970) 'Theology and Philosophy in early Greek thought' (*Philosophical Quarterly* II (1952), 97–123), in Furley and Allen 1970, pp. 92–129.
(1953) 'Isonomia', *American Journal of Philology* LXXIV, 337–66.
(1953/1975) Review of Raven 1948 (*Gnomon* XXV (1953), 29–35) in Allen and Furely 1975, pp. 166–76.
(1955/1970) Review of Cornford 1952 (*Gnomon* XXVII (1955), 65–76) in Furley and Allen 1970, pp. 42–55.
(1959/1975) 'A note on Zeno B 1' (from *Gnomon* XXXI (1959), 193–204) in Allen and Furley 1975, pp. 177–83.
(1964) 'Ἰσονομία πολιτική', in *Isonomia*, edd. J. Mau and E. G. Schmidt (Berlin), pp. 1–35.
(1966*a*/1975) 'A note on Zeno's arrow' (*Phronesis* XI (1966), 3–18) in Allen and Furley 1975, pp. 184–200.
(1966*b*/1975) 'Zeno's Race Course' (*Journal of the History of Philosophy* IV (1966), 95–108), in Allen and Furley 1975, pp. 201–20.
(1975) *Plato's Universe* (Oxford).
(forthcoming) 'The role of observation in Plato's conception of astronomy', in *Science and Sciences in Plato*, ed. J. P. Anton (New York).
Vogel, C. J. de (1966) *Pythagoras and early Pythagoreanism* (Assen).
Vogel, K. (1936) *Beiträge zur griechischen Logistik* (Sitzungsberichte der mathematisch-naturwissenschaftlichen Abteilung der bayerischen Akademie der Wissenschaften zu München, Jahrgang 1936, München, pp. 357–472).
Vogt, H. (1908–9) 'Die Geometrie des Pythagoras', *Bibliotheca Mathematica*, Dritte Folge IX, 15–54.
(1909–10) 'Die Entdeckungsgeschichte des Irrationalen nach Plato und anderen Quellen des 4 Jahrhunderts', *Bibliotheca Mathematica*, Dritte Folge X, 97–155.
(1915) 'Zur Entdeckungsgeschichte des Irrationalen', *Bibliotheca Mathematica*, Dritte Folge XIV, 1913–14 (1915), 9–29.
(1925) 'Versuch einer Wiederherstellung von Hipparchs Fixsternverzeichnis', *Astronomische Nachrichten* CCXXIV, 17–54.
Wächter, T. (1910) *Reinheitsvorschriften im griechischen Kult* (Religionsgeschichtliche Versuche und Vorarbeiten IX, 1 (1910–11), Giessen).
Waerden, B. L. van der (1940–1) 'Zenon und die Grundlagenkrise der griechischen Mathematik', *Mathematische Annalen* CXVII, 141–61.
(1947–9) 'Die Arithmetik der Pythagoreer', *Mathematische Annalen* CXX, 127–53 and 676–700.
(1951) *Die Astronomie der Pythagoreer* (Verhandelingen der koninklijke Nederlandse Akademie van Wetenschappen, Afdeling Natuurkunde, XX, 1, Amsterdam).
(1954) *Science Awakening* (trans. A. Dresden) (Groningen).
(1958) 'Drei umstrittene Mondfinsternisse bei Ptolemaios', *Museum Helveticum* XV, 106–9.
(1960) 'Greek astronomical calendars and their relation to the Athenian civil calendar', *Journal of Hellenic Studies* LXXX, 168–80.

(1965–7) 'Vergleich der mittleren Bewegungen in der babylonischen, griechischen und indischen Astronomie', *Centaurus* XI, 1–18.

(1970) 'Berichtigung zu meine Arbeit "Vergleich der mittleren Bewegungen in der babylonischen, griechischen und indischen Astronomie"', *Centaurus* XV, 21–5.

Waschkies, H.-J. (1970–1) 'Eine neue Hypothese zur Entdeckung der inkommensurablen Grössen durch die Griechen', *Archive for History of the Exact Sciences* VII, 325–53.

Wasserstein, A. (1958) 'Theaetetus and the history of the theory of numbers', *Classical Quarterly* NS VIII, 165–79.

(1962) 'Greek scientific thought', *Proceedings of the Cambridge Philological Society* NS VIII, 51–63.

(1972) 'Le rôle des hypothèses dans la médecine grecque', *Revue Philosophique* CLXII, 3–14.

Watkins, J. W. N. (1957) 'Between analytic and empirical', *Philosophy* XXXII, 112–31.

Wedberg, A. (1955) *Plato's Philosophy of Mathematics* (Stockholm).

Weidauer, K. (1954) *Thukydides und die hippokratischen Schriften* (Heidelberger Forschungen 2, Heidelberg).

Weil, E. (1951/1975) 'The place of logic in Aristotle's thought' (originally 'La place de la logique dans la pensée aristotélicienne', *Revue de métaphysique et de morale* LVI (1951), 283–315) in Barnes, Schofield, Sorabji 1975, pp. 88–112.

Weinreich, O. (1909) *Antike Heilungswunder* (Religionsgeschichtlichen Versuchen und Vorarbeiten VIII, 1 (1909–10), Giessen, 1909).

Weiss, H. (1942) *Kausalität und Zufall in der Philosophie des Aristoteles* (Basel).

Wellmann, M. (1895) *Die pneumatische Schule* (Philol. Untersuch. 14) (Berlin).

(1901) *Die Fragmente der sikelischen Ärzte Akron, Philistion und des Diokles von Karystos* (Berlin).

(1929) 'Die Schrift περὶ ἱρῆς νούσου des Corpus Hippocraticum', *Sudhoffs Archiv für Geschichte der Medizin* XXII, 290–312.

West, M. L. (1963) 'Three Presocratic cosmologies', *Classical Quarterly* NS XIII, 154–76.

(1967) 'Alcman and Pythagoras', *Classical Quarterly* NS XVII, 1–15.

(1971) *Early Greek Philosophy and the Orient* (Oxford).

Whorf, B. L. (1956) *Language, Thought, and Reality* (M.I.T. Press and New York and London).

Wieland, W. (1962) *Die aristotelische Physik* (Göttingen).

(1972) 'Zeitliche Kausalstrukturen in der aristotelischen Logik', *Archiv für Geschichte der Philosophie* LIV, 229–37.

Wilamowitz-Moellendorff, U. von (1901) 'Die hippokratische Schrift περὶ ἱρῆς νούσου', *Sitzungsberichte der königlich preussischen Akademie der Wissenschaften zu Berlin* (Berlin, 1901), 2–23 (reprinted in Vol. III of *Kleine Schriften*, Berlin, 1969, pp. 278–302).

(1929) 'Die καθαρμοί des Empedokles', *Sitzungsberichte der preussischen Akademie der Wissenschaften, phil.-hist. Kl.*, Jahrgang 1929 (Berlin), 626–61 (reprinted in Vol. I of *Kleine Schriften*, Berlin, 1935, pp. 473–521).

Wilcox, S. (1942) 'The scope of early rhetorical instruction', *Harvard Studies in Classical Philology* LIII, 121–55.

Willetts, R. F. (1967) *The Law Code of Gortyn* (Kadmos Suppl. 1, Berlin).

Wilpert, P. (1956–7) 'Aristoteles und die Dialektik', *Kant-Studien* XLVIII, 247–57.

Wilson, B. R. (ed.) (1970) *Rationality* (Oxford).

Wilson, J. A. (1949) 'Egypt', in Frankfort 1949, pp. 39–133.

(1962) 'Medicine in ancient Egypt', *Bulletin of the History of Medicine* xxxvi, 114–23.

Winch, P. (1958) *The Idea of a Social Science* (London).

(1964/1970) 'Understanding a primitive society' (*American Philosophical Quarterly* I (1964), 307–24) in B. R. Wilson 1970, pp. 78–111.

Wolff, H. J. (1970) '*Normenkontrolle*' *und Gesetzesbegriff in der attischen Demokratie. Untersuchungen zur* γραφή παρανόμων (Sitzungsberichte der Heidelberger Akademie der Wissenschaften, phil.-hist. Kl., Jahrgang 1970, 2, Heidelberg).

Woodbury, L. (1965) 'The date and atheism of Diagoras of Melos', *Phoenix* xix, 178–211.

(1976) 'Aristophanes' *Frogs* and Athenian literacy: *Ran.* 52–53, 1114', *Transactions of the American Philological Association* cvi, 349–57.

Wright, L. (1973–4) 'The astronomy of Eudoxus. Geometry or physics?' *Studies in History and Philosophy of Science* iv, 165–72.

Wussing, H. (1974) 'Zur Grundlagenkrisis der griechischen Mathematik', in *Hellenische Poleis*, Vol. iv, ed. E. C. Welskopf (Berlin), pp. 1872–95.

Zaehner, R. C. (1966) *Hindu Scriptures* (London and New York).

Zeller, E. and Mondolfo, R. (1932–74) *La Filosofia dei Greci nel suo sviluppo storico* (*Die Philosophie der Griechen*, translated, edited and enlarged by R. Mondolfo) 3 Parts (Firenze).

Zeuthen, H.-G. (1896) 'Die geometrische Construction als "Existenzbeweis" in der antiken Geometrie', *Mathematische Annalen* xlvii, 222–8.

(1910) 'Notes sur l'histoire des mathématiques, VIII. Sur la constitution des livres arithmétiques des Eléments d'Euclide et leur rapport à la question de l'irrationalité', *Oversigt over det kongelige Danske Videnskabernes Selskabs Forhandlinger*, 395–435.

(1913) 'Sur les connaissances géométriques des grecs avant la réforme platonicienne de la géométrie', *Oversigt over det kongelige Danske Videnskabernes Selskabs Forhandlinger*, 431–73.

(1915) 'Sur l'origine historique de la connaissance des quantités irrationelles', *Oversigt over det kongelige Danske Videnskabernes Selskabs Forhandlinger*, 333–62.

Zilsel, E. (1941–2) 'The sociological roots of science', *American Journal of Sociology* xlvii, 544–62.

(1942) 'The genesis of the concept of physical law', *Philosophical Review* li, 245–79.

(1945) 'The genesis of the concept of scientific progress', *Journal of the History of Ideas* vi, 325–49.

Zinner, E. (1950) 'Cl. Ptolemaeus und das Astrolab', *Isis* xli, 286–7.

Zubov, V. P. (1959) 'Beobachtung und Experiment in der antiken Wissenschaft', *Das Altertum* v, 223–32.

Zuntz, G. (1971) *Persephone* (Oxford).

INDEX OF PASSAGES REFERRED TO

ARISTOTLE (*continued*)

209 n. 414; (383a32ff), 209 n. 415; (383b11ff), 209 n. 414; (383b13ff), 209;
(383b20), 142 n. 86; (383b20ff), 210; (383b23), 209 n. 414; (384a11ff), 209
n. 416; (384a20ff), 210; (384a26ff), 209 n. 417; (384a29ff), 209 n. 417;
(385b12ff), 210 n. 419; (387b9ff), 210 n. 418; (388a33ff), 210 n. 418;
(388b11), 209; (388b12), 209 n. 414; (389a9ff), 209 n. 416; (389a19ff),
209 n. 417; (389a22f), 209 n. 416

de An. (407b27ff), 253 n. 120; (409a6), 112 n. 288; (411a8), 11 n. 9; (417b22f),
137 n. 62; (420a14ff), 164 n. 201; (III 3), 136 n. 61; (427b3), 135 n. 53;
(428a11ff), 136 n. 61; (428a16ff), 136 n. 61; (428b2ff), 136 n. 61; (428b18f),
136 n. 61; (428b23ff), 136 n. 61

Sens. (437a17–b1), 97 n. 200; (2, 437a19ff), 161 n. 187; (438b2), 164 n. 201;
(438b13f), 164 n. 201; (444a10ff), 165 n. 202

Somn. Vig. (458a15ff), 161 n. 186

Div. Somn. (463a4–b11), 43 n. 177; (463b13ff), 43 n. 177

Long. (464b32f), 97 n. 204

Juv. (468a13ff), 213; (468b28ff), 216; (469a10ff), 162 n. 191

Resp. (470b8f), 220 n. 483; (470b9f), 205 n. 389; (471b23ff), 220 n. 483;
(475b7ff), 214 n. 442; (476b30ff), 214 n. 442; (477a20f), 215 n. 448;
(480b22ff), 97 n. 200

HA (I 2–3), 213; (488b29ff), 213 n. 438; (489a8ff), 213 n. 438; (491a9ff), 137
n. 64, 212 n. 436; (491a34ff) 165 n. 202; (491b2ff) 215 n. 452; (491b20ff),
164 n. 201; (492a14ff), 156 n. 160; (493b14ff), 212 n. 429; (494b21ff), 163
n. 194; (494b27ff), 214 n. 441; (494b29ff), 165 n. 202; (494b33ff), 165
n. 202; (495a4ff), 165 n. 202; (495a11ff), 164 n. 201; (I 17), 161 n. 186;
(496a9ff), 164; (496a11), 157 n. 164; (496a13), 161 n. 186; (496b4ff), 157
n. 164; (496b24ff), 157 n. 165; (497a32), 164 n. 198; (501b19ff), 215
n. 453; (510a21ff), 164; (510b3f), 216 n. 461; (511b13ff), 157, 163 n. 194;
(III 2f, 511b23–513a7), 157 n. 167; (511b20ff), 157 n. 164; (512a4ff),
21 n. 63; (512a9ff), 21 n. 63; (512a29ff), 21 n. 63; (512a30f), 158 n. :58;
(III 3), 161 n. 186; (512b17ff), 158 n. 168; (512b24ff), 158 n. 168; (512b32ff),
21 n. 63; (513a12ff), 163 n. 194; (513b26ff), 127 n. 8; (514a18ff), 165
n. 202; (514a32ff), 21 n. 63; (514b3ff), 21 n. 63; (515a28ff), 161 n. 186;
(516a18f), 215 n. 452; (516a19f), 212; (520b23ff), 209 n. 417; (521a2ff),
215 n. 448; (521a26f), 215 n. 450; (523a18ff), 209 n. 416; (523a25f), 223
n. 493; (IV 1–7), 213, 214 n. 441; (524a5ff), 212 n. 432; (524b4), 214 n. 441;
(524b21f), 214 n. 442; (524b32), 214 n. 441; (525a8f), 164 n. 198; (527b1ff),
213 n. 440; (530a2f), 213 n. 439; (531a12ff), 213 n. 439; (531b8ff), 213
n. 439; (532b7f), 213 n. 440; (538a22ff), 215 n. 458; (538b15ff), 215
n. 456; (540b15f), 215 n. 458; (541b8ff), 212 n. 432; (544a12f), 212 n. 432;
(552b15ff), 212 n. 431; (561a4ff), 158 n. 172; (561a6ff), 216 n. 463;
(561a11f), 216 n. 463; (VI 10, 565b1ff), 144 n. 94; (566a6–8), 212 n. 427;
(566a14ff), 164 n. 198; (572b29), 44 n. 191; (573a11ff), 212 n. 427;
(574b4), 44 n. 191; (574b15ff), 212 n. 427; (579b2ff), 212 n. 430; (579b5ff),
212 n. 435; (580a19–22), 212 n. 429; (582b28ff), 215 n. 450; (583a4ff),
215 n. 450; (583a14ff), 223 n. 493; (583b2ff), 215 n. 459, 217 n. 470;
(583b5ff), 215 n. 459, 217 n. 470; (583b14ff), 160 n. 181, 163 n. 194;
(584a12ff), 215 n. 459; (585b35ff), 219 n. 477; (587b1), 44 n. 190;

GAUDENTIUS
GELLIUS
GORGIAS

HERACLITUS

HIPPOCRATIC CORPUS (*continued*)

(6, 44.11ff), 94 n. 186, 150 n. 126; (6, 44.15ff), 150 n. 126; (7, 46.9ff), 94 n. 184; (7, 46.11), 149 n. 124; (7, 46.11ff), 152 n. 137; (7, 46.17ff), 94 n. 186, 150 n. 127; (7, 48.10ff), 94 n. 186, 150 n. 127; (7, 50.3), 103 n. 244; (7. 50.9), 149 n. 124; (7, 50.9ff), 94 n. 186, 150 n. 128; (8, 50.19), 103 n. 244; (11), 157 n. 167; (11, 60.1ff), 21 n. 63
Nat. Mul. (περὶ γυναικείης φύσιος) (1, L VII 312.1ff), 41 n. 164; (1, 312.9), 41 n. 164; (10, 326.3ff), 91 n. 172; (96, 412.19ff), 223 n. 493
Nat. Puer. (περὶ φύσιος παιδίου) (13, L VII 488.22ff), 160 n. 181; (29, 530.10ff), 158 n. 172
Off. (κατ᾽ ἰητρεῖον) (1, L III 272.1–5), 135 n. 48
Oss. (περὶ ὀστέων φύσιος) (8, L IX), 157 n. 167; (9) 157 n. 167
Praec. (παραγγελίαι) (1, *CMG* I, 1 30.3ff), 135 n. 48; (1, 30.18ff), 39 n. 151; (2, 31.6ff), 91 n. 172; (5, 31.26ff), 90 n. 163; (7, 32.22ff), 91 n. 174; (8, 33.5ff), 91 n. 174; (10, 33.32ff), 89 n. 158; (12, 34.5ff), 90 n. 166
Prog. (προγνωστικόν) (1, L II 110.2ff), 45 n. 194, 90 n. 169, 151 n. 134; (1, 112.5f), 41 n. 164; (1, 112.6ff), 90 n. 169; (1, 112.10f), 48 n. 209; (2, 112.12ff), 152 n. 135; (3ff, 118.7ff), 152 n. 136; (9, 132.6ff), 152 n. 137; (11–14, 134.13–146.15), 152 n. 138; (12, 138.15ff), 152 n. 140; (12, 142.12ff), 152 n. 140; (20, 168.16ff), 121 n. 326, 155 n. 153; (25, 190.6ff), 90 n. 164
Prorrh. II (προρρητικόν) (1f, L IX 6.1ff), 45 n. 195, 89 n. 160; (1, 8.2ff), 45 n. 195; (2, 8.11), 45 n. 195; (3, 10.23ff), 45 n. 195; (27, 60.1ff), 91 n. 172; (34, 66.8ff), 91 n. 172; (41, 70.22ff), 91 n. 172; (42, 74.4ff), 91 n. 172
Steril. (περὶ ἀφόρων) (213, L VIII 410.14f), 91 n. 172; (214, 414.17ff), 223 n. 493; (215, 416.8ff), 223 n. 493; (215, 416.13ff), 223 n. 493; (219, 422.23ff), 223 n. 493; (230, 440.13f), 91 n. 172
Superf. (περὶ ἐπικυήσιος) (25, *CMG* I, 2, 280.28ff), 223 n. 493
VM (περὶ ἀρχαίης ἰητρικῆς) (1, *CMG* I, 1 36.2), 95 n. 193; (1, 36.2ff), 135 n. 51, 140 n. 73, 147 n. 109; (1, 36.15ff), 135 n. 51, 140 n. 73, 147 n. 106; (1, 36.16), 147 n. 108; (1, 36.18–21), 95 n. 194; (2, 37.1ff), 135 n. 49; (2, 37.7ff), 39 n. 152; (2, 37.9ff), 95 n. 195, 228 n. 7, 260 n. 149; (2, 37.17ff), 39 n. 152, 228 n. 7; (2, 37.17–19), 95 n. 196; (3, 37.20ff), 135 n. 49; (4, 38.27ff), 39 n. 152; (4, 39.2ff), 39 n. 152; (5, 39.21ff), 148 n. 114; (6, 39.27ff), 148 n. 115; (8, 41.8f), 135 n. 49; (9–12), 121 n. 326; (9, 41.20ff), 121 n. 326, 135 n. 52; (9, 41.25ff), 39 n. 152; (9, 42.6ff), 39 n. 152; (13–17, 44.8ff), 147 n. 109; (13ff), 148 n. 114; (13, 44.8), 147 n. 108; (13, 44.18ff), 148 n. 115; (13, 44.27f), 95 n. 197; (14, 45.17f), 41 n. 164; (14, 45.18ff), 248 n. 97; (14, 45.26ff), 147 n. 110; (14, 45.28ff), 147 n. 111; (15, 46.18ff), 54 n. 228; (15, 46.20ff), 148 n. 115; (15, 46.22ff), 148 n. 114; (15, 46.26ff), 91 n. 173, 95 n. 197; (15, 47.5ff), 148 n. 116; (15, 47.6), 148 n. 118; (16ff), 54 n. 228; (16, 48.10f), 54 n. 228; (17, 48.21), 95 n. 198; (17, 48.21ff), 54 n. 228; (17, 49.2), 54 n. 228; (18, 49.10f), 152 n. 137; (18, 49.16ff), 54 n. 228; (19, 50.2ff), 147 n. 113; (19, 50.7ff), 54 n. 229, 147 n. 112; (19, 50.9ff), 54 n. 228; (19, 50.14ff), 147 n. 113; (20, 51.6ff), 147 n. 107; (20, 51.10f), 34 n. 119; (20, 51.12f), 95 n. 193; (20, 51.17ff), 135 n. 49; (20, 51.24), 54 n. 229; (20, 52.3), 54 n. 229; (21, 52.17ff), 39 n. 152, 54 n. 227; (22, 53.1ff), 158 n. 174; (22, 53.12f), 134 n. 47; (22, 53.12ff), 148 n. 117, 158 n. 174; (24, 55.4ff), 148 n. 117; (24, 55.12), 148 n. 118

PLATO (*continued*)

131 n. 24; (83a), 131 n. 26; (91a), 100 n. 219; (92d), 116; (97b ff), 36 n. 131; (99ab), 54 n. 230; (99b), 140 n. 72; (102d f), 132 n. 27; (108de), 102 n. 237; (115a), 44 n. 184

Phdr. (244a ff), 227 n. 4; (244d–245a), 29 n. 98; (245c), 102 n. 237; (246a ff), 102 n. 237; (259e ff), 100 n. 219; (263d), 101 n. 225; (265de), 102 n. 240; (266d ff), 81 n. 109; (266e–267a), 81 n. 110; (270c ff), 101 n. 225; (272d ff), 100 n. 219; (273a), 81 n. 110; (274cd), 230 n. 13; (274e), 44 n. 185; (277b6), 101 n. 225

Phlb. (28c), 248 n. 94; (33c ff), 131 n. 23; (58a ff), 84 n. 134, 100 n. 219; (58a–c), 102 n. 241; (59a), 119 n. 317; (59a–c), 102 n. 241

Plt. (280e), 29 n. 98; (285ab), 102 n. 240; (296bc), 259 n. 145; (297e ff), 254 n. 127; (298c2 f), 254 n. 127; (299a), 253 n. 120; (303c), 99 n. 209

Prm. (127ab), 262 n. 156; (128cd), 72; (128d5 f), 111 n. 284; (136a4 f), 111 n. 284; (137e), 105 n. 255

Prt. (315c), 87 n. 146; (318d ff), 87 n. 146; (319b f), 254 n. 127; (320c ff), 244 n. 73; (328d ff), 100 n. 220; (329b), 100 n. 220; (333a), 101 n. 233; (334e–335d), 100 n. 220; (338de), 101 n. 226; (339b–d), 101 n. 233; (343a), 249 n. 99; (348a), 101 n. 226; (349a), 80 n. 106; (361a–c), 101 n. 233

R. (336c), 100 n. 224; (336e), 101 n. 227; (337a b), 101 n. 231; (343a ff), 101 n. 228; (350c–e), 101 n. 226; (358b), 99 n. 209; (364b ff), 29 n. 98, 227 n. 4; (364e f), 44 n. 189; (426b), 29 n. 98; (436b8 ff), 124 n. 331; (476d ff), 119 n. 317; (510c ff), 113–14; (511b), 114 n. 292; (521cd), 132 n. 31; (523ab), 132 n. 31; (523b), 119 n. 317, 131 n. 26, 132 n. 33; (523c), 132 n. 31; (524d ff), 103 n. 248; (525b–e), 132 n. 31; (525de), 117 n. 306; (526ab), 132 n. 31; (529b7–c1), 132 n. 33; (529c4 ff), 132 n. 32; (529cd), 131; (529d7 ff), 133 n. 35; (529e3 ff), 132 n. 28; (530a7–b4), 132 n. 34; (530b1–4), 132 n. 28; (530b6 ff), 132 n. 29; (530b8 f), 132 n. 31; (530c2 f), 132 n. 32; (530d), 145; (530d ff), 119 n. 320, 133 n. 36; (530d–531c), 145 n. 100; (531a1–3), 133 n. 36; (531b7), 145 n. 101; (531c), 133 n. 36; (531cd), 102 n. 239; (531e), 102 n. 238; (532d ff), 102 n. 239; (533c), 102 n. 238; (534ab), 102 n. 239; (534b), 102 n. 238; (534d), 102 n. 238; (537cd), 102 n. 239; (539b–d), 101 n. 227; (557b), 245 n. 81; (561de), 245 n. 81; (562b–563b), 245 n. 81; (606e f), 65 n. 37; (616de), 174 n.251; (617a), 174 n. 249; (617ab), 174 n. 250

Smp. (186d ff), 248 n. 97; (202e–203a), 29 n. 98

Sph. (226d ff), 44 n. 190; (234e f), 99 n. 209; (248a ff), 131 n. 23; (253b–e), 102 n. 240; (261d ff), 124 n. 332

Tht. (147d), 106 nn. 260 and 265; (147d ff), 111 n. 283; (149cd), 29 n. 98; (151e), 135 n. 54; (152a), 135 n. 54; (162e), 116; (164e), 116 n. 299; (166a ff), 135 n. 54; (166d), 135 n. 54; (167e), 101 n. 227; (168e), 116 n. 299; (184bc), 131 n. 23; (201e ff), 108 n. 270

Ti. (20d ff), 238 n. 40; (21e ff), 238 n. 39; (24c), 227 n. 4; (27d ff), 201 n. 377; (29bc), 102 n. 237, 201 n. 377; (36b ff), 174 n. 249; (36b–d), 174 n. 251; (36c7), 174 n. 249; (36d), 174 n. 250; (38de), 174 nn. 250 and 252; (38e–39b), 174 n. 250; (39cd), 174–5; (40c), 175 n. 254; (40cd), 174 n. 252, 181 n. 293; (47ab), 133 n. 35; (47e f), 248 n. 94; (51d ff), 119 n. 317;

GENERAL INDEX